Advancing VLSI through Machine Learning

This book explores the synergy between very large-scale integration (VLSI) and machine learning (ML) and its applications across various domains. It investigates how ML techniques can enhance the design and testing of VLSI circuits, improve power efficiency, optimize layouts, and enable novel architectures.

This book bridges the gap between VLSI and ML, showcasing the potential of this integration in creating innovative electronic systems, advancing computing capabilities, and paving the way for a new era of intelligent devices and technologies. Additionally, it covers how VLSI technologies can accelerate ML algorithms, enabling more efficient and powerful data processing and inference engines. It explores both hardware and software aspects, covering topics like hardware accelerators, custom hardware for specific ML tasks, and ML-driven optimization techniques for chip design and testing.

This book will be helpful for academicians, researchers, postgraduate students, and those working in ML-driven VLSI.

Materials, Devices, and Circuits: Design and Reliability

Series Editor:
Shubham Tayal, K. K. Paliwal, Amit Kumar Jainy

This series is designed to illustrate the many new and exciting challenges in the expansive and interdisciplinary field of materials science, device engineering, reliability, device-circuit co-design and their applications. The scope of this series is broad, with reference works and textbooks offering insight into all aspects of materials development, device fabrication, circuit analysis and their reliability. The titles in this series deliver authoritative information to professionals, researchers, and students across material and device engineering as well as other scientific disciplines. Each volume offers a comprehensive approach covering fundamentals, technologies, their evolution advancements and applications and includes real-world examples where appropriate.

Tunneling Field Effect Transistors: Design, Modeling and Applications
Edited by T.S. Arun Samuel, Young Suh Song, Shubham Tayal, P. Vimala, and Shiromani Balmukund Rahi

Quantum-Dot Cellular Automata Circuits for Nanocomputing Applications
Edited by Trailokya Sasamal, Hari Mohan Gaur, Ashutosh Kumar Singh, and Xiaoqing Wen

Device Circuit Co-Design Issues in FETs
Edited by Shubham Tayal, Billel Smaani, Shiromani Balmukund Rahi, Samir Labiod, and Zeinab Ramezani

Human-Machine Interface Technology Advancements and Applications
Edited By Ravichander Janapati, Usha Desai, Shrirang Ambaji Kulkarni, and Shubham Tayal

Negative Capacitance Field Effect Transistors: Physics, Design, Modeling and Applications
Edited by Young Suh Song, Shubham Tayal, Shiromani Balmukund Rahi, and Abhishek Kumar Upadhyay

Laser Applications in Manufacturing
Edited by Pankaj Kumar, Manowar Hussain, Amit Kumar Jain, and Sunil Pathak

5G and Beyond Wireless Networks: Technology, Network Deployments and Materials for Antenna Design
Edited by Indrasen Singh, Shubham Tayal, Niraj Pratap Singh, Vijay Shankar Tripathi, and Ghanshyam Singh

5G and Beyond Wireless Communications: Fundamentals, Applications, and Challenges
Edited by Abhijit Bhowmick, Yogesh Kumar Choukiker, Indrasen Singh, and Arumugam Nallanathan

For more information about this series, please visit: https://www.routledge.com/Materials-Devices-and-Circuits/book-series/MDCDR

Advancing VLSI through Machine Learning

Innovations and Research Perspectives

Edited by
Abhishek Narayan Tripathi,
Jagana Bihari Padhy, Indrasen Singh,
Shubham Tayal, and Ghanshyam Singh

CRC Press
Taylor & Francis Group
Boca Raton London New York

CRC Press is an imprint of the
Taylor & Francis Group, an **informa** business

Designed cover image: Abhishek Narayan Tripathi

First edition published 2025
by CRC Press
2385 NW Executive Center Drive, Suite 320, Boca Raton FL 33431

and by CRC Press
4 Park Square, Milton Park, Abingdon, Oxon, OX14 4RN

CRC Press is an imprint of Taylor & Francis Group, LLC

ISBN: 978-1-032-77428-2 (hbk)
ISBN: 978-1-032-77429-9 (pbk)
ISBN: 978-1-003-48303-8 (ebk)

DOI: 10.1201/9781003483038

Typeset in Times
by codeMantra

Contents

Preface vii
About the Editors viii
List of Contributors ix

Chapter 1 Optimizing Circuit Synthesis: Integrating Neural Networks and Evolutionary Algorithms for Increased Design Efficiency 1

Swagata Devi, Naushad Manzoor Laskar, Shreyas Nirgude, and Saanvi Srivastava

Chapter 2 Study of Physical Process Analysis and Phenomena of Insights of Trapping in Performance Degradation in AlGaN/GaN HEMTs 14

Rayabarapu Venkateswarlu, G.P.S.C. Mishra, and B. Acharya

Chapter 3 Framework for Design and Performance Evaluation of Memory Using a Memristor 36

Jalpaben Pandya, Ramji Gupta, and Alpana Pandey

Chapter 4 Innovative Design and Optimization of High-Power Amplifiers: A Comparative Study with GaN HEMT and CMOS Technologies 52

Manju Yadav, Ramji Gupta, Abhishek Narayan Tripathi, and Alpana Pandey

Chapter 5 Exploring FPGA Architecture Designs for Matrix Multiplication in Machine Learning 71

Jyoti Kori

Chapter 6 Silicon Chip Design and Testing 86

Shasanka Sekhar Rout and Jagana Bihari Padhy

Chapter 7 A Novel Deep Learning Approach for Early Brain Tumor Detection 101

Ankit Singh, Ajay Singh Raghuvanshi, B. Ganesh, and K.V.S. Harsha Vardhan

Chapter 8 TCAD-Augmented Machine Learning for the Prediction of Device Behaviour and Failure Analysis 113

Sandeep Moparthi and Gopi Krishna Saramekala

Chapter 9 Opportunities and Challenges for ML-Based FPGA Backend Flow 127

Saif Ahmad, Mohd Javed Khan, Indrasen Singh, and Zeenat Fatima

Chapter 10 Role of Machine Learning Applications in VLSI Design 149

Mohd Javed Khan, Pooja Gaur, Indrasen Singh, and Saif Ahmad

Chapter 11 Application of Artificial Intelligence/Machine Learning in VLSI Design 184

Abhay Pratap Singh, Vimal Kumar Mishra, and Shamim Akhter

Chapter 12 FinFET-Based 9T SRAM for Enhanced Performance in AI/ML Applications 198

Yagnadatt Sarangi, Kartik Singh, and Shelja Kaushal

Chapter 13 Power Consumption and SNM Analysis of 6T and 7T SRAM Using 90 nm Technology 215

Devanshi Tripathi, Vedha Tripathi, Siddhant Bhushan, and Prachi Sharma

Chapter 14 Transforming Electronics: An Extensive Analysis of Hyper-FET Technological Developments and Utilisation 224

Debashish Dash and Manwendra Chauhan

Chapter 15 VLSI Realization of Smart Systems Using Blockchain and Fog Computing 233

R. Hanumantharaju, K. N. Shreenath, Sagarika Choudhury, Swagata Devi, and Naushad Manzoor Laskar

Index 251

Preface

In recent years, the fields of very large-scale integration (VLSI) and machine learning (ML) have witnessed an unprecedented convergence, sparking a new era of innovation in electronic system design. This book, *Advancing VLSI through Machine Learning: Innovations and Research Perspectives*, is a culmination of efforts to bridge the gap between these two dynamic disciplines, exploring their synergistic potential to revolutionize computing capabilities and shape the future of intelligent devices and technologies.

At the heart of this book lies a comprehensive exploration of foundational principles and cutting-edge applications in VLSI-ML integration. From optimizing circuit synthesis using neural networks and evolutionary algorithms to innovative designs of high-power amplifiers utilizing GaN HEMT and CMOS technologies, each chapter delves into critical aspects that define the forefront of research and development.

The journey begins with an in-depth analysis of physical processes and phenomena in AlGaN/GaN HEMTs, providing crucial insights into performance degradation and reliability—a foundational understanding essential for robust circuit design. Subsequently, the framework for memory design and evaluation using Memristors offers a glimpse into future memory technologies, showcasing the transformative potential of emerging electronic components.

Throughout the book, readers will explore FPGA architectures tailored for matrix multiplication in ML applications, paving the way for efficient computation in data-intensive tasks. Furthermore, innovative approaches in deep learning for early brain tumour detection underscore the transformative impact of ML in healthcare, highlighting its application beyond traditional electronic systems.

The integration of technology computer-aided design (TCAD) augmented ML offers a predictive framework for device behaviour and failure analysis, empowering designers with enhanced tools for performance optimization and reliability assurance. Additionally, chapters dedicated to the role of ML in FPGA backend flows and the application of artificial intelligence (AI)/ML algorithms in VLSI design underscore the transformative impact of intelligent algorithms in streamlining design processes and optimizing system performance.

As we embark on this journey through the realms of FinFET-based SRAM for AI/ML applications, power consumption analysis in SRAM technologies, and the technological advancements of hyper-FET, the book culminates in a comprehensive analysis of smart system realization using blockchain and fog computing—an exploration into the future of distributed, secure, and intelligent electronic systems.

This book is intended to serve as a definitive resource for researchers, engineers, and students alike, offering insights into the latest advancements, challenges, and opportunities in VLSI-ML integration. It is our hope that the chapters within will inspire innovation, foster collaboration across disciplines, and propel the development of next-generation electronic systems that redefine the boundaries of possibility.

About the Editors

Dr. Abhishek Narayan Tripathi is currently an Assistant Professor in the Department of Micro and Nanoelectronics, School of Electronics Engineering, Vellore Institute of Technology, Vellore, Tamil Nadu, India. He holds a Ph.D. in ECE with a specialization in VLSI Design and Embedded Technology from MANIT, Bhopal. His research work includes the development of methodologies for dynamic power and leakage power estimation in FPGA and ASIC-based implementations, VLSI system design, AI, deep learning, and microprocessor architecture.

Dr. Jagana Bihari Padhy is an Assistant Professor in the Department of Embedded Technology, School of Electronics Engineering, Vellore Institute of Technology, Vellore, Tamil Nadu, India. He holds a Ph.D. in ECE with a specialization in optical wireless system design from IIIT Bhubaneswar. His research work includes the development of optical system design both in wired and wireless methodologies for the next generation of communication 5G and beyond.

Dr. Indrasen Singh is an Assistant Professor (Sr. Grade-2) in the Department of Embedded Technology, School of Electronics Engineering, Vellore Institute of Technology, Vellore, Tamil Nadu, India. His research interests are in the areas of cooperative communication, stochastic geometry, modelling of wireless networks, heterogeneous networks, millimetre wave communications, device-to-device communication, and 5G/6G communication.

Dr. Shubham Tayal is an Assistant Professor in the Department of Electronics and Communication Engineering, SR University, Warangal, India. He has more than 6 years of academic/research experience in teaching at the UG and PG levels. He received his Ph.D. in Microelectronics and VLSI Design from the National Institute of Technology, Kurukshetra; M.Tech. (VLSI Design) from YMCA University of Science and Technology, Faridabad; and B.Tech. (Electronics and Communication Engineering) from MDU, Rohtak. His research interests include simulation and modelling of multi-gate semiconductor devices, device-circuit co-design in digital/analogue domain, ML, and Internet of Things.

Prof. Ghanshyam Singh received a Ph.D. degree in Electronics Engineering from the Indian Institute of Technology, Banaras Hindu University, Varanasi, India, in 2000. At present, he is a full Professor with the Department of Electrical and Electronics Engineering, APK Campus, University of Johannesburg, South Africa. His research and teaching interests include RF/microwave engineering, millimetre/THz wave antennas and their applications in communication and imaging, next-generation communication systems (OFDM and cognitive radio), and nanophotonics. He has more than 19 years of teaching and research experience in electromagnetic/microwave engineering, wireless communication, and nanophotonics.

Contributors

B. Acharya
Department of Electronics and Communication Engineering
National Institute of Technology Raipur
Chhattisgarh, India

Saif Ahmad
Department of Electronics and Communication Engineering
Integral University Lucknow
Lucknow, India

Shamim Akhter
Department of Electronics and Communication Engineering
Jaypee Institute of Information Technology
Noida, India

Siddhant Bhushan
Department of Electronics and communication
School of Electronics Engineering
Vellore Institute of Technology
Tamil Nadu, India

Jagana Bihari Padhy
School of Electronics Engineering
Vellore Institute of Technology
Tamil Nadu, India

Manwendra Chauhan
Department of Micro and Nanoelectronics
School of Electronics Engineering
Vellore Institute of Technology
Tamil Nadu, India

Sagarika Choudhury
From Industry Tejas Networks Ltd
Bangalore, India

Debashish Dash
Department of Electronics and communication
School of Electronics Engineering
Vellore Institute of Technology
Tamil Nadu, India

Swagata Devi
Department of ECE
Faculty of Engineering
Assam Down Town University
Assam, India

Zeenat Fatima
Department of ME
Integral University Lucknow
Uttar Pradesh, India

B. Ganesh
Department of Electronics and Communication
National Institute of Technology Raipur
Chhattisgarh, India

Pooja Gaur
Department of Electronics and Communication Engineering
Integral University Lucknow
Uttar Pradesh, India

Ramji Gupta
Department of Electronics and Communication Engineering
Parul University
Gujarat, India

R. Hanumantharaju
Department of ECE
Siddaganga Institute of Technology
Tumakuru, India

K. V. S. Harsha Vardhan
Department of Electronics and Communication
National Institute of Technology Raipur
Chhattisgarh, India

Mohd Javed Khan
Department of Electronics and Communication Engineering
Integral University Lucknow
Uttar Pradesh, India

Shelja Kaushal
Department of Micro and Nanoelectronics
School of Electronics Engineering
Vellore Institute of Technology
Tamil Nadu, India

Jyoti Kori
Department of Electronics and Computer Science
Thakur College of Engineering
Mumbai, India

Naushad Manzoor Laskar
Department of Micro and Nanoelectronics
School of Electronics Engineering
Vellore Institute of Technology
Tamil Nadu, India

G. P. S. C. Mishra
Department of Electronics and Communication Engineering
National Institute of Technology Raipur
Chhattisgarh, India

Sandeep Moparthi
Department of Micro & Nanoelectronics
School of Electronics Engineering
Vellore Institute of Technology
Tamil Nadu, India

Shreyas Nirgude
School of Electronics Engineering
Vellore Institute of Technology
Tamil Nadu, India

Alpana Pandey
Department of Electronics and Communication Engineering
National Institute of Technology Bhopal
Madhya Pradesh, India

Jalpaben Pandya
Parul University
Gujarat, India

Abhay Pratap Singh
Department of Electronics and Communication Engineering
Jaypee Institute of Information Technology
Noida, India

Ajay Singh Raghuvanshi
Department of Electronics and Communication
National Institute of Technology Raipur
Chhattisgarh, India

Gopi Krishna Saramekala
Department of Electronics and Communication Engineering
National Institute of Technology Calicut
Kerala, India

Shasanka Sekhar Rout
Department of Electronics & Communication
GLA University
Mathura, India

Prachi Sharma
Department of Micro and Nanoelectronics, School of Electronics
School of Electronics Engineering
Vellore Institute of Technology
Tamil Nadu, India

K. N. Shreenath
Department of ECE
Siddaganga Institute of Technology
Tumakuru, India

Ankit Singh
Department of Electronics and Communication
National Institute of Technology Raipur
Chhattisgarh, India

Indrasen Singh
School of Electronics Engineering
Vellore Institute of Technology
Tamil Nadu, India

Kartik Singh
Department of Electronics and communication
School of Electronics Engineering
Vellore Institute of Technology
Tamil Nadu, India

Saanvi Srivastava
School of Electronics Engineering
Vellore Institute of Technology
Tamil Nadu, India

Abhishek N. Tripathi
School of Electronics Engineering
Vellore Institute of Technology
Tamil Nadu, India

Devanshi Tripathi
Department of Electronics and communication
School of Electronics Engineering
Vellore Institute of Technology
Tamil Nadu, India

Vedha Tripathi
Department of Electronics and communication
School of Electronics Engineering
Vellore Institute of Technology
Tamil Nadu, India

Rayabarapu Venkateswarlu
Department of Electronics and Communication Engineering
National Institute of Technology Raipur
Chhattisgarh, India

Vimal Kumar Mishra
Department of Electronics and Communication Engineering
Jaypee Institute of Information Technology
Noida, India

Manju Yadav
Department of Electronics and Communication Engineering
Parul University
Gujarat, India

Sarangi Yagnadatt
Department of Electronics and communication
School of Electronics Engineering
Vellore Institute of Technology
Tamil Nadu, India

1 Optimizing Circuit Synthesis

Integrating Neural Networks and Evolutionary Algorithms for Increased Design Efficiency

Swagata Devi, Naushad Manzoor Laskar, Shreyas Nirgude, and Saanvi Srivastava

1.1 INTRODUCTION

The integration of neural networks (NNs) with evolutionary algorithms represents a paradigm shift in analog designs and simulation tools and offers a compelling solution to the challenges in circuit design. This approach not only accelerates the circuit synthesis process but also lays the foundation for future advancements in the field of circuit design automation [1]. Traditionally, designers relied on manual methods to optimize circuit configurations, which proved to be labor-intensive and time-consuming. With the advent of computational techniques, evolutionary algorithms emerged as powerful tools for automating the circuit synthesis process. These algorithms enable designers to explore a vast search space efficiently, identifying optimal circuit configurations based on specified performance criteria [2]. In traditional simulation-driven circuit synthesizers, data generated by previous generation is often disregarded, leading to inefficiencies in the optimization process. In contrast, the power of artificial neural networks (ANNs) within the optimization loop uses data from preceding generations to train NNs, by establishing a valuable mechanism for capturing the underlying relationships between circuit configurations and performance metrics [3]. Once trained, these NNs serve as efficient surrogates for the simulator, enabling rapid estimation of performance parameters for new candidate solutions. This paradigm shifts from simulation-based evaluation to NN-based prediction results in a considerable minimization in execution time, significantly increasing the efficiency of the circuit synthesis process [4,5].

Over the past decade, focused research was carried out to improve artificial intelligence (AI), applying optimization techniques. Abdolrasol et al. [6] used ANNs optimized using various algorithms, including genetic algorithms (GAs), particle swarm

DOI: 10.1201/9781003483038-1

optimization (PSO) method, artificial bee colony (ABC) method, and backtracking search algorithm (BSA) method, alongside advanced techniques like lightning search optimization (LSA) and whale optimization (WOA). ANNs via optimization algorithms are used to tune parameters for optimal structure. The results demonstrate improved ANN performance by the above algorithms in solving energy management problems for systems. Devi et al. [7] optimized parameters for a self-biased amplifier in seizure diagnosis preamplifiers, using metaheuristics and an ANN method, targeting a minimal preamplifier area. The ANN method outperformed other methods, including the efficient whale optimization technique, achieving a 50% smaller preamplifier area. This optimization is vital for developing compact, wearable, or implantable neurodiagnostic devices, tested in a Cadence 180 nm Semi-Conductor Laboratory (SCL) environment. Wang et al. [8] introduced a novel analog circuit optimization method combining a GA-based global optimization engine and an ANN-based local optimization engine, significantly speeding up automated analog circuit sizing by utilizing parallel computation for both global optimization via SPICE simulations and local minimum searches through ANN model predictions, and this system achieves a fourfold increase in optimization speed. It successfully designed complex analog circuits like operational amplifiers and Chebyshev filters, offering results comparable to traditional methods but much faster. In similar studies, Rosa et al. [9] used an ANN-based methodology for the automatic circuit sizing and layout of analog ICs, employing deep learning to learn from optimized design patterns. The methodology features two ANN architectures for direct mapping from specifications to device sizes and includes topology selection, demonstrating a significant efficiency in circuit sizing. While Chakraborty et al. [10] discussed the application of multilayer feed-forward NNs in optimizing CMOS analog circuits, we emphasize on the NN model's ability to replace empirical and numerical modeling methods for enhanced accuracy and computational efficiency. Harsha et al. [11] proposed an ANN model for optimizing the design of 2-stage op-amps, addressing the challenge of multidimensional trade-offs in analog circuit design. The model, validated against SPICE simulations, demonstrates the potential of ANNs to achieve design optimization swiftly and accurately. Budak et al. [12] presented an efficient machine learning-based global optimization method for analog circuit sizing, named ESSAB. This method tackles the challenges of prediction errors and high ML costs in the face of stringent design specifications, through innovations in design ranking and ANN model construction for circuit performance. The integration of soft computing methods in analog IC design automation represents a paradigm shift toward more efficient, accurate, and cost-effective methods. By addressing critical challenges, such as computational expense, prediction accuracy, and multidimensional optimization, these AI-driven approaches hold the promise of significantly advancing the field of electronic design automation (EDA).

Therefore, despite the significant progress in AI-assisted electronic circuit design utilizing methods like Bayesian optimization and GAs, a gap persists in designing implantable circuits with cloud connectivity. This niche demands optimization approaches tailored for the dual challenges of minimizing size while ensuring functionality. The presented research innovatively targets this by optimizing a high-swing self-biased folded cascode amplifier for biomedical applications, using advanced techniques, including metaheuristic algorithms and ANNs. It stands out by conducting both quantitative and qualitative analyses to evaluate optimization strategies, particularly their speed and efficiency in avoiding local optima, thus addressing the gap and enriching methodological approaches for designing constrained implantable electronics.

1.2 OPTIMIZATION OF HIGH-SWING SELF-BIASED FOLDED CASCODE (HSSFC) AMPLIFIER

Figure 1.1 depicts the HSSFC amplifier, an enhanced variant of the traditional folded cascode amplifier, incorporating source-degeneration-based current sources, M_3–$M_{4,}$ rather than conventional ones. This leads to an enhancement in input noise. It is facilitated by using current scale resistors, R_1 and R_2, which reduce the current flow through the folded area of M_7–M_{10}, where just a fraction of differential pair M_1–M_2 current flows. The choice of device sizes, resistance values, and other design parameters ensures high overall transconductance, leading to lower noise. Furthermore, to enhance swing, an additional transistor, M_{13}, is introduced as shown in Figure 1.1 [13]. In the HSSFC, hybrid whale particle swarm optimization (HWPSO) and other bioinspired algorithms are tailored for low-frequency/biomedical applications, which is optimized to minimize the circuit area while satisfying specific constraints as shown in Table 1.1. The objective function for the design is stated in equation (1.1), which is minimized subject to equations (1.2)–(1.11).

$$\text{Minimize CF} = \sum_{k=0}^{12} W_k L_k \tag{1.1}$$

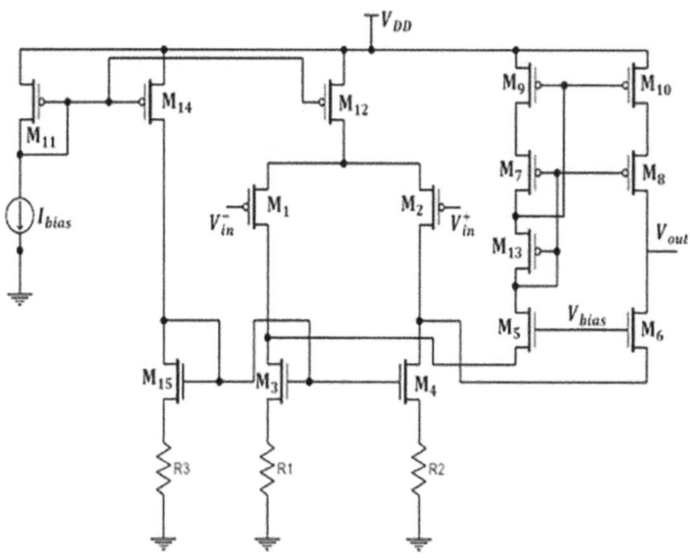

FIGURE 1.1 High-swing self-biased folded cascode (HSSC) amplifier [13].

TABLE 1.1
Specifications for Amplifier Design

Design Criteria	Specifications
Slew rate, SR (V/μs)	≥10
Differential gain, A_v (dB)	>40
Cutoff frequency, f_{-3dB} (kHz)	≤15
$V_{CM(min)}$ (V)	≥−1.6
$V_{CM(max)}$ (V)	≤1.6
CMRR (dB)	>60
PSRR (dB)	>50
Power dissipation, P (μW)	≤7
Input referred noise, S_n (f) (V/√Hz)	≤500n

Subject to

$$I_{12} = (\text{SR})C_L \tag{1.2}$$

$$\left(\frac{W}{L}\right)_4 = \frac{2I_4}{K_n V_{\text{DS4}}{}^2} = \left(\frac{W}{L}\right)_3 \tag{1.3}$$

$$\left(\frac{W}{L}\right)_6 = \frac{2I_6}{K_n V_{\text{DS6}}{}^2} = \left(\frac{W}{L}\right)_5 \tag{1.4}$$

where

$$V_{\text{DS4}a}(\text{sat}) = V_{\text{DS6}}(\text{sat}) = \frac{V_{\text{out(min)}} - |V_{\text{SS}}|}{2}$$

$$\left(\frac{W}{L}\right)_{10} = \frac{2I_{10}}{K_P V_{\text{SD10}}{}^2} \text{ and } \left(\frac{W}{L}\right)_8 = \frac{2I_8}{K_P V_{\text{SD8}}{}^2} \tag{1.5}$$

where

$$V_{\text{SD8}}(\text{sat}) = V_{\text{SD10}}(\text{sat}) = \frac{V_{\text{DD}} - V_{\text{out (max)}}}{2}$$

$$\left(\frac{W}{L}\right)_1 = \left(\frac{W}{L}\right)_2 = \frac{g^2{}_{m1}}{\dfrac{K_p 2I_3}{K+1}} = \frac{f_\mu^2 C_L{}^2}{\dfrac{K_p 2I_3}{K+1}} \tag{1.6}$$

$$\left(\frac{W}{L}\right)_{12} = \frac{2I_{12}}{K_p\left[V_{\text{DD}} - V_{\text{CM(max)}} - \sqrt{\dfrac{2I_3}{K_p\left(\frac{W}{L}\right)_1}} - V_{\text{tp}}\right]^2} \tag{1.7}$$

TABLE 1.2
Technology Constants Used

Specification	Values Used
Supply V_{DD} (V)	1.8
Supply V_{SS} (V)/GND	0
Threshold PMOS, V_{tp} (V)	−0.42
Threshold NMOS, V_{tn} (V)	0.42
Process parameter NMOS, K_n (μA/V²)	355
Process parameter PMOS, K_p (μA/V²)	75
Technology node	180 nm

$$\left(\frac{W}{L}\right)_4 = \frac{2I_4}{K_n\left[V_{CM(min)} - |V_{SS}| + V_{tp}\right]^2} = \left(\frac{W}{L}\right)_3 \tag{1.8}$$

$$A_v = g_{m2}g_{m6}r_{06}\left(r_{02}\|r_{04} \| R2)\|\right) | \left(g_{m8}r_{08}r_{010}\right) \tag{1.9}$$

$$S_n(f) = \frac{8KT}{g_{m1}^2}\left(\frac{g_{m1}}{3} + \frac{1}{R1} + \frac{2g_{m9}}{3} + \frac{2g_{m13}}{3}\right) \tag{1.10}$$

$$P = \left(V_{DD} + |V_{SS}|\right)\left(I_{12} + I_9 + I_{10}\right) \tag{1.11}$$

The transistor channel lengths are specified as 1 μm for all transistors, M_1–M_{13}. This configuration results in a six-dimensional optimization challenge shown in equation (1.12) similar to Ref. [14], aimed at minimizing the circuit area while adhering to the optimal specifications required for low-frequency applications [15,16]. The design constant values used are shown in Table 1.2.

$$X_{\text{self biased}} = \left[\text{SR},\ f_\mu,\ V_{CM(min)}, V_{CM(max)}, V_{out(min)},\ V_{out(max)}\ \right] \tag{1.12}$$

The circuit parameters after employing HWPSO and bioinspired algorithm simulations are shown in Table 1.3. HWPSO proved more consistent and statistically superior to other algorithms in over 20 trials, as evidenced by lower standard deviations and rigorous statistical tests as shown in Table 1.2 and Figure 1.2. Validation in Cadence Virtuoso confirmed HWPSO's effectiveness, achieving a high gain and suitable cutoff frequency for biomedical amplifiers, with a minor trade-off in power dissipation. Simulations closely matched algorithmic predictions, with minimal deviations attributed to unmodeled parasitics and higher-order effects, demonstrating HWPSO's robustness in circuit design optimization over other approaches [17].

TABLE 1.3
Comparison of Design Parameters and Best Results Achieved Through HWPSO

Design Criteria	PSO	HBPSO	DE	DEPSO	DA	WOA	GOA	SSA	HWPSO
Slew rate (V/μs)	10.02	12.07	12.07	12.07	15.7	15.7	10.02	10.02	15.7
UGB (MHz)	8	8.31	8.31	8.31	2.7	2.7	8	8	2.7
Gain (dB)	56.1	55.8	55.8	55.8	40.01	40.01	56.1	56.1	40.01
$V_{CM(min)}$ (V)	–0.608	–0.688	–0.688	–0.688	–1.77	–1.77	–0.608	–0.608	–1.77
$V_{CM(max)}$ (V)	1.36	1.42	1.42	1.42	1.49	1.49	1.36	1.36	1.49
Power dissipation (μW)	4.97	4.97	4.97	4.97	1.76	1.76	4.97	4.97	1.76
CMRR (dB)	70.8	69.7	69.7	69.7	67.8	67.8	70.8	70.8	67.8
PSRR (dB)	58.7	58.7	58.7	58.7	57.2	59.5	58.7	58.7	59.5
Input referred noise ($nV/\sqrt{Hz}$)	179.8	177.9	177.9	177.9	238.02	238.02	179.8	179.8	238.02
Phase margin (degrees)	84.1	85.2	85.2	85.2	89.5	89.5	84.1	84.1	89.5
Cutoff frequency (kHz)	12.1	12.7	12.7	12.7	13.8	13.8	12.1	12.1	13.8
Total area (m^2)	3.04e-10	2.94e-10	2.94e-10	2.94e-10	2.80e-10	2.80e-10	3.04e-10	3.04e-10	**2.80e-10**

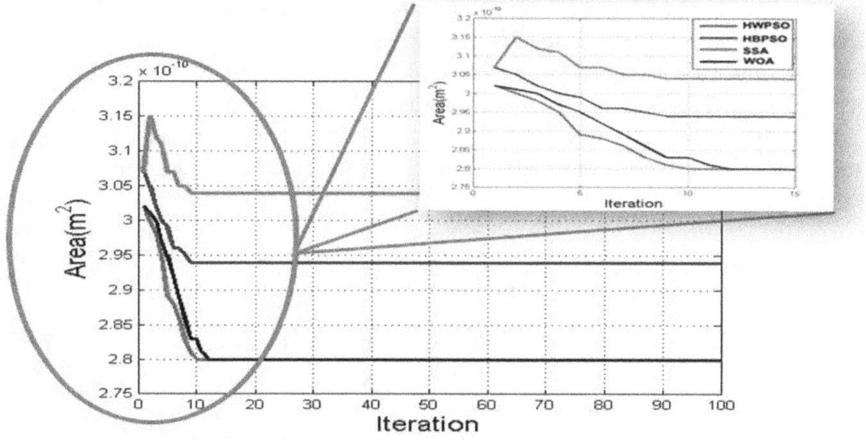

FIGURE 1.2 Convergence plot of optimization algorithms.

1.3 ARTIFICIAL NEURAL NETWORK METHODS (ASSISTED OBJECTIVE ATTAINMENT METHOD)

The methodology for determining optimal circuit parameters integrates the fitting of NNs with the objective attainment method, as illustrated in a comprehensive flowchart. This approach begins with the supervised training of an ANN, utilizing a dataset generated from around 150 approximated simulations within the Cadence 180 nm SCL environment and HWPSO-based best results. This dataset encompasses a range of circuit specifications, including Slew Rate (SR), A_v, and Unity-Gain Bandwidth (UGB), as mentioned in Table 1.1. These parameters are crucial for calculating circuit area and for conducting further simulations to assess performance metrics like gain, phase, noise, and power. A discriminative NN is developed to predict circuit area from six input parameters, offering a swift alternative to extensive numerical simulations. This is particularly advantageous for complex circuit models, where ANNs expedite the calculation process across dense parameter spaces. The ANN employed is a shallow feed-forward-based backward propagation network (FFBN), potentially expanding to a deep network structure for larger datasets. This network uses the sigmoid symmetric transfer function for its activation function, tested with ten nodes in the hidden layer.

Data preparation involves dividing the initial dataset into training, validation, and testing sets, with stochastic partitioning to facilitate diverse training scenarios. The training process adjusts the ANN's weights and biases through iterations until the output closely matches target values derived from Cadence simulations. The fitting quality is assessed using mean squared error (MSE) and regression analysis. On establishing an effective ANN model, the next phase is optimization through the goal attainment method. This method, part of nonlinear programming techniques, like sequential quadratic programming (SQP), aims to minimize the deviation from set design goals, considering various constraints. The methodology's merit lies in its ability to achieve precise optimization results, guiding the design toward minimal circuit area while satisfying specified performance criteria. Figure 1.3 presents a flowchart illustrating the protocol employed for estimating optimal circuit parameters through NN fitting and the goal attainment method.

This study presents a novel approach to reduce circuit synthesis time by integrating ANNs with bioinspired algorithms. Traditional reliance on simulations for

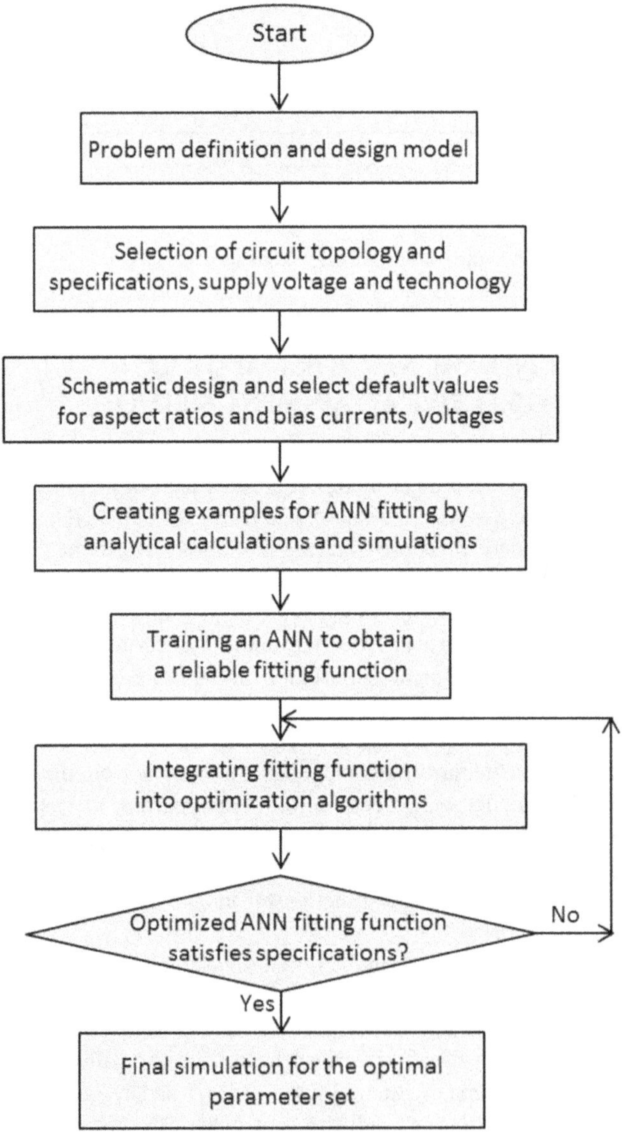

FIGURE 1.3 Flowchart calculating the best set of circuit parameters.

performance evaluation is minimized by using ANNs to predict outcomes based on circuit configurations. The process involves initializing solutions, evaluating them, training ANNs with historical data, and using these ANNs for performance prediction. This results in efficiently exploring the search area, ultimately speeding up the identification of optimal circuit configurations through iterative optimization until convergence criteria are met.

1.4 RESULTS AND DISCUSSION

Simulation results demonstrate the effectiveness of integrating ANNs with bioinspired algorithms in enhancing the efficiency of circuit synthesis by replacing simulations with ANNs for performance prediction, and significant reductions in execution times can be achieved without compromising the quality of the synthesized circuits. Comparative analyses between traditional simulation-based approaches and our proposed methodology highlight the superiority of the latter in terms of computational efficiency and design optimization. The results illustrated are determined in Octave, version 2.0, and MathWorks MATLAB, version R2019a. The ANN employed for relation fitting between the overall area of circuit and the six design criteria was a shallow FFBN shown in Figure 1.4.

The relationship between the total circuit size and the eight associated parameters using an ANN is a shallow FFBN, the topology of which is depicted in Figure 1.4. With only one hidden layer, the suggested ANN is shallow due to the dataset's total of 150 samples. A deep structure — one with multiple hidden layers — might be employed if there are more datasets available [18]. Considering a given number of circuit variables, the i/p's and o/p's, the optimal ANN design can be found by varying the total neuron numbers present in the hidden and output layers, introducing additional hidden layers (and utilizing methods from deep learning [19]), or altering the activation function.

In the described work, the authors utilized a shallow FFBN NN with a sigmoid symmetric transfer function as the activation function. This NN was employed to fit the relationship between the overall circuit area and seven related parameters. The NN architecture is illustrated in Figure 1.4, featuring dual hidden layers. It is noted that with a dataset comprising 150 examples, the chosen architecture is shallow. However, the authors suggest that in scenarios where a greater volume of data is available, a deep NN architecture might be more appropriate. A deep structure

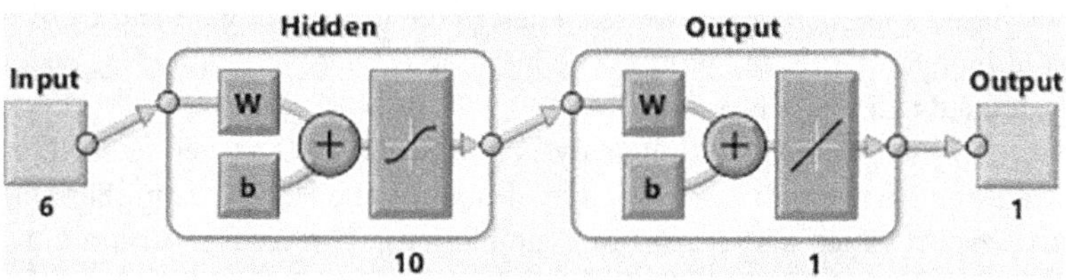

FIGURE 1.4 Artificial neural network (ANN) architecture employed in circuit area fitting.

typically involves several hidden layers. The optimization of the ANN configuration can involve various strategies as follows:

a. **Adding More Hidden Layers**: Deep learning techniques can be employed to create a network with multiple hidden layers, enabling the network to capture more complex patterns in the data.
b. **Adjusting Neurons Present in Hidden and Output Layers:** The total neurons present in each layer can significantly impact the network's capacity to learn and generalize. This parameter can be tuned based on the problem's complexity and the available data.
c. **Changing the Activation Function:** The activation function plays a crucial role in introducing nonlinearity to the network. While the sigmoid symmetric transfer function was used in the described work, alternative activation functions like ReLU (rectified linear unit), Tanh (hyperbolic tangent), or others could be explored to potentially improve the model's performance.

The sigmoid symmetric transfer function was used as activation function in this chapter in equation (1.13):

$$y = \frac{2}{1 + e^{-2x}} \tag{1.13}$$

The parameters employed during training were 1000 for maximum epochs, 0 for goal, 10^{-7} for the minimum gradient of performance, and μ begins with 0.001 and increases on factor of 0.001 and a maximum of 10^{10}. The number of failures allowed during checking validation is six at maximum. The constants and variables for the ANN-assisted goal attainment method are the same as the HWPSO to find the optimized value of the objective function. The ANN fitting functions are obtained by using the Levenberg–Marquardt (LM) algorithm for nonlinear least-squares curve-fitting problems on a ratio of 70:15:15, which implied that out of 150 examples in a dataset, 70% (104) were employed for training, 15% (23) in validating, and rest 15% (23) for the purpose of testing. Using 150×7 matrix as an input and a column of 150 numerical values as an output, the training for 150 examples for feeding the ANN was conducted. The process of ANN fitting function training with ten neurons needed one second to meet training criteria by reaching six validations successfully, and the drop in gradient is shown in Figure 1.5.

The optimized values obtained are shown in Table 1.4. The results indicate that the ANN-based approach demonstrated a design having higher gain and UGB but at the cost of increased area and power. The increase is, however, within the specifications as defined in Table 1.1.

Table 1.5 presents the validation analysis for the HWPSO as well as ANN methods in Cadence Virtuoso. It is achieved by designing the HSSFC amplifier with the respective aspect ratios and performing simulations. The results illustrate a close agreement with both ANN and HWPSO methods.

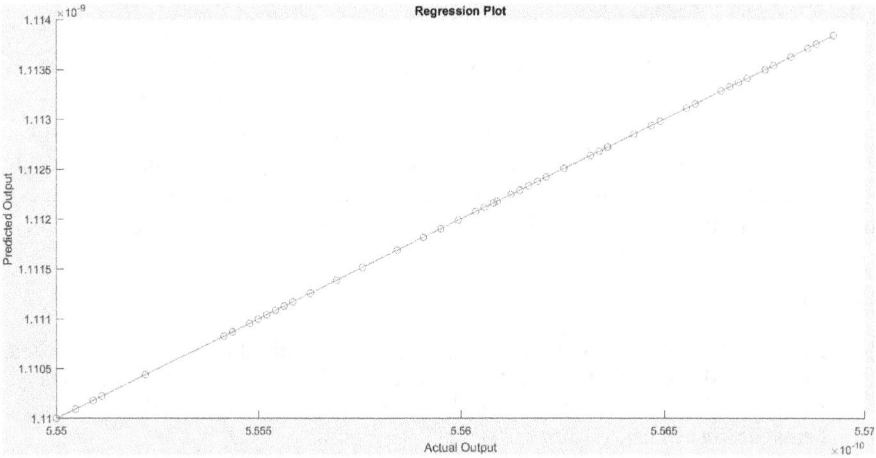

FIGURE 1.5 Data regression for ANN fitting function employed.

TABLE 1.4
Results for Optimization and ANN-Based Approaches

Parameters	ANN-Based Goal Attainment Method [10 Nodes]	HWPSO
Slew rate (V/μs)	13.7374	15.7
Gain (dB)	79.45	40.01
UGB (MHz)	5.783	2.7
ICMR min (V)	−0.2440	−1.77
ICMR max (V)	0.9454	1.49
Power (μW)	4.125	1.76
Area (mm^2)	$5.55e^{-10}$	2.80e-10

TABLE 1.5
Comparison of Cadence Simulation Results for Optimization and ANN-Based FFBP Methods

Approach	Gain (dB)	Power (μW)	Noise (μV)	Phase (degree)	BW (kHz)	Area (μm^2)
HWPSO	40.01	1.76	238.02	89.5	13.8	280
ANN-based goal attainment method [10 nodes]	79.5	4.12	226.69	89.7	10.7	555

1.5 CONCLUSION

To conclude, the integration of ANNs with bioinspired algorithms offers a promising approach to enhance the efficiency of circuit synthesis. By leveraging ANNs as surrogate models for performance prediction, the reliance on time-consuming simulations can be mitigated, thereby accelerating the design process. Future research directions may explore further refinements to the proposed methodology and investigate its applicability to various circuit synthesis problems across different domains.

REFERENCES

1. İslamoğlu G., Çakici T. O., Afacan E., & Dündar G. Artificial neural network assisted analog IC sizing tool. In *16th International Conference on Synthesis, Modeling, Analysis and Simulation Methods and Applications to Circuit Design (SMACD)*, IEEE, Lausanne, Switzerland (2019); pp. 9–12.
2. Laskar N. M., Guha K., Paul P. K., & Baishnab K. L. Random offset minimization in low frequency front-end amplifiers using swarm intelligence-based techniques. *Evolutionary Intelligence* (2021); 14:1317–1335.
3. Joshi D., Dash S., Reddy S., Manigilla R., & Trivedi G. Multi-objective hybrid particle swarm optimization and its application to analog and RF circuit optimization. *Circuits, Systems, and Signal Processing* (2023); 42(8):4443–4469.
4. Laskar N. M., Guha K., Nath S., Baishnab K. L., & Paul P. K. Optimal sizing of recycling folded cascode amplifier for low-frequency applications using new hybrid swarm intelligence-based technique. *Applied Artificial Intelligence* (2020); 34(12):880–897.
5. Afacan E., Lourenço N., Martins R., & Dündar G. Machine learning techniques in analog/RF integrated circuit design, synthesis, layout, and test. *Integration* (2021); 77:113–130.
6. Abdolrasol M. G., Hussain S. S., Ustun T. S., Sarker M. R., Hannan M. A., Mohamed R., et al. Artificial neural networks-based optimization techniques: a review. *Electronics* (2021); 10(21):2689.
7. Devi S., Guha K., Jakšić O., Baishnab K. L., & Jakšić Z. Optimized design of a self-biased amplifier for seizure detection supplied by piezoelectric nanogenerator: metaheuristic algorithms versus ANN-assisted goal attainment method. *Micromachines* (2022); 13(7):1104.
8. Li Y., Wang Y., Li Y., Zhou R., & Lin Z. An artificial neural network assisted optimization system for analog design space exploration. *IEEE Transactions on Computer-Aided Design of Integrated Circuits and Systems* (2019); 39(10):2640–2653.
9. Rosa J. P., Guerra D. J., Horta N. C., Martins F. M., & Lourenço N. C. *Using Artificial Neural Networks for Analog Integrated Circuit Design Automation* (Vol. 1). Berlin: Springer (2020).
10. Chakraborty M. Artificial neural network for performance modeling and optimization of CMOS analog circuits. arXiv preprint arXiv: (2012);1212.0215.
11. Harsha M. V., & Harish B. P. Artificial neural network model for design optimization of 2-stage op-amp. In *2020 24th International Symposium on VLSI Design and Test (VDAT)*, IEEE, Bhubaneswar, India: Indian Institute of Technology (IIT) (2020); 1–5.
12. Budak A. F., Gandara M., Shi W., Pan D. Z., Sun N., & Liu B. An efficient analog circuit sizing method based on machine learning assisted global optimization. *IEEE Transactions on Computer-Aided Design of Integrated Circuits and Systems* (2021); 41(5):1209–1221.

13. Laskar N. M., Guha K., Nath S., Chanda S., Baishnab K. L., Paul P. K., & Rao K. S. Design of high gain, high bandwidth neural amplifier IC considering noise-power trade-off. *Microsystems Technologies* (2021); 27(2):585–599.
14. Laskar N. M., Guha K., Nath S., Baishnab K. L., & Paul P. K. Optimal sizing of recycling folded cascode amplifier for low-frequency applications using new hybrid swarm intelligence-based technique. *Applied Artificial Intelligence* (2020); 34(12):880–897.
15. Mirjalili S., & Lewis A. The whale optimization algorithm. *Advances in Engineering Software* (2016); 95:51–67.
16. Saremi S., Mirjalili S., & Lewis A. Grasshopper optimisation algorithm: theory and application. *Advances in Engineering Software* (2017); 105:30–47.
17. Mirjalili S. Dragonfly algorithm: a new meta-heuristic optimization technique for solving single-objective, discrete, and multi-objective problems. *Neural Computing & Applications* (2016); 27(4):1053–1073.
18. Mohammadzaheri M., Tafreshi R., Khan Z., Ghodsi M., Franchek M., & Grigoriadis K. Modelling of petroleum multiphase flow in electrical submersible pumps with shallow artificial neural networks. *Ships and Offshore Structures* (2020); 15(2):174–183.
19. Hinton G. Deep learning: a technology with the potential to transform health care. *JAMA* (2018); 320(11):1101–1102.

2 Study of Physical Process Analysis and Phenomena of Insights of Trapping in Performance Degradation in AlGaN/GaN HEMTs

Rayabarapu Venkateswarlu, G.P.S.C. Mishra, and B. Acharya

2.1 INTRODUCTION

Wide bandgap materials such as gallium nitride (GaN)-based power devices grown on silicon, sapphire, and silicon carbide are essentially used for large power, microwave, and millimetre (mm)-wave applications. Due to its peculiar properties, it can deliver high breakdown voltages and high switching speed to metal-oxide-semiconductor field-effect transistor (MOSFETs) with low dynamic on-resistance (Dynamic-R_{ON}). The AlGaN/GaN heterojunction consists of piezoelectric and spontaneous polarizations that induce a high electron charge density called 2-dimensional (2D) electron gas (2-DEG). Heterojunction properties of these III-nitride materials form the device GaN-based AlGaN/GaN heterostructure in high electron mobility transistor. Because of excellent properties of the high electron mobility, these are used in advanced mm-wave, radar, avionics, power switching, and power amplification applications [1]. For mm-wave communications, high-power and high-efficiency power amplifiers are needed to meet the demands of high data rate transmission. Fifth-generation mobile communications applications use solid-state devices. Traditional silicon-based devices cannot fulfil the requirement of high power and high frequency. The design of radio frequency (RF) circuits developed on GaN employing metal-organic chemical vapour deposition has recently made significant advancements [2–6]. Silicon nitride is a popular passivation material that has been tried on GaN HEMTs using a variety of production techniques, including PECVD (plasma-enhanced chemical vapour deposition) [7]. The name itself suggests that the devices' speed (on/off ratio) is remarkably fast, ranging from 10^6 to 10^{10}, due to their possession of the 2-DEG. The foremost important characteristic of the high electron mobility devices is its threshold voltage (V_{TH}) and saturated drain current (I_D). The threshold voltage shift occurs due to the negatively charged traps below the gate electrode. The rise in drain current is observed due to pinch-off voltage (V_P) shift, resulting in a decrease in gate

DOI: 10.1201/9781003483038-2

current I_g and I_{gdo} [8]. In addition to that, the hot electron carrier traps will present in epitaxial for long-time stresses, degrading the device DC characteristics and RF performance [9].

2.2 ENERGY STATES IN CONDUCTION BAND DESCRIPTION

Theoretical background of AlGaN/GaN HEMT: HEMT is a type of field effect transistor (FET) which employs heterojunction structures of III–V group elements for the applications with high switching speed and high-power requirements. GaN is a group III-nitride direct bandgap semiconductor that has the hexagonal crystal lattice structure. GaN compound is a very hard material having bilayers of gallium and nitride molecules in wurtzite crystal structure. The bilayers consist of crystal lattice mismatch that causes the spontaneous polarization (P_{SP}) and the piezoelectric polarization (P_{PE}). As the concentration of aluminium (Al) in the strained AlGaN increases, so does the piezoelectric polarization (PPE) and the total polarization-induced electron charge density. The existence of P_{SP} and P_{PE} (spontaneous and piezoelectric polarization) charges is responsible for 2-DEG (2-DEG of the order, i.e., $10^{13}cm^{-2}$) sheet charge carrier densities. The charge density and bandgap discontinuity occur due to the Al mole content. The usage of Al content gives the proper optimization of the devices. These devices work on the modulation doping principle; hence, these devices are also termed as modulation-doped FETs or heterojunction FETs. This is achieved by the transportation of the carriers from donor ions in an adjacent barrier layer of $Al_xGa_{(1-x)}As/Al_xGa_{(1-x)}N$, where x denotes the Al concentration into high purity GaAs/GaN with the electrons and ions being separated by the built-in heterojunction potential [10] (Figures 2.1 and 2.2).

Some advantages of HEMTs are as follows:

i. Wide bandwidth
ii. High linearity
iii. High electron charge motility

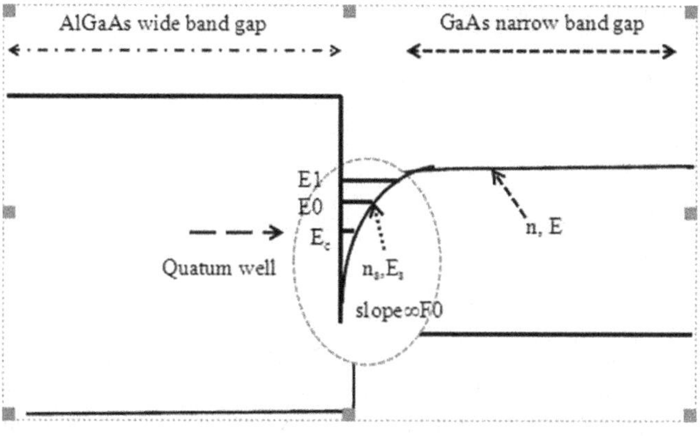

FIGURE 2.1 Energy levels in heterojunction quantum well and its properties [10].

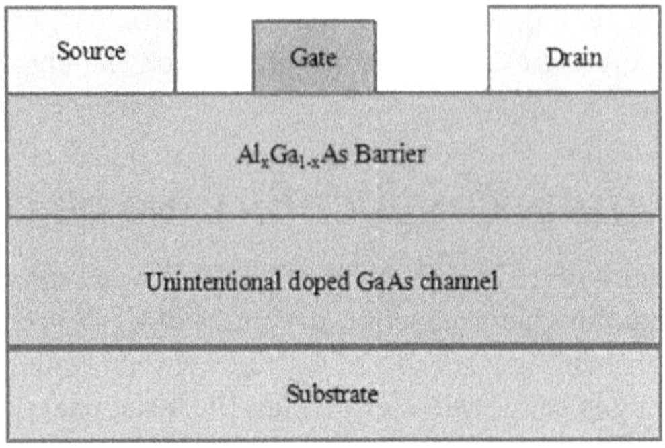

FIGURE 2.2 Cross-sectional view of $Al_xGa_{1-x}As/GaAs$ HEMT where x denotes aluminium-percentage content [10].

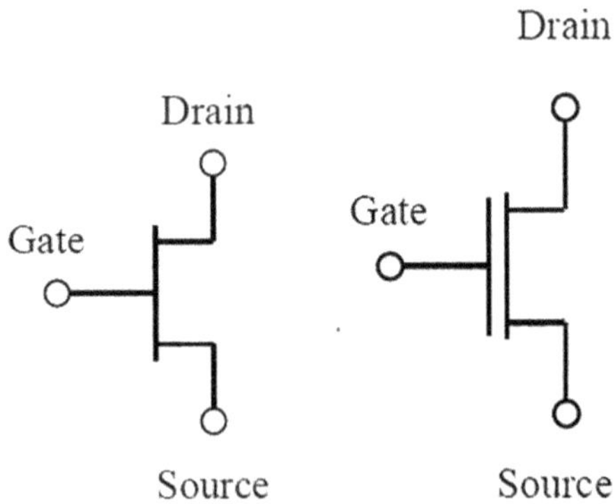

FIGURE 2.3 HEMT device symbols: (a) basic HEMT and (b) MOSHEMT.

iv. High-frequency (switching) performance
v. Low noise
vi. High power charge density
vii. High thermal stabilities

Applications of GaN transistors (Figure 2.3) are as follows:

i. Fifth-generation and mm-wave communications cellular base stations.
ii. Power supplies.
iii. Space and radar surveillance applications.
iv. Power switching.

The HEMT transistor low-frequency model equivalent circuit is shown in Figure 2.4.

Despite these potential characteristics, GaN HEMTs have several challenges that make the transistor unreliable. The safe operation (SOP) of the device under external pressures is a major difficulty in high frequency and power-switching applications due to the device's complexity. Defects are the localized states in the semiconductor

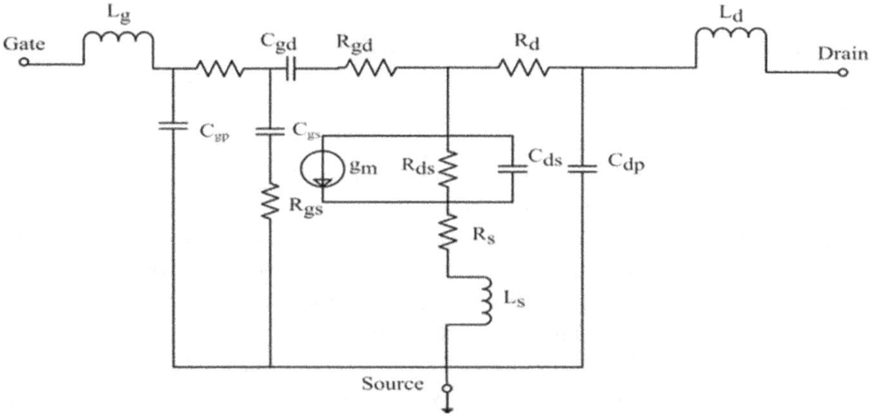

FIGURE 2.4 Electrical equivalent circuit of the HEMT [13].

materials that can capture and release the charge carriers (electrons and holes). Traps can affect the device performance by influencing the carrier mobility, charge distribution, and other electrical characteristics. When the traps are captured or released, the current fluctuates, which causes the devices to behave nonlinearly and with excessive noise. The speed and high-power performance of the device may be affected by the traps close to the channel because they can disperse charge carriers and deplete their mobility. Consequently, the transconductance, G_m, may be reduced, and hence, rise in Dynamic-R_{ON}. Traps are localized states that can emit or capture the electrons/holes by the process called trapping/de-trapping. The possible traps are surface traps present at the passivation and barrier interface, the traps present in the barrier layer, buffer traps present in the buffer layer, and interface traps involved at the AlGaN/GaN interface. Accumulation of holes at the passivation/AlGaN causes surface traps. Buffer traps/deep-level traps are mainly responsible for the DC–RF performance degradation [11]. Temperature variation results in traps because of threshold voltage shift [12]. Short channel effects (SCEs) are also responsible for GaN HEMT high-frequency performance degradation. Possible techniques to reduce the traps are by dielectric passivation, gate region insertion into the barrier called gate recess techniques, proper surface fabrication preparation, and the growth of doping profile in epitaxial layer.

Various drain current and threshold voltage-induced traps in AlGaN/GaN HEMT are studied and presented. Different measurement techniques like the pulsed-RF excitation, drain-lag, pulsed I_{DS}–V_{DS}, gate-lag, transient response, and frequency dispersion are used.

2.3 THE CHALLENGING PARAMETERS

The challenging parameters that affect the device characteristics include:

Capturing and releasing of carriers: When the trapped charge carriers are released, they can contribute to current fluctuations, prone to noise and nonideal behaviour of the device. There was an increase in GaN channel 2-DEG concentration, when the trapped electrons emitted. Traps can capture charge carriers (electrons in n-HEMTs) as they move through the 2D electron gas channel, altering the overall mobility and current flow. The increase in the device's internal temperature leads to a rise of gate leakage current and an increase in the voltage drop across the gate terminal. The trapping effect is primarily due to the presence of the high field in the drain terminal at the gate edge side between gate and drain.

Carrier mobility: Traps can restrict the charge carrier's ability to move, which lowers their mobility. Carriers scatter as they encounter the traps, which results in a less effective flow of current. The speed, general effectiveness, and amplification capabilities of the device are impacted by the traps.

Threshold voltage shift (V_{TH} variation): The threshold voltage determines the state at which the device is switched to on state. When a device switching takes place from off to on and vice-versa, it can be altered because of the traps. The large gate negative bias at the gate edge causes a threshold voltage shift because it leaves more electrons below the gate terminal, which causes surface donor traps to generate virtual gates. The virtual gate formation is due to decaying of surface net positive charge. This has happened due to electron traps present at the surface donor-like states [14]. These fluctuations can result in variations in device behaviour and performance, making it challenging to control and predict its operational characteristics.

Degradation and ageing effect: As charges formed inside the transistor in the form of traps, the performance characteristics of the device may gradually change over the time. As a result, there is a decrease in the device performance that includes decrease in gain and a shorter operational life span of the device.

Noise and instabilities: Traps bring unpredictability into the capture and release of charge carriers, which causes noise and instability. The electrical behaviour of the device fluctuates because of the randomness, which adds to noise, worse signal quality, and unstable output.

2.4 TRAPPING MECHANISMS

Different types of traps are as follows:

2.4.1 Gate Leakage Traps

Traps located close to the gate-channel interface can promote the generation of undesirable leakage currents. Power usage and energy efficiency were impacted. According to the charge neutrality condition, in the absence of external electric field, the field magnitudes thus formed have equal polarization charges. As the charges of opposite signs accumulate and the fixed positive charge polarization increases, the ionization of acceptor traps also rises, leading to a decrease in electron concentration [15]. The threshold voltage (shift) variation is due to strongly influencing the field under the gate, mainly whenever the devices applied to bias, i.e., quiescent during off-state [8,16]. The gate leakage current dominates mainly when the voltage applied to the Schottky gate electrode exceeds the critical field value [17]. The essential phenomenological models used to improve the reduction in the gate leakage is by gate electrons trapped into barrier surface [9,17], lattice mismatch, defects in AlGaN barrier [9], and inverse piezoelectric polarization effect in the barrier layer [18]. The combination of better quality SiO_2/AlGaN and Si_3N_4/AlGaN interfaces decreases the gate leakage current to few orders in magnitude below those of normal frequently used high electron mobility transistor devices, good linearity characteristics, and high saturation currents. Double layers also allow for a large positive threshold voltage shift. Incorporation of indium (In) surfactants and small concentration of carbon (C) improves the surface and interface roughness [19]. If deep acceptors are introduced by plasma treatment into the AlGaN barrier layer's surface, they compensate for the high-density acceptable charge, and then gate depletion layer thickness increases. Due to electron tunnelling, the gate leakage current decreases [20,21].

2.4.2 Interface Traps

Usually, interface traps are located between semiconductor and the insulating material or passivation layer. These traps arise due to defects introduced during device fabrication or due to inherent nature of the interface. The drain current variation trends due to concentration variations of the ionized acceptor-like interface traps. Interface traps can affect the charge transport characteristics across the interface and alter the characteristics of the devices [15]. In comparison with metal-oxide semiconductor devices, the interconnection between the dielectric material and the semiconductor at the channel, the buried-channel group III-nitride metal–insulator–semiconductor heterostructure consists of two interfaces one at the passivation/group III-nitride and the other at the barrier to channel layer interface. This complicates the extraction of interface traps in III-nitride heterostructures [22] (Figure 2.5).

Trapping and de-trapping effects are the defects that present in the device due to a mismatch in lattice structures and current transportation process issues. The traps are either a donor trap or an acceptor trap. Traps are formed if one or more electrons capture or emit to or from the conduction band (CB) and holes to or from the valence band (VB). For instance, a trapping is empty, and those that are negatively charged get ionized when one or more energy levels are occupied by the electrons. Similar to this, a donor trap is positively charged and ionized if one or more energy levels are vacant, and

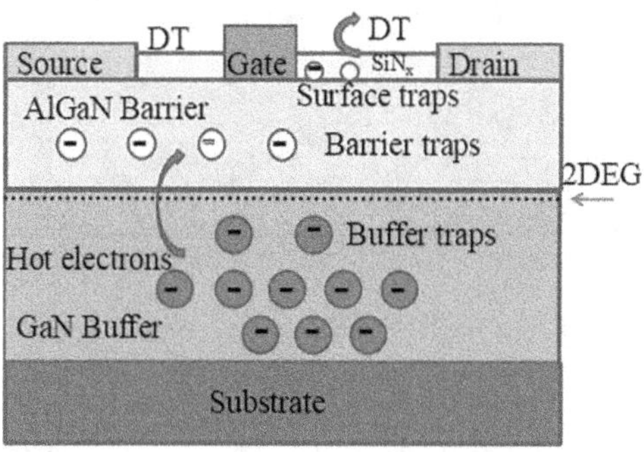

FIGURE 2.5 Various trapping phenomena of the GaN HEMT devices at different drain biasing conditions. DT is referred to as direct tunnelling [23].

all the levels have been filled with electrons. By the recombination process, the emission energy released is due to radiative recombination (photon), nonradiative recombination (phonon), or auger recombination. If a carrier captured in a confined state can combine with a carrier of the opposing sign, traps will be created if the opposite sign carrier is also trapped locally. Trapping and de-trapping phenomenon degrades the transistor performance, and hence, the reliability of the GaN/AlGaN transistors. The interfacial traps may present at the passivation/AlGaN interface, AlGaN/GaN interface, and buffer/GaN interface [23]. Different trap measuring methods are (i) transient drain-lag and gate-lag, (ii) pulsed I_{DS}–V_{DS} technique, (iii) frequency dispersion, and (iv) pulsed-RF excitation. Reduction in current will take place in GaN devices before and after the execution of high drain bias, which is termed as current collapse (CC), current slump, or current dispersion [24]. The initial CC is more in highly doped buffer layer, but a faster recovery process is observed due to emission of larger proportion of trapped electrons. On set, a slower recovery process depends on the lower content of ferrous concentration. Negative effects of traps in the buffer can be minimized by the iron (Fe)-buffer doping profile. The shift in threshold voltage or drift (ΔV_{TH}) is due to which traps fall under the gate, also affecting the I_{ds}–V_{DS} saturation curve. Given that type T3 traps are in proximity with the 2D electron gas, it is anticipated that they will primarily affect the DC current–voltage (*I*–*V*) characteristics [25]. In semi-on state conditions, the hot electron trappings accelerate the traps present at the passivation/barrier interface and the AlGaN high-resistivity buffer layer. This is the strong reason why the 2-DEG gets depleted at the barrier/channel layer (Figure 2.6).

The probability of an electron hopping toward a drain is when a strong electric field is present, but it reaches a threshold value, indicating that it becomes saturated. Under both conditions for gate-lag and drain-lag measurements, the 2D electron gas depletion eventually increases the Dynamic-R_{ON} that is provoked by tunnelled gate electrons caught in surface traps [26]. The extrinsic electron traps or deep-level traps in the buffer layer are acceptor traps (when electrons capture) and donor traps (when electrons emit). The intrinsic surface states induce carrier's capture/release of electrons that further increase the CC. These phenomena are usually originated from the buffer-doping profile, dislocations, internal crystal defects, and GaN process technology.

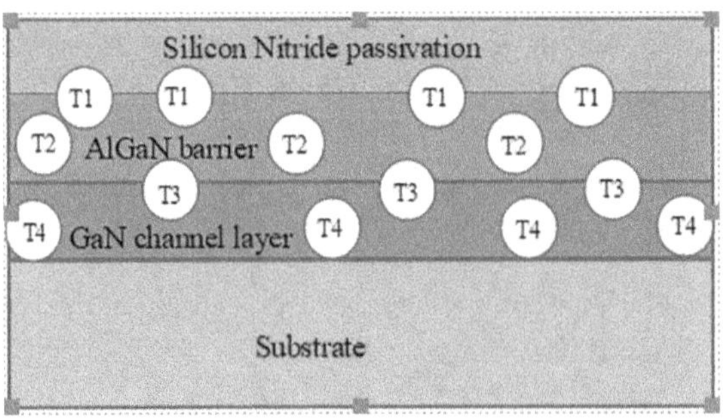

FIGURE 2.6 HEMT cross-sectional view and trap locations: (I) surface traps type T1, (II) AlGaN barrier traps type T2, (III) AlGaN/GaN interface type T3 traps, and (IV) GaN buffer type T4 traps [25].

2.4.3 Bulk Traps

Bulk traps are found within the semiconductor materials. These may result from crystal defects, impurities, and other imperfections. The AlGaN/GaN layer lattice mismatch causes significant spontaneous and strain-induced piezoelectric polarization fields, which further raises the density of 2-DEG. An increase in the Al concentration percentage in the barrier layer causes the discontinuities in conduction band, and the doping density interface is most responsible for the high charge carrier density. As the temperature increases, a drop in drain current is observed because of the AlGaN/GaN lattice mismatch at the interface layer. But the threshold voltage varies with the doping concentration, thickness of the barrier layer, and mole concentration in AlGaN barrier. The threshold voltage (V_{TH}) is increased to a more negative direction with the increase in buffer doping concentration and AlGaN thickness. For actual operation of the device, pinch-off was essentially governed by the gate leakage currents. The predominant off-state leakage current is due to vertical tunnelling of electrons from the gate metal to semiconductor. For the better pinch-off characteristics, low Al concentration and high Schottky barrier height are preferred. High-power microwave devices can be optimized with high Al content by achieving low trap centres, ohmic contacts with low resistance, and good quality crystals [27].

Under high drain bias conditions (V_DS > 0), electrons are generated through side leakage and pass through the surface states of the active region, eventually leading to dislocation. Oxidizing localized dislocations greatly influence the CC [28]. In an off-state condition, the vertical and lateral field distribution in the buffer is more complicated, and dispersed traps were identified. The 2D electron gas progressively depleted while rising in v_{ds}. Under the drain field plate (FP), a strong vertical electric field is produced [29,30]. The method by which electrons move from the barrier to the passivation causes a strong band tilt when it is depleted [31]. The drain saturation current maintains at a minimum value for a duration of time until the traps start to release the caught electrons due to the traps established in the AlGaN layer that collect charges from the channel [32]. The strain in the inter layers and additional operational strain can be removed by choosing the lattice matched barrier layer.

2.4.4 Deep Level Traps

The deep levels are the energy states that an electron was given to conduction band, or it can accept from it. The traps present in the bandgap of the semiconductor with different energy states are separated by VB and CB. These traps can capture and release carriers with relatively high activation energies adversely affecting carrier lifetime and recombination dynamics. Buffer is a semi-insulating material used to reduce hot electron injection traps in deep levels. Buffers doped with Fe have small CC. The buffer layer is used to insulate the device and carriers channel confinement. To confine the channel carriers, the GaN buffer layer must be incorporated sufficiently to have high deep acceptors. To maximize the RF power performance of the transistor, a good buffer engineering compensation technique is needed to keep the position at the Fermi level close to conduction band in the upper half of the energy band [33]. Intrinsic or extrinsic deep donor or acceptor grown defects are the two types of doping methods that are employed, resulting in the Fermi level being stuck in the top region of the bandgap [34]. The Fe-doping concentration influences the deep levels in buffer [35]. Trap levels are examined by increasing the Fe-doping in the buffer. It is found that CC significantly depends on trap locations in the buffer. Fe doping causes an increase in the concentration of native GaN defects [36]. In RF operation conditions, GaN HEMTs are subjected to the most challenging conditions, including high-current charge density, strong electric field, and channel temperatures. RF operation impels clinker modes that did not observe during thermal and DC tests [37]. A high-resistive semi-insulating buffer enables high output power, improves breakdown voltages, and minimizes SCEs and buffer leakage [38]. Utilizing intrinsic growth defects to create deep donors and acceptors allows for buffer isolation [39]. High gate leakage current was observed for low-doped C (Low-C) compared to stepped-doped C and high-doped C (stepped-C and high-C) devices, hence, increase in AlGaN/GaN interface energy band slope. Electron tunnelling is responsible for gate leakage. Due to high-C, RF performance of the devices severely degraded, lowering output power and drain efficiency [24]. A nominal C-doping trap level is to be chosen [14]. A high-resistive GaN can be achieved by introducing a deep-level acceptor trap of Fe or C. Carbon doping is more dominant over Fe because it can be organized in step fashion great control over impurities in the channel [39,40]. High-C concentration enables high breakdown voltage but also forms floating p-n junction with 2D electron gas (between 2-DEG and buffer), leading to more CC [34,41,42]. At large electrical fields, GaN HEMT microwave output power is limited by the localized states in C-doped buffers. C concentration enables the traps by forming dislocations and trappings at the GaN/aluminium nitride (AlN) nucleation interface layer. By focussing more on the interaction of Carbon, the dislocations can be minimized for GaN HEMTs used in microwave applications. Dislocation densities can be minimized by using semi-insulating GaN substrates, and the C concentration can be controlled by the addition of C-carrying gas, rather changing growth parameters [43] (Figure 2.7).

To achieve high breakdown voltage without affecting its RF performance is to minimize acceptor trapping concentration. Leakage current is due to the following: (i) if the concentration of Si doping is higher than or comparable to that of acceptor

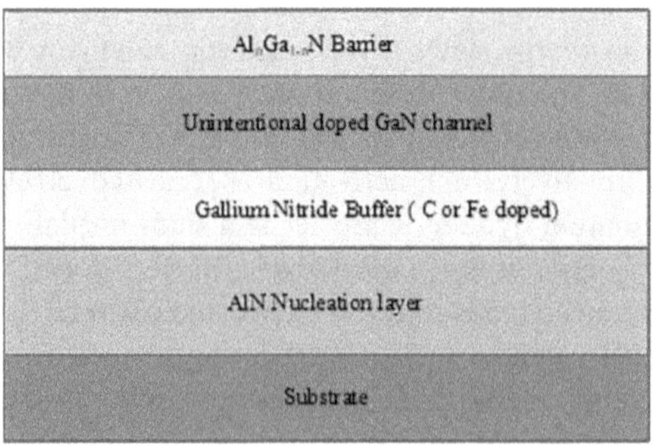

FIGURE 2.7 HEMT structure (a) with thick buffer layer (b) with thin channel layer, where n-defines aluminium concentration [43].

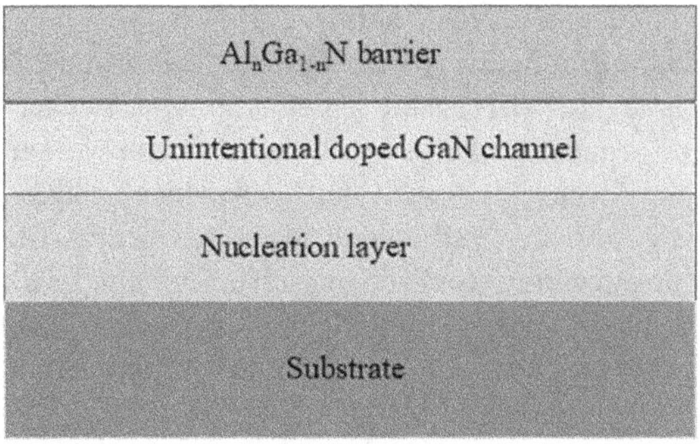

FIGURE 2.8 Graded-AlGaN buffer layer AlGaN/GaN HEMT [49].

doping and (ii) if acceptor concentration is high as that of Si doping. In the first case, leakage current is because of the Si-doping-induced excess electrons. In the second case, leakage current is because of the holes induced by acceptor traps. The holes-induced drain electrode is going to function in a strong electric field combined with the holes induced by acceptor doping that cumulatively increases leakage current [44]. High-frequency HEMT without buffer is grown on an epitaxial AlN nucleation layer [45] (Figure 2.8).

A linear graded negative polarization charges with Al concentration in AlGaN layer has been engineered from 25% to 16%. By introducing negative charges uniformly with a charge of −0.2505 C/cm^3 and a GaN/AlGaN interface polarization charge of −1.33 μC/cm^2, background ionized donors are compensated, which are positively charged, causing the CB to stretch away from the Fermi energy level [46]. But deep-level traps suppress the leakage current in buffer and SCEs, and these unwanted buffer traps/deep-level traps would be minimized by exclusion of thick buffer.

2.4.5 Surface Traps

These defects are present on the surface of the semiconductors. These traps can lead to charge carrier recombination and scattering, affecting the speed and efficiency of the devices [47]. CC happens when a GaN HEMT device is overdriven because of the surface negative charge present between the gate and drain [48]. This surface traps behave as though there is a virtual gate in addition to the intrinsic gate by extending the depletion region at the access region [11,12,48]. The virtual gate effect is caused by hot electrons from a channel, or electrons pushed from a gate surface trapping [47,49]. Hot electrons induce lattice defects, or traps will degrade the reliability of GaN transistors. The electrons generated due to high drain bias (hot electrons) enhance the trapping effects and cause to form new trapping/de-trapping process since these trapped charges influence the electric field (hot electron equivalent temperature), leading to affect the static and RF performance of AlGaN/GaN HEMTs. Electroluminescence (EL) measurement at various gate-bias values and drain bias reveals the effects of hot electrons [8]. Hot electrons limit the output power delivered at RF frequencies and may eventually lead to the phenomenon called negative differential output conductance (NDC) at low drain bias. The threshold voltage positive shift and decreased drain current were observed in on-state. In off-state breakdown, the electron-hole pairs (EHPs) are generated due to impact ionization. In the holes flowing toward negatively charged gate, few of them were trapped in the surface passivation layer. Because of discontinuity at valence band in the SiN*x*/AlGaN/GaN interface, transfer of holes to silicon nitride consequently rises in drain current and results in more negative threshold voltage shift. The high field is concentrated at gate edge of the drain side when the drain bias is larger than the pinch-off voltage. This strong electric field produces hot electrons due to high potential, and these were extended into barrier layer captured by AlGaN surface traps. With the increase of drain stress voltage and pulse time, the virtual gate gradually extends and degrades in drain current. This CC is eliminated by the following: (a) surface passivation may be used to reduce AlGaN surface trap density; (b) adding donor doping at AlGaN surface to compensate surface traps; and (c) the high electric field present at drain side can be suppressed by using different types of FP configurations. The FPs may be source FP, gate FP, double gate FP, drain FP, and extension of these FPs toward the access region. Unfortunately, the FPs incorporate additional internal capacitances, affecting the RF performance of the devices. Surface traps are the dominating factor in producing CC, and the upper channel gate to drain region has many mobile electrons to screen lower channel potential fluctuations.

2.4.6 Dislocation Traps

The localized energy states that function as traps can be produced by crystal defects and dislocations in the material lattice structure. An increase in Al doping between the discontinuities in conduction band and doping densities interface is most responsible for the high charge carrier density. As the temperature increases, there is a drop in drain current observed due to lattice mismatch at the AlGaN/GaN interface layer. But the threshold voltage varies with the thickness of the barrier layer and Al-mole concentration in the AlGaN barrier. For actual operation of the device, pinch-off characteristics are mainly governed by the gate leakage currents. The vertical tunnelling of electrons from the gate metal to semiconductor causes predominant off-state gate leakage current. For the better pinch-off characteristics, low Al concentration and high Schottky barrier height are preferred. High power microwave devices can be optimized with high Al content by achieving low trap centres, ohmic contacts with low resistance, and good quality crystals [27]. Multiple FPs were implemented to further improve the robustness and increased power density. The FP integrated with the gate electrode is used for reducing the electron traps and gate resistance. The FPs integrated with drain terminal are used for tailoring the high electric field and for attaining high breakdown voltages [29,30].

2.4.7 Radiation-Induced Traps

Cosmic rays and X-rays can introduce defects within the semiconductor material, which can lead to traps. The devices degrade rapidly under DC operating conditions than when there is no DC bias [48]. Neutron radiation introduces defects, which

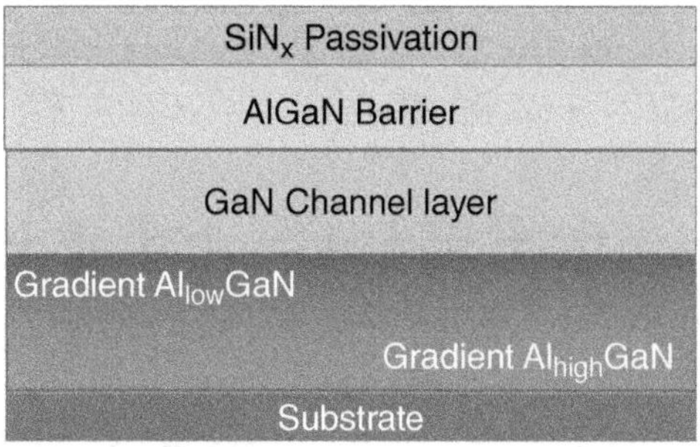

FIGURE 2.9 Spontaneous (P_{sp}) and piezoelectric (P_{pe}) charges of HEMT [51].

can decrease the 2D electron gas and mobility at the semiconductor/semiconductor, semiconductor/insulator, spacer, and buffer layers [50]. The space charge region is altered in the AlGaN layer by the introduction of acceptor-like charged defects. The deep-level defects below the Fermi level are negatively charged. This makes band tilting in the high electron mobility transistor devices and bands goes up with respect to the Fermi level and hence results in reduced 2D electron gas density, and this reduces drain current [51] (Figure 2.9).

An increase in gate current may result from traps acting as tunnelling and are located within the bandgap and close to the AlGaN/metal interface. The thermionic emission (TE) and electron-hole generation-recombination (G-R) process have negligible impact on gate leakage current. A consistent distribution of point defects like vacancies, interstitials, and Frenkel pairs as well as certain deeper damage cascades are produced throughout the epitaxial layers by irradiation with 2 MeV protons [52]. In particular, when proton irradiation of 1.8 MeV with fluences up to 10^{14}cm^{-2} is used under the gate region, the traps result in both permanent v_{th}-shifts and dynamic v_{th}-dispersion that can have a considerable impact on the properties of the device. The balanced trapping/de-trapping rates during off-state conditions determine the overall increase in Dynamic-R_{ON}. The devices prove a substantial reduction in Dynamic-R_{ON} at higher proton irradiation fluences higher than 10^{13}cm^{-2}. The Dynamic-R_{ON} is completely suppressed at a drain bias voltage (V_{DS}) of 600 V/150 °C at highest fluences (1.5×10^{14}cm^{-2}) [53,54]. Physical defects incorporated during the fabrication process result in surface donor traps. The intentional Fe-doped (i-Fe) deep traps are used to provide deep acceptors to capture free electrons and to provide semi-insulating behaviour to the buffer layer [54] (Figure 2.10).

HEMT space charges $\pm\sigma_{\text{polarization}}$, $\sigma_{\text{2-D electron gas}}$, σ_{holes} and $\sigma_{\text{donor trap ionization}}$ [55].

where

$\pm\,\sigma_{\text{polarization}}$ is polarization-induced dipole charges.
$\sigma_{\text{2D-electron gas}}$ is charge due to the 2D electron gas.
σ_{holes} is charge due to holes accumulation at the AlGaN surface.
$\Sigma_{\text{donor trap ionization}}$ is the charge due to donor trap ionization.

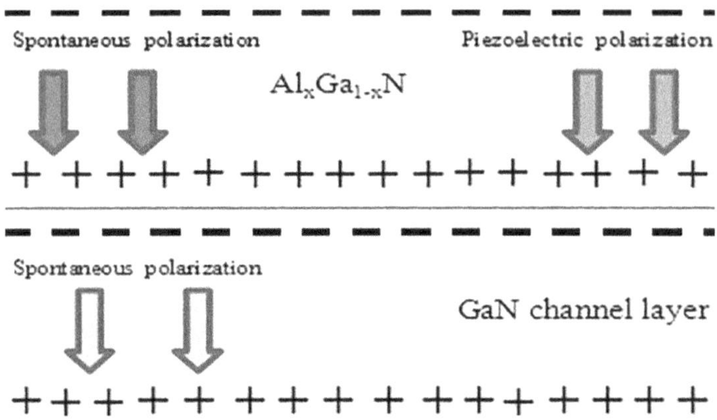

FIGURE 2.10 Possible interface and charge-induced polarization charges in AlGaN/GaN system.

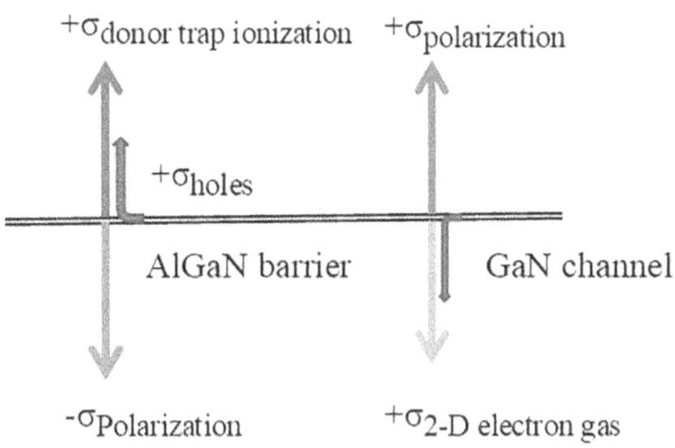

FIGURE 2.11 Various trapping mechanisms in the AlGaN/GaN system [56].

A large drain current is likely to result in the kink effect, the trapping/de-trapping of electrons captured by deep traps that are present in the buffer at the interface. The kink effect is due to the traps situating between drain edges of the gate electrode in the GaN buffer layer. The kink effect is generated by hot electron trappings of donor-like ionized deep level traps in the buffer layer [56]. Several traps are located above the channel, in the AlGaN barrier or the surface, and in the buffer. Buffer traps are temperature independent, but AlGaN traps are thermally activated [57]. The presence of passivation effectively eliminates the condition of a significant transconductance decrement. The high energetic electrons cause an impact on ionization. The hot electron traps have been characterized by examining the gate current due to huge gathering of holes by the impact of the ionization phenomenon [8,57]. The electron charge sheet density decreases in the channel because of the trapped electrons injected into the buffer layer. A current reduction takes place because of the slow response of deep level traps in the semi-insulating buffer (Figure 2.11).

Different types of interface charges are as follows: (a) positive polarization charges, (b) negative ionized acceptor-like traps, (c) negatively charged electrons. The group III-nitride metal–insulator–semiconductor heterostructures wide bandgap properties lead to moderately deep interface traps with lengthy emission time constant (τ_e), which restrict the measurable energy depth at room temperature.

2.5 MEASUREMENT TECHNIQUES OF TRAPS

Different trap measurement methods are listed below:

i. Pulsed *I–V* measurement.
ii. Drain-lag and gate-lag measurement.
iii. The deep-level transient spectroscopy (DLTS) measurement.
iv. The capacitance–voltage (*C–V*) measurement.

2.5.1 Pulsed Current–Voltage Measurement

This method characterizes the drain current by applying the quiescent bias conditions (v_{gsq}, v_{dsq}). By measuring the rate of change of the drain current, it can be accomplished to determine the rate of the electron-emitting charge (releasing) and capturing time, e and C_{tun} (electron tunnelling). The temperature and the trap level's position with respect to the CB's bottom influence the emission probability. The triangular-shaped AlGaN barrier layer is responsible for large gate electron trap tunnelling, whereas the trap-assisted tunnelling depends on the surface trap location, the energy levels of traps, and the field applied. Multiple forms rely on the measurement of the functional temperature dependency of the release and capture rates of transient spectroscopy, from which the trap activation energy and its capture cross-section can be determined [56,58,59].

Pulsed measurement eliminates device self-heating that is used to distinguish between the trapping and thermal effects. Higher V_{DS} under pulse bias avoids the device to get permanent degrade condition.

2.5.2 Gate-Lag and Drain-Lag Measurement

Gate-lag measurement is the study of the gate voltage variation produced by delayed drain current characteristics of the device [8]. Gate lag is used to analyse the transient drain current consequences. A delay in the channel current's response to a change in the gate potential is a result of charge trapping in the device's structure. Gate-lag studies give vital details on the traps that are necessary for identifying the active flaws. Gate lag is measured by a gate pulse of short duration that is applied from pinch-off to on-state with drain voltage at fixed voltage [9]. In gate-lag measurement technique, a pulse of a certain period with a duration is applied at the gate electrode by keeping V_{DS} constant (fixed V_{DS}). Similarly, a drain lag is a measurement technique where a pulse is applied to the drain electrode, keeping the gate bias V_{GS} at a fixed value. Gate lag is defined as the rate of change in drain current that produces a current delay. Because the hole accumulation along the surface cancels, the traps are responsible for the gate lag, and furthermore, the device is unaffected by acceptor traps [60]. The density of

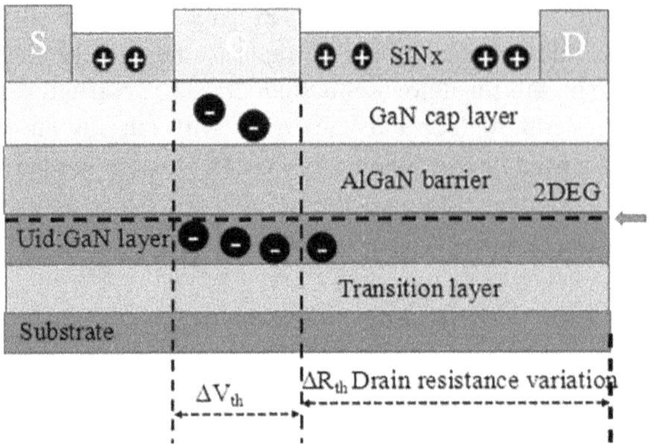

FIGURE 2.12 Circuit diagram of transient V_{DS} to get drain response [55].

2D electron gas can be altered by surface donor-like traps that are filled and removed. Gate lag happens since the surface trapping affect acts as an electron trap. The dipole charges originate from the polarization-induced charges, and 2D electron gas charges are present in the space charge components, charges due to donor-like trap ionization [61]. The charge is driven on by a collection of holes close to the AlGaN surface. The electrons that immediately tunnel into the Schottky gate contact when a new device undergoes stress can occupy the donor-like traps and make them neutral; when the total number of donor-like traps decreases; 2D electron gas density decreases as well, leading to a decrease in initial drain current (Figure 2.12).

The Poole–Frenkel (PF) effect emission model is used to analyse the field-dependent reverse gate leakage characteristics. Photon-assisted tunnelling (PAT) and direct tunnelling (DT) emission are a slow process, owing to activation emission process [60–63]. By keeping the gate bias fixed and the channel open, the turn-on pulsating mode of the drain current tends to drop with time, and a decreased drain current is measured in the turn-off pulsating mode after the drain voltage returns to 0 V [14]. The surface traps leading to drastic decrease in drain current in turn affect the microwave characteristics, which lowers output power [62]. A small number of the electrons may tunnel through the barrier during the on-state and get captured by the deep-level trap. The electron traps in the channel causes the drain current reduction. Owing to the filling process enabled by hot electrons, a portion of the channel's electrons is drained into the buffer layer when a high drain voltage is applied, making it easier for the deep-level traps to capture electrons [63,64]. The trapping process is achieved by empty acceptor traps that have the ability to emit holes or capture electrons, and the de-trapping process is achieved by filled acceptor traps that have the ability to emit electrons or capture holes [65]. The interface charges produce polarization that is responsible for hot electron buffer traps [66]. By the increase in drain bias, the hot electrons (heated up) in the channel are injected into the buffer due to large gate edge electric field. Drain lag is a decay in drain current with slow current transient response when V_{DS} is applied. If the voltage within the pulse is higher when compared with the quiescent bias point, traps capture free electrons and traps release their charges when this voltage is lower than the quiescent bias (Figure 2.13).

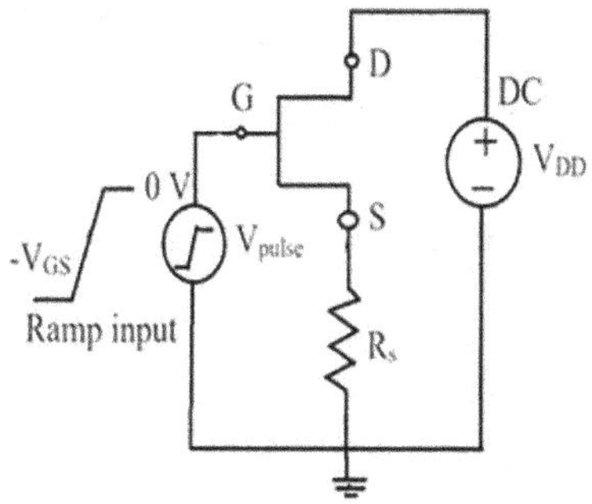

FIGURE 2.13 Transient current measurement of AlGaN/GaN HEMT [55].

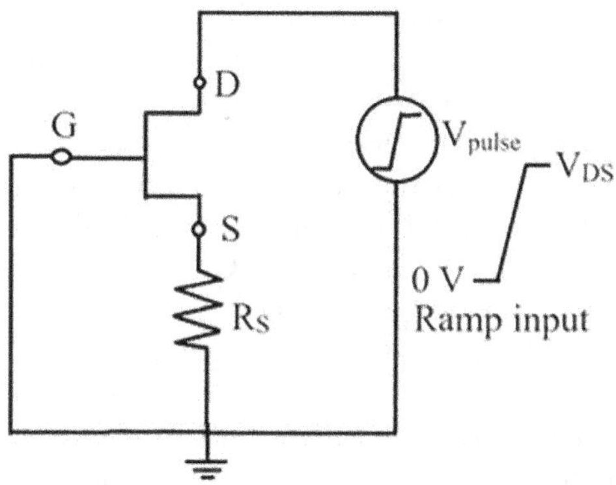

FIGURE 2.14 Drain current variation in device (a) turn-on state and (b) turn-off state [62].

The surface traps leading to drastic decrease in drain current in turn affect the microwave characteristics, which lowers the output power of the device. The DLTS measurement: the defects are most likely caused by a lattice mismatch parameter and strain-induced co-efficiency between different GaN epitaxial layers and the substrate when GaN is grown epitaxial. These defects may tend to propagate into the AlGaN layer [67]. To analyse the origin of these defects, the DLTS technique is used. A constant gate voltage with a pulse width is applied to measure the drain current and v_{ds} [68]. Current-DLTS measurement is based on the channel temperature-varied transient drain current response. Trap activation energy and cross-sectional information are precisely achieved by using this technique. Current-DLTS measurement is used to characterize the traps under semi-on stress state. When the device is stressed from the state-off to semi-on stress state, the generated hot electrons in the conducting channel being forced to buffer (AlGaN) get trapped by deep traps present in buffer. Capacitance-DLTS measurement is used to analyse the sudden change in voltage produced by transient gate capacitance. This method provides the information of variations of trap properties under the gate terminal. This method does not provide the precise measurements for the small gate area devices [8] (Figure 2.14).

TABLE 2.1
Semiconductor materials and its properties

Materials/ Properties	Mobility (μ)	Permittivity (ε)	Energy Band gap (E_g)	Figure-of Merit (FOM)	Temperature (°C)
Silicon	1300	11.4	1.1	1.0	300
Gallium Arsenide	5000	13.1	1.4	9.6	300
Silicon Carbide	260	9.7	2.9	3.1	600
Gallium Nitride	1500	9.5	3.4	24.6	700

2.5.3 The Capacitance–Voltage Measurement

The *C–V* measurement is used to measure the *C–V* characteristics by applying a gate voltage with the frequency of operation. A drain pulse with small time duration is applied to reach the specified final value. After stabilizing the pulse, the gate voltage is forced to move to the RB (reverse bias) in deep pinch-off and then pushed to FB condition. Keeping the longer transition duration allows to gradually shift and capture the response of the device. If the gate voltage is less than the threshold voltage, the 2D electron gas at the ALGaN barrier/GaN channel interface is pinch-off with channel density. By keeping the vertical depletion region constant, the barrier is extended into the buffer. When the device is turned on, channel charge density starts to grow. The gate capacitance starts rising as the channel density starts building from pinch-off and causes a reduced effective barrier. The depletion width edge gradually shifts to the AlGaN barrier/GaN channel interface. A point is reached where this is equal to the threshold voltage. This is the point where the depletion width is equal to the background doping. If the gate capacitance reaches the peak value, it indicates a strong channel conduction. A further increase in gate voltage, exhibiting a slight decay in gate capacitance, indicates that the channel gets saturated, and the depletion region is growing toward source terminal. The *C–V* curves exhibit a sharp change in variation of drain bias. This causes an extension of the depletion layer in 2D electron gas regions [69].

2.6 CONCLUSION

The promising properties of GaN devices motivating the device technology led to large signal applications. The GaN high electron mobility transistor power devices are useful for high switching speed and high-power requirement. In this discussion, we studied the origin for the trap generation. The DC and RF characteristics are disturbed due to unwanted traps. Traps are produced due to the physical technology process. To remove the traps, different engineering techniques have been implemented. Gate engineering techniques are used to control the gate leakage and control of electrons within the device. By the implementation of various techniques, the linearity and noise performance can be improved; hence, the devices are used in communication systems and a wide range of radar applications. A good buffer design can control the traps by a proper doping design. Surface traps can be removed by proper passivation. Barrier traps can be removed with barrier layer engineering. Strain engineering has been used to increase transistor speed while reducing the number of traps by employing buffer layers. A trade-off should be done in a transistor design to increase the reliability by reducing traps. Finally, we discussed trap measurement techniques.

REFERENCES

1. F. Iucolano and T. Boles, "GaN-on-Si HEMTs for wireless base stations," *Mater. Sci. Semicond. Process.*, vol. 98, pp. 100–105, 2019, doi: 10.1016/j.mssp.2019.03.032.
2. U. K. Mishra, P. Parikh, and Y. Wu, "AlGaN / GaN HEMTs: An overview of device operation and applications," *Proc. IEEE*, vol. 90, no. 6, pp. 1022–1031, 2002, doi: 10.1109/JPROC.2002.1021567.
3. U. K. Mishra, Y. Wu, B. P. Keller, S. Keller, and S. P. Denbaars, "GaN microwave electronics," *IEEE Trans. Microw. Theory Tech.*, vol. 46, no. 6, pp. 756–761, 1998.
4. E. Santi, S. Member, K. Peng, S. Member, H. A. Mantooth, and J. L. Hudgins, "Modeling of wide-bandgap power semiconductor devices: Part II," *IEEE Trans. Electron Devices*, vol. 62, no. 2, pp. 434–442, 2015, doi: 10.1109/TED.2014.2373373.
5. T. Palacios et al., "High-Power AlGaN / GaN HEMTs for Ka-band applications," *IEEE Trans. Electron Devices*, vol. 26, no. 11, pp. 781–783, 2005.
6. B. J. Baliga, "Gallium nitride devices for power electronic applications," *Semicond. Sci. Technol.*, vol. 28, no. 7, 2013, doi: 10.1088/0268-1242/28/7/074011.
7. G. Koley, V. Tilak, L. F. Eastman, and M. G. Spencer, "Slow transients observed in AlGaN/GaN HFETs: Effects of SiNx passivation and UV illumination," *IEEE Trans. Electron Devices*, vol. 50, no. 4, pp. 886–893, 2003, doi: 10.1109/TED.2003.812489.
8. M. Meneghini et al., "Investigation of trapping and hot-electron effects in GaN HEMTs by means of a combined electrooptical method," *IEEE Trans. Electron Devices*, vol. 58, no. 9, pp. 2996–3003, 2011, doi: 10.1109/TED.2011.2160547.
9. M. Meneghini et al., "Time-dependent degradation of AlGaN/GaN high electron mobility transistors under reverse bias," *Appl. Phys. Lett.*, vol. 100, no. 3, pp. 2012–2015, 2012, doi: 10.1063/1.3678041.
10. J. J. Coleman, Two-dimensional channel flows over rough surfaces. *J. Hydraul. Eng.*, vol. 109, pp. 175–188, 1983.
11. B. M. Green, V. Tilak, V. S. Kaper, J. A. Smart, J. R. Shealy, and L. F. Eastman, "Microwave power limits of AlGaN/GaN HEMTs under pulsed-bias conditions," *IEEE Trans. Microw. Theory Tech.*, vol. 51, no. 2, pp. 618–623, 2003, doi: 10.1109/TMTT.2002.807680.
12. Y. Murao, M. Hayakawa, K. Ohgami, and T. Kaneko, "A study of GaN HEMTs current collapse impacts on doherty multistage PA linearity," *2015 IEEE Compound Semiconductor Integrated Circuit Symposium CSICS 2015*, 11–14 October 2015, doi: 10.1109/CSICS.2015.7314483.
13. P. C. Chao, S. C. Palmateer, P. M. Smith, U. K. Mishra, K. H. G. Duh, and J. C. M. Hwang, "Millimeterwave low-noise high electron mobility transistors," *IEEE Electron Dev. Lett.*, vol. EDL-6, pp. 531–533, Oct. 1985.
14. A. Chini et al., "Experimental and numerical analysis of hole emission process from carbon-related traps in GaN buffer layers," *IEEE Trans. Electron Devices*, vol. 63, no. 9, pp. 3473–3478, 2016, doi: 10.1109/TED.2016.2593791.
15. W. Zhang, Y. Zhang, W. Mao, X. Ma, J. Zhang, and Y. Hao, "Influence of the interface acceptor-like traps on the transient response of AlGaN/GaN HEMTs," *IEEE Electron Device Lett.*, vol. 34, no. 1, pp. 45–47, 2013, doi: 10.1109/LED.2012.2227235.

16. D. Bisi et al., "Deep-level characterization in GaN HEMTs-Part I: advantages and limitations of drain current transient measurements," *IEEE Trans. Electron Devices*, vol. 60, no. 10, pp. 3166–3175, 2013, doi: 10.1109/TED.2013.2279021.
17. R. J. Trew, Y. Liu, W. W. Kuang, and G. L. Bilbro, "The physics of reliability for high voltage AlGaN/GaN HFET's," *2006 IEEE Compound Semiconductor Integrated Circuit Symposium CSIC*, pp. 103–106, 2006, doi: 10.1109/CSICS.2006.319925.
18. R. Sun, Y. C. Liang, Y. C. Yeo, Y. H. Wang, and C. Zhao, "Realistic trap configuration scheme with fabrication processes in consideration for the simulations of AlGaN/GaN MIS-HEMT devices," *IEEE J. Emerg. Sel. Top. Power Electron.*, vol. 4, no. 3, pp. 720–729, 2016, doi: 10.1109/JESTPE.2016.2549959.
19. M. A. Khan et al., "Insulating gate III-N heterostructure field-effect transistors for high-power microwave and switching applications," *IEEE Trans. Microw. Theory Tech.*, vol. 51, no. 2, pp. 624–633, 2003, doi: 10.1109/TMTT.2002.807681.
20. M. A. Khan, S. Member, G. Simin, J. Yang, and J. Zhang, "Insulating gate III-N Heterostructure field-effect transistors for high-power microwave and switching applications," *IEEE Trans. Microw. Theory Tech.*, vol. 51, no. 2, pp. 624–633, 2003, doi: 10.1109/TMTT.2002.807681.
21. S. Mizuno, Y. Ohno, S. Kishimoto, K. Maezawa, and T. Mizutani, "Large gate leakage current in AlGaN/GaN high electron mobility transistors," *Jpn. J. Appl. Phys., Part 1 Regul. Pap. Short Notes Rev. Pap.*, vol. 41, no. 8, pp. 5125–5126, 2002, doi: 10.1143/jjap.41.5125.
22. S. Yang, S. Liu, Y. Lu, C. Liu, and K. J. Chen, "AC-capacitance techniques for interface trap analysis in GaN-based buried-channel MIS-HEMTs," *IEEE Trans. Electron Devices*, vol. 62, no. 6, pp. 1870–1878, 2015, doi: 10.1109/TED.2015.2420690.
23. R. Ye et al., "An overview on analyses and suppression methods of trapping effects in AlGaN/GaN HEMTs," *IEEE Access*, vol. 10, pp. 21759–21773, 2022, doi: 10.1109/ACCESS.2021.3139443.
24. S. Gustafsson et al., "Dispersive effects in microwave AlGaN/AlN/GaN HEMTs with carbon-doped buffer," *IEEE Trans. Electron Devices*, vol. 62, no. 7, pp. 2162–2169, 2015, doi: 10.1109/TED.2015.2428613.
25. B. Syamal, X. Zhou, S. Ben Chiah, A. M. Jesudas, S. Arulkumaran, and G. I. Ng, "A comprehensive compact model for GaN HEMTs, including quasi-steady-state and transient trap-charge effects," *IEEE Trans. Electron Devices*, vol. 63, no. 4, pp. 1478–1485, 2016, doi: 10.1109/TED.2016.2533165.
26. A. K. Rashmi, S. Haldar, M. Gupta, and R. S. Gupta, "Comprehensive analysis of small-signal parameters of fully strained and partially relaxed high Al-content lattice mismatched AlmGa1-m N/GaN HEMTs," *IEEE Trans. Microw. Theory Tech.*, vol. 51, no. 2 II, pp. 607–617, 2003, doi: 10.1109/TMTT.2002.807678.
27. S. Ghosh et al., "OFF-state leakage and current collapse in AlGaN/GaN HEMTs: a virtual gate induced by dislocations," *IEEE Trans. Electron Devices*, vol. 65, no. 4, pp. 1333–1339, 2018, doi: 10.1109/TED.2018.2808334.
28. Y.-F. Wu, M. Moore, A. Saxler, T. Wisleder, and P. Parikh, "40-W / mm Double Field-plated GaN HEMTs," *Device Research Conference*, Pennsylvania, pp. 151–152, 26 June 2006.
29. R. S. Pengelly, S. M. Wood, J. W. Milligan, S. T. Sheppard, and W. L. Pribble, "A review of GaN on SiC high electron-mobility power transistors and MMICs," *IEEE Trans. Microw. Theory Tech.*, vol. 60, no. 6 PART 2, pp. 1764–1783, 2012, doi: 10.1109/TMTT.2012.2187535.
30. M. Wespel et al., "Trapping effects at the drain edge in 600 v GaN-on-Si HEMTs," *IEEE Trans. Electron Devices*, vol. 63, no. 2, pp. 598–605, 2016, doi: 10.1109/TED.2015.2506904.

31. N. K. Subramani, J. Couvidat, A. Al Hajjar, J. C. Nallatamby, R. Sommet, and R. Quere, "Identification of GaN buffer traps in microwave power AlGaN/GaN HEMTs through low frequency S-parameters measurements and TCAD-based physical device simulations," *IEEE J. Electron Devices Soc.*, vol. 5, no. 3, pp. 175–181, 2017, doi: 10.1109/JEDS.2017.2672685.
32. A. Shanbhag, M. P. Sruthi, A. Chakravorty, N. Dasgupta, and A. Dasgupta, "Compact modeling of static and transient effects of buffer traps in GaN HEMTs," *IEEE Trans. Electron Devices*, vol. 69, no. 3, pp. 999–1005, 2022, doi: 10.1109/TED.2022.3145334.
33. A. G. Hfets et al., "Punch-through in short-channel AlGaN/GaN HFETs," *IEEE Trans. Electron Devices*, vol. 53, no. 2, pp. 395–398, 2006.
34. M. J. Uren, J. Moreke, and M. Kuball, "Buffer design to minimize current collapse in GaN/AlGaN HFETs," *IEEE Trans. Electron Devices*, vol. 59, no. 12, pp. 3327–3333, 2012, doi: 10.1109/TED.2012.2216535.
35. M. Silvestri, M. J. Uren, and M. Kuball, "Iron-induced deep-level acceptor center in GaN/AlGaN high electron mobility transistors: energy level and cross section," *Appl. Phys. Lett.*, vol. 102, no. 7, pp. 1–5, 2013, doi: 10.1063/1.4793196.
36. M. Meneghini et al., "Buffer traps in Fe-doped AlGaN/GaN HEMTs: investigation of the physical properties based on pulsed and transient measurements," *IEEE Trans. Electron Devices*, vol. 61, no. 12, pp. 4070–4077, 2014, doi: 10.1109/TED.2014.2364855.
37. D. Bisi et al., "Hot-electron degradation of AlGaN/GaN high-electron mobility transistors during RF operation: correlation with GaN buffer design," *IEEE Electron Device Lett.*, vol. 36, no. 10, pp. 1011–1014, 2015, doi: 10.1109/LED.2015.2474116.
38. M. J. Uren et al., "Control of short-channel effects in GaN/AlGaN HFETs," *Proceedings of 1st 2006 European Microwave Integrated Circuits Conference EuMIC 2006*, pp. 65–68, 2006, doi: 10.1109/EMICC.2006.282751.
39. T. Suski et al., "Mechanism of yellow luminescence in GaN," *Appl. Phys. Lett.*, vol. 67, no. 5, p. 2188, 1995, doi: 10.1063/1.115098.
40. S. Heikman, S. Keller, S. P. Denbaars, and U. K. Mishra, "Growth of Fe doped semi-insulating GaN by metalorganic chemical vapor deposition," *Appl. Phys. Lett.*, vol. 81, no. 3, pp. 439–441, 2002, doi: 10.1063/1.1490396.
41. A. Fariza et al., "Leakage currents and Fermi-level shifts in GaN layers upon iron and carbon-doping," *J. Appl. Phys.*, vol. 122, no. 2, 2017, doi: 10.1063/1.4993180.
42. M. J. Uren et al., "'leaky dielectric' model for the suppression of dynamic RON in carbon-doped AlGaN/GaN HEMTs," *IEEE Trans. Electron Devices*, vol. 64, no. 7, pp. 2826–2834, 2017, doi: 10.1109/TED.2017.2706090.
43. J. Bergsten et al., "Electron trapping in extended defects in microwave AlGaN/GaN HEMTs with carbon-doped buffers," *IEEE Trans. Electron Devices*, vol. 65, no. 6, pp. 2446–2453, 2018, doi: 10.1109/TED.2018.2828410.
44. V. Joshi, S. P. Tiwari, and M. Shrivastava, "Part II: proposals to independently engineer donor and acceptor trap concentrations in gan buffer for ultrahigh breakdown AlGaN/GaN HEMTs," *IEEE Trans. Electron Devices*, vol. 66, no. 1, pp. 570–577, 2019, doi: 10.1109/TED.2018.2878787.
45. D. Y. Chen et al., "Microwave performance of 'buffer-free' GaN-on-SiC high electron mobility transistors," *IEEE Electron Device Lett.*, vol. 41, no. 6, pp. 828–831, 2020, doi: 10.1109/LED.2020.2988074.
46. A. Gowrisankar et al., "Compensation dopant-free GaN-on-Si HEMTs with a polarization engineered buffer for RF applications," *IEEE Trans. Electron Devices*, vol. 70, no. 4, pp. 1622–1627, 2023, doi: 10.1109/TED.2023.3244514.
47. R. Vetury, N. Q. Zhang, S. Keller, and U. K. Misha, "The impact of surface states on the DC and RF characteristics of AlGaN/GaN HFETs," *IEEE Trans. Electron Devices*, vol. 48, no. 3, pp. 560–566, 2001, doi: 10.1109/16.906451.

48. K. Matsuzaki et al., "Radiation hardness of pseudomorphic HEMT and dual-gate GaAs MES-FET MMICs," *RADECS 97: Fourth European Conference on Radiation and its Effects on Components and Systems (Cat. No.97TH8294)*, pp. 114–119, 1998, doi: 10.1109/radecs.1997.698865.
49. M. A. Kahn, Q. C. Chen, J. N. Kuznia, and M. S. Shur, "Current/voltage characteristic collapse in AlGaN/GaN heterostructure insulated gate field effect transistors at high drain bias," *Electron. Lett.*, vol. 30, no. 25, pp. 2175–2176, 1994, doi: 10.1049/el:19941461.
50. M. Papastamatiou, N. Arpatzanis, G. J. Papaioannou, C. Papastergiou, and A. Christou, "Neutron radiation effects in high electron mobility transistors," *IEEE Trans. Electron Devices*, vol. 44, no. 3, pp. 364–372, 1997, doi: 10.1109/16.556145.
51. A. Kalavagunta et al., "Electrostatic mechanisms responsible for device degradation in AlGaN/AlN/GaN HEMTs," *2007 9th European Conference on Radiation and Its Effects on Components and Systems*, pp. 0–6, 2007, doi: 10.1109/RADECS.2007.5205538.
52. A. D. Koehler et al., "Proton radiation-induced void formation in Ni/Au-gated AlGaN/GaN HEMTs," *IEEE Electron Device Lett.*, vol. 35, no. 12, pp. 1194–1196, 2014, doi: 10.1109/LED.2014.2363433.
53. M. Meneghini et al., "Total suppression of dynamic-Ron in AlGaN/GaN- HEMTs through proton irradiation," *IEEE Trans. Electron Devices*, vol. 4, no. c, pp. 753–756, 2017.
54. T. Nelson et al., "Examination of trapping effects on single-event transients in GaN HEMTs," *IEEE Trans. Nucl. Sci.*, vol. 70, no. 1, pp. 328–335, 2023, doi: 10.1109/TNS.2022.3220235.
55. J. M. Tirado, J. L. Sánchez-Rojas, and J. I. Izpura, "Trapping effects in the transient response of AlGaN/GaN HEMT devices," *IEEE Trans. Electron Devices*, vol. 54, no. 3, pp. 410–417, 2007, doi: 10.1109/TED.2006.890592.
56. G. Meneghesso et al., "Surface-related drain current dispersion effects in AlGaN-GaN HEMTs," *IEEE Trans. Electron Devices*, vol. 51, no. 10, pp. 1554–1561, 2004, doi: 10.1109/TED.2004.835025.
57. A. Wakejima, A. F. Wilson, S. Mase, T. Joka, and T. Egawa, "Evaluation of electron trapping speed of AlGaN/GaN HEMT with real-time electroluminescence and pulsed I-V measurements," *IEEE Trans. Electron Devices*, vol. 60, no. 10, pp. 3183–3189, 2013, doi: 10.1109/TED.2013.2273796.
58. S. Mollah et al., "Current collapse in high-Al channel AlGaN HFETs," *Appl. Phys. Express*, vol. 12, no. 7, 2019, doi: 10.7567/1882-0786/ab24b1.
59. A. Santarelli et al., "GaN FET nonlinear modeling based on double pulse I/V characteristics," *IEEE Trans. Microw. Theory Tech.*, vol. 62, no. 12, pp. 3262–3273, 2014, doi: 10.1109/TMTT.2014.2364236.
60. J. M. Tirado, J. L. Sanchez-Rojas, and J. I. Izpura, "2D simulation of static surface states in AlGaN/GaN HEMT and GaN MESFET devices," *Semicond. Sci. Technol.*, vol. 20, no. 8, pp. 864–869, 2005, doi: 10.1088/0268–1242/20/8/042.
61. Y. Kazami, D. Kasai, and K. Horio, "Numerical analysis of slow current transients and power compression in GaAs FETs," *IEEE Trans. Electron Devices*, vol. 51, no. 11, pp. 1760–1764, 2004, doi: 10.1109/TED.2004.837383.
62. W. D. Hu, X. S. Chen, and W. Lu, "Intrinsic mechanism of drain-lag and current collapse in GaN-based HEMTs," *2009 IEEE Workshop on Microelectronics and Electron Devices, WMED 2009*, pp. 35–37, 2009, doi: 10.1109/WMED.2009.4816142.
63. S. Faramehr, K. Kalna, and P. Igić, "Drift-diffusion and hydrodynamic modeling of current collapse in GaN HEMTs for RF power application," *Semicond. Sci. Technol.*, vol. 29, no. 2, 2014, doi: 10.1088/0268-1242/29/2/025007.

64. K. Horio and A. Nakajima, "Physical mechanism of buffer-related current transients and current slump in AlGaN/GaN high electron mobility transistors," *Jpn. J. Appl. Phys.*, vol. 47, no. 5 PART 1, pp. 3428–3433, 2008, doi: 10.1143/JJAP.47.3428.
65. W. D. Hu, X. S. Chen, and W. Lu, "Intrinsic mechanism of drain-lag and current collapse in GaN-based HEMTs," *IEEE Trans. Microw. Theory Tech.*, pp. 9–11, 2009, doi: 10.1109/WMED.2009.4816142.
66. W. D. Hu, X. S. Chen, F. Yin, J. B. Zhang, and W. Lu, "Two-dimensional transient simulations of drain lag and current collapse in GaN-based high-electron-mobility transistors," *J. Appl. Phys.*, vol. 105, no. 8, 2009, doi: 10.1063/1.3106603.
67. X. A. Cao, J. M. Teetsov, M. P. D'Evelyn, D. W. Merfeld, and C. H. Van, "Electrical characteristics of InGaN/GaN light-emitting diodes grown on GaN and sapphire substrates," *Appl. Phys. Lett.*, vol. 85, no. 1, pp. 7–9, 2004, doi: 10.1063/1.1767280.
68. H. Mosbahi, M. Gassoumi, M. Charfeddine, C. Gaquiere, M. A. Zaidi, and H. Maaref, "Investigation of deep levels in AlGaN/GaN HEMTs on silicon substrate by conductance deep level transient spectroscopy," *5th International Conference on Design & Technology of Integrated Systems in Nanoscale Era, DTIS* 2010, pp. 10–12, 2010, doi: 10.1109/DTIS.2010.5487570.
69. S. Kumar et al., "Temperature and bias dependent trap capture cross section in AlGaN / GaN HEMT on 6-in silicon with carbon-doped buffer," *IEEE Trans. Electron Devices*, vol. 64, no. 12, pp. 4868–4874, 2017.

3 Framework for Design and Performance Evaluation of Memory Using a Memristor

Jalpaben Pandya, Ramji Gupta, and Alpana Pandey

3.1 INTRODUCTION

The electronics sector is experiencing rapid growth due to the emergence of flexible and intelligent electronics. Flexible and smart electronics are distinguished by their elevated circuit complexity and substantial data storage capacity [1–4] Historically, data storage has been achieved via scaling CMOS technology to keep in pace with Moore's law. However, when attempting to scale processors below 22-nm CMOS technology, quantum effects arise, and the chips become more prone to process, voltage, and temperature (PVT) changes [5,6]. This leads to a decline in the performance of the CMOS technology when it comes to storing large amounts of data while using less power and occupying a larger space. Therefore, it is necessary to construct nonvolatile memories employing the new memristor technology, which effectively addresses all the current problems associated with the CMOS technology [7–12].

The research on memory design with memristors has focused on exploring the potential benefits of memristor-based memories, including its high density, low power consumption, and nonvolatile nature. Memristors, also known as memory resistors, are electronic devices with two terminals that control the electric current and store the charge last passed through it in a circuit [5,11–13]. Memristor is a component which links flux and charge as given in equation (3.1) [11,14].

$$d\varnothing = M\ dq \tag{3.1}$$

where

$\varnothing$ – flux
M – memristance
q – charge

This device was merely a conceptualization until it was physically fabricated in 2008. This first model of memristor consisted of two layers, namely TiO_2 layer (undoped)

 DOI: 10.1201/9781003483038-3

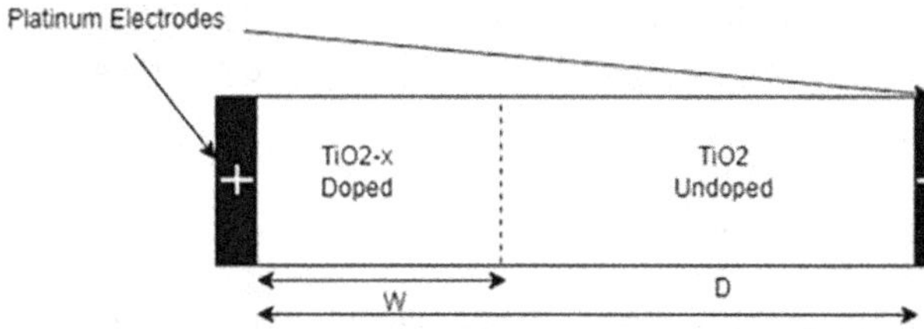

FIGURE 3.1 Vertical cross-sectional structure of HP model memristor where W is known as the state variable that corresponds to the TiO_{2-x} region width and D symbolizes the memristor device width.

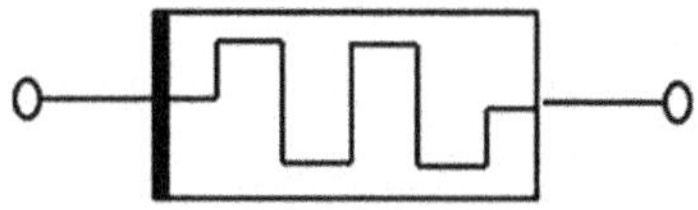

FIGURE 3.2 Symbolic representation of a memristor [7].

which is highly resistive and TiO_{2-x} layer (doped with oxygen vacancies) which is highly conductive in nature [7,13] Platinum electrodes are connected on both the sides wherein the TiO_2 behaves as a negative terminal, whereas the TiO_{2-x} layer behaves as a positive terminal as illustrated in Figure 3.1 along with its symbol as shown in Figure 3.2.

If a positive supply is applied to the (TiO_{2-x}) layer and the negative supply to the TiO_2 layer, the positive oxygen ions that are present in the metallic layer experience a force of repulsion due to which these ions are pushed deep into the insulating layer of TiO_2. This creates a boundary between the two materials shifting towards the insulating layer. Thus, the overall percentage of conducting TiO_{2-x} increases. Consequently, the overall memristor resistance decreases corresponding to an "ON" state.

In contrast, if a negative supply is given to the (TiO_{2-x}) layer and the positive supply is given to the TiO_2 layer, under this condition, the positively charged oxygen ions that are present in the insulator layer experience a force of attraction due to which it pulls the positively charged oxygen ions from the TiO_2 layer. This results in the boundary between the two materials shifting towards the conducting layer. Hence, the width of the insulating layer TiO_2 increases, thereby making the device less conductive [15]. The resistance returns to its initial state, which corresponds to an "OFF" state. By applying voltage to the electrodes, which causes the resistive switching layer to undergo a resistance shift between the high-resistance state R_{OFF} and the low-resistance state R_{ON}, the device's resistance may be adjusted. The SET process and RESET process are the terms used to describe the transitions between high- and low-resistance states, respectively.

Even though memristors have well-defined mathematical formulas, simulating a circuit using them is difficult. The absence of mathematical models for simulating memristor circuits is the reason for this. Therefore, in an effort to mimic a memristor's behaviour, the three fingerprints, namely "Bow Tied *V–I* curve, *V–I* Curve Area inversely related to frequency, and V–I curve attaining a single value at infinite frequency" need to be met [14,16].

3.2 MEMRISTOR MODELS

The behaviour of the physical memristor developed was quite similar to the mathematical formulas suggested by Leon Chua [11]. After the development of this model, numerous efforts have been undertaken to create an appropriate and uniform memristor model that can capture the accurate memristor behaviour suggested by Chua [11]. The different memristor models available are as follows:

1. Linear ionic drift model
2. Nonlinear ionic drift model
3. Yakopcic neuromorphic memristor model
4. Simmons tunnel barrier model
5. ThrEshold adaptive memristor model (TEAM)
6. Voltage TEAM (V-TEAM) model
7. Elwakil model for double-loop hysteresis
8. Model for complementary resistive switches (CRS)

The linear ionic drift model [12] provides basic equations for modelling an ideal memristor. Nevertheless, it is plagued by boundary effect and diminished nonlinearity. To address the boundary effect, several window functions have been suggested for memristor modelling, including Strukov, Joglekar, Biolek, Prodromakis, Piecewise, and Zha. Each of these functions has its own advantages and disadvantages. Table 3.1 shows various window functions together with their respective drawbacks.

In addition, other window functions have been developed to address the boundary lock problem by introducing nonlinearity and emulating the properties of an actual memristor device. Jeetendra Singh [24] proposed a window function that considers both the linear and nonlinear dopant drift of dopants within a memristor. This function demonstrates a significant nonlinearity at the frontiers and effectively addresses other boundary concerns, making it valuable for nonlinear memristor models. Therefore, it is crucial that the window design incorporates the capability to introduce controlled nonlinearity, addresses border impact and boundary lock issues, and allows for scalability.

The nonlinear ionic drift model [25] yields more precise outcomes in relation to memristive devices when compared to the linear ionic drift model. Nevertheless, it demonstrates asymmetrical switching behaviour without any threshold being taken into account. The Yakopcic neuromorphic memristor model [26] enables the incorporation of memristors with threshold characteristics to construct intricate neuromorphic memristor-based systems [5,26]. The Simmons tunnel barrier model [27] is another memristor model created using SPICE software. This model accurately represents the device as a resistor coupled with an electron tunnel barrier. This model demonstrates symmetric switching without any explicit correlation between voltage and current and is a complex model. The threshold adaptive memristor model (TEAM) [28] represents a device that adjusts its resistance based

TABLE 3.1
Different Window Functions along with its Disadvantages

Sr. No	Window	Window Function	Disadvantages
1.	Strukov's window function [12]	$f(w)=\dfrac{w\,(1-w)}{D^2}$ D and w are the parameters as stated in Figure 1.1	Boundary lock issue present and cannot capture nonlinear effects
2.	Joglekar's window function [17]	$f(x)=1-(2x-1)^{2p}$ where x – internal state variable and $x=\dfrac{w}{D}$	Boundary lock issue present
3.	Biolek's window function [18]	$f(x)=1-\left(x-\text{sgn}\,(-i)\right)^{2p}$ where p – positive integer $\text{sgn}\,(-i)$ – current through memristor	Discontinuities at the boundaries
4.	Prodromakis' window function [19]	$f(w)=j\left\{1-\left[\left(\dfrac{w}{D}-0.5\right)^2+0.75\right]^p\right\}$ where p, j – control parameters	At low frequencies, it fails to retain the pinched hysteresis loop
5.	Piecewise window function [20]	$f(x)=\left(1+\left(\dfrac{x-0.25}{a}\right)^{2b}\right)^{-1};$ $x_0\le x\le 1-x$ $f(x)=kx\,(1-x);\quad$ otherwise where $x=\dfrac{w}{D}$	Needs careful parameter selection to guarantee precision
6.	Zha's window function [21]	$f(x)=j\left(1-\left\{0.25\left[x-\text{sgn}(-i)\right]^2+0.75\right\}^p\right)$ where p, j are the control parameters	Boundary lock issues present
7.	Window function suggested by T. Anusudha [22]	$f(x)=J\left[1-2\left(x^3-x+1\right)^p\right]$ where x – internal state variable p – positive integer J – control parameter	Needs careful parameter selection to guarantee precision
8.	Window function suggested by Junrui [23]	$f(x)=j\left\{1-\begin{bmatrix}\alpha x^3+a^2\left[x-\text{stp}(-i)\right]^2\\+\left(1-a^2\right)+\beta x^2+\gamma x\end{bmatrix}^p\right\}$ where α,β,γ – of nonlinear drift parameters j – control parameter p – positive integer	Complicated function

on threshold currents and changes in resistance when the input current surpasses a certain level. This model is characterized by its intricate nature and high level of precision, making it particularly valuable for applications involving switching or digital processes. The V-TEAM model [29] is a developed model that considers the threshold voltage. The V-TEAM model demonstrates superior accuracy in comparison to the existing memristor models that are currently in use [29,30]. Elwakil devised a model for double-loop hysteresis [31] that offers representations for both current- and voltage-driven memristors that display asymmetrical switching behaviour. Memristor-based memory circuits are affected by sneak path currents, which significantly impair their performance [32,33]. To prevent sneak path current in the circuit, the model for CRS [33] is employed. This model consists of two memristive components that are coupled in an anti-serial arrangement. This model is effective in addressing the problems related to sneak path currents. However, when dealing with bigger memory array sizes, it experiences slight convergence challenges.

In theory, the behaviour of memristor devices can be accurately predicted using the stated memristor device models. However, there has been limited study conducted on memristor devices in real-world contexts, where their usage differs dramatically from theoretical studies [15]. Insufficient attention is given to the interaction between memristive circuits and the real world when they are utilized in an uncontrolled context, leading to potentially inaccurate conclusions.

The previously described models only partially consider nonlinearity and kinetics, which impacts the device speed. However, for modelling accurate speed at the device level, the performance of memory cells built with memristors is compromised because design considerations for various parameters, such as leakage current, state drift, noise, device variability, temperature, and parasitic effects, are omitted [34–37]. Additionally, other emulator models tailored for certain applications have been developed [34–43] to mimic and verify the functionality of memristors in those applications. Therefore, it may be inferred that there is no one memristor model that fulfils all objectives.

3.3 RELATED WORK

Memristors have been used extensively in both analogue and digital applications. The recent developments for memristors in analogue domains include memory circuit for artificial intelligence and machine learning [45,45], memristive biosensors [46], artificial neural networks [44,47,48], neuromorphic applications [35,44,49,50], speech encryption [43], spiking neural network [30,44], convolutional neural network [41], associative memories [51], memory cell design [7,9,52–56], and physical unclonable function (PUF) [2,45,57–59].

Recent developments of memristor-based circuits in digital domain are random number generator [60], ternary memory application [61], in-memory computation [62], ternary memristor decode, logic gates [63,64], flip flop, Advanced Encryption Standard (AES) encryption [44,65], and key generation for encryption [44,66].

Figure 3.3 depicts the investigation conducted on the research done using memristors in both the domains. For using memristors, it is necessary to have a model that accurately replicates the behaviour of a memristor, in addition to providing mathematical models, circuit designs, and simulation tests. Since 2008, many Simulation Program with Integrated Circuit Emphasis (SPICE) models and emulators of memristors have been developed to facilitate the design and simulation of memristor circuits. Although there have been some advancements in SPICE models and practical

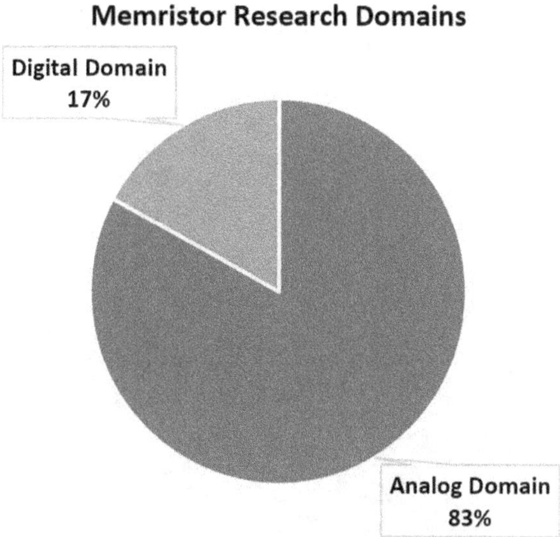

FIGURE 3.3 Memristor research domains.

emulators, the progress in developing a digital circuit simulator for memristors is falling behind. Nevertheless, the accessibility of simulation tools poses a challenge when designing a comprehensive memristor-based system. Figure 3.4 depicts the various software applications utilized for simulating circuits based on memristors.

Memristors have a lot of potential for use in a wide range of applications, and looking into its basic nature, it is an obvious choice to be used for memory design. Various memory circuits, such as 1R, 1T1R, 2T2R, 2T2M, 3T2R, 5CNTFET2R, 6T1M, 7T1R, 7T2M, 8T2R, and 9T2R, are designed and compared in Table 3.2.

Referring to all the above circuits, the memory cells are majorly evaluated for read delay, write delay, hold state power consumption, and static noise margin. Thus, there remains a scope for a better memory cell design using a memristor.

3.4 DESIGN CONSIDERATION FOR MEMRISTOR-BASED MEMORY DESIGN

Referring to all the above circuit designs, for memory circuits as shown in Table 3.2, the parameters evaluated are read delay, write delay, hold power, and area. For an efficient memory design using a memristor, the desirable parameters which need to be considered [70–72] are shown in Table 3.3.

An increased $\frac{R_{OFF}}{R_{ON}}$ ratio is advantageous for facilitating a correct read operation. However, during the read operation, state drift occurs in the memristor, necessitating memory refreshing. When designing, it is important to evaluate the trade-off between speed and robustness, taking into account the state drift issue.

Endurance is measured by the frequency of write operations on a device without causing any impact on the stored data, and it is intended that this value should be maximized.

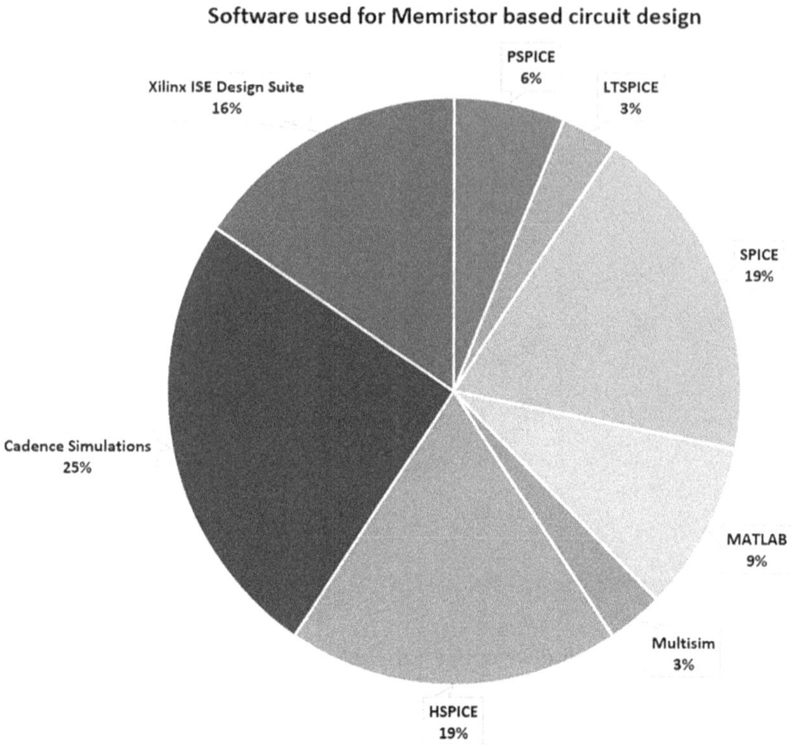

FIGURE 3.4 Software used for memristor-based circuit design.

TABLE 3.2
Comparison of Memristor-based Memory Circuits

Memory Cell	Number of Access Transistors	Number of Data Storage Elements	Circuit Design Issues
6T SRAM cell [67]	Two transistors	Four transistors	Affected by process, voltage, and temperature (PVT) variations and increased read and write delays
2T2M [7]	Two transistors in the form of transmission gate	Two memristors connected in differential configuration	Immune to PVT variation during read operation
2T2M [68]	Two transistors	Two memristors connected in differential configuration	Increased write delay

(*Continued*)

TABLE 3.2 (*Continued*)
Comparison of Memristor-based Memory Circuits

Memory Cell	Number of Access Transistors	Number of Data Storage Elements	Circuit Design Issues
1T1M [8]	One transistor	One memristor connected in single-ended configuration	1. Optimization of reading voltage, pulse width, height, and speed for correct read operation 2. For writing, high voltage is required to achieve fast switching
5CNFET2M [53]	Four carbon nano tube field effect transistors (CNTFET) in the form of transmission gates	Two memristors connected in differential configuration	Increased area
6T1M [69]	Two along with four control transistors	One memristor	Increased leakage power during read operation
7T2M [54]	Two transistors	Four transistors along with two control memristors and one control transistor	Increased area

TABLE 3.3
Design Consideration for a Memory

Sr. No	Parameters	Desired Value	Typical Value
1.	$\frac{R_{OFF}}{R_{ON}}$ ratio	As high as possible	Around 200
2.	Endurance (Years)	As high as possible	>10
3.	Data retention (s)	As high as possible	10^5 s
4.	Device speed	As high as possible	–
5.	Sneak path current	As low as possible	–
6.	Data refreshment (number of cycles)	As high as possible	$\sim 14 \times 10^3$
7.	Read delay (ns)	As low as possible	Few ns
8.	Write delay (ns)	As low as possible	< 50 ns
9.	Area (μm^2)	As low as possible	–

During the read mechanism, state drift occurs which can read wrong data. Thus, a high data retention rate is desirable. One of the ways to achieve this is using bidirectional storage mechanism [69], which will disconnect the circuit from the bit lines during read operation. Data refreshment circuit also needs to be designed to overcome the erroneous value due to the state drift condition. One way to accomplish this is to design multiple programs/erase circuit, and the second way is to design a circuit such that the read access time is less than the switching rate of the memristor [53].

Additionally, the memory circuit design considerations also need to be considered. This basically involves the selection of a memory architecture, the design of read and write operation mechanisms, the choice of a sensing mechanism, and the selection of suitable peripheral circuitry for connecting with memristor arrays.

The memory architectural options include crossbar arrays and grid Resistive Random Access Memory (RRAM) architecture [56]. The grid RRAM architecture exhibits higher energy efficiency and superior access time. Nevertheless, its style exhibits larger dimensions, resulting in increased expenses. As no access transistors are required for designing the memory cell using the crossbar structure, this architecture is quite appealing as it occupies a compact area. While the crossbar RRAM topology offers advantages in terms of compactness, it also presents challenges in switching time, access time, and reliability [56]. Furthermore, it is crucial to establish a proficient and efficient read and write mechanism for the cell, ensuring optimal performance in terms of efficiency, speed, and reliability. This entails taking into account many parameters, such as voltage and current levels, pulse lengths, and sensing mechanisms.

3.5 SIMULATION OF THE MEMORY CELL USING MEMRISTORS

The operations and architecture of 2T2M RRAM have been described in this section. As shown in Figure 3.5, the circuit uses transmission gates comprising of transistors M1 (PMOS) and M2 (NMOS) as compared to the traditionally used NMOS access transistor. The transmission gate allows for both strong one and strong zero logic without any signal degradation. The memristors M3 and M4 act as the storage devices and are connected in differential configuration so that it is less susceptible to

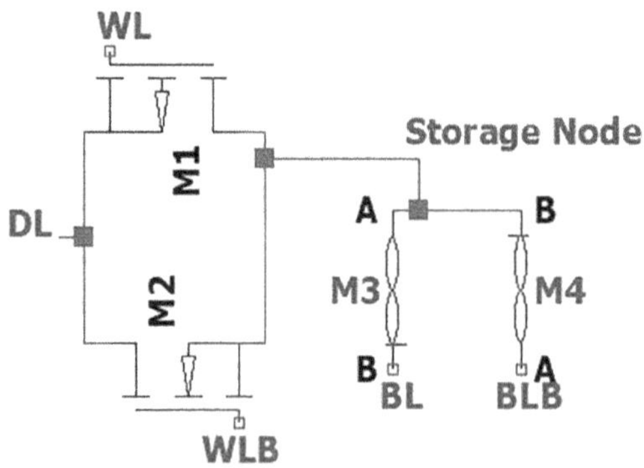

FIGURE 3.5 2T2M memory cell.

PVT variations as opposed to single-ended connection. The linear ionic drift model using the Joglekar window function is employed for the memristor, which addresses the challenges associated with nonlinear dopant drift. The simulation is conducted using LTspice® software, an open-source simulator that is compatible with SPICE and permits both netlist programming and waveform analysis. Memristor parameters selected are $R_{ON} = 100\ \Omega$, $R_{OFF} = 16\ \mathrm{K}\Omega$, $R_{INIT} = 11\ \mathrm{K}\Omega$, $D = 10$, $p = 10$.

3.5.1 For Write Operation

For storing any bit in a cell, the word line (WL) should be connected to logic high and word line bar (WLB) to a logic low. The data to be stored is applied at data line (DL), and the complement of applied data is provided to bit line (BL) and bit line bar (BLB). As the memristors are connected in a differential configuration, either of the memristor will be ON, and the data will be stored at the storage node. The output of the write '1' operation is illustrated in Figure 3.6.

In order to read the stored data value, BL and BLB must be discharged prior to the read operation. WL is then given a logic high, followed by WLB, receiving a logic low. As soon as the transistors of the transmission gate become active, BL and BLB begin to charge at a rate proportional to their resistance levels. The data will be sensed by the sense amplifier when the voltage difference exceeds 50 mV.

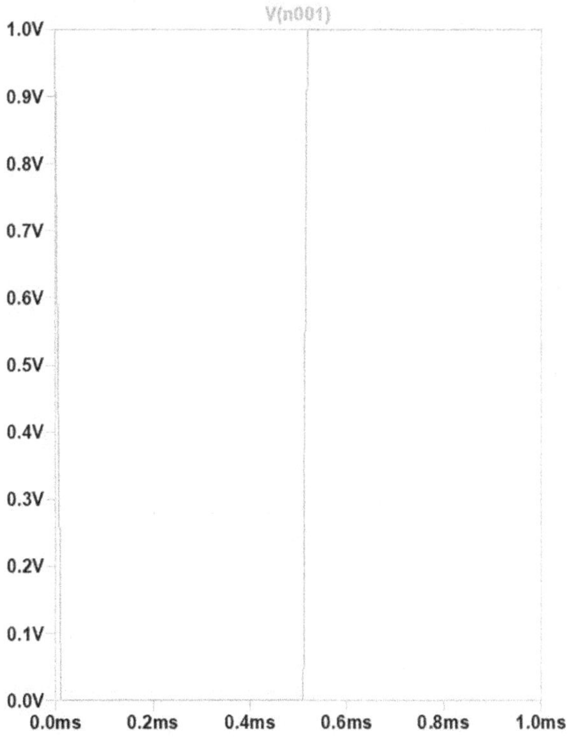

FIGURE 3.6 Output for write operation.

TABLE 3.4
Comparison of Write Delay

Memory cell	1R [55]	1T1R [52]	2T2M [68]	2T2M [7]	5CNTFET2M [9]	6T1M [69]	7T2M [54]	Proposed circuit
Write delay	–	100 ns	1 ns	25 ns	2.4 ns	3.3 ns	8.66 ps	0.39 μs

3.6 RESULTS AND DISCUSSION

For simulating our proposed device, firstly, the HP Lab's linear drift model of the memristor was implemented. Every Metal Oxide Semiconductor Field Effect Transistor (MOSFET) component was simulated utilizing a 45-nm technology node. The results obtained for write delay using the proposed RRAM cell is compared with the other memory circuits. The write delay of a memristor in the proposed circuit is determined by averaging the write delays of write-1 and write-0. The write latency for the proposed 2T2M is 0.39 μs. As shown in Table 3.4, the write delay of this circuit is greater than the other memory cell designs described in Ref. [7]. This is contributed to the fact that the linear ionic drift model used in the memristor assumes the following three things: a homogeneous electric field, average ion mobility, and a linear connection between applied voltage and current, and hence, it is not a typical indication of memristive features. The technology node and the operating voltages are different for each memory cell. This design is a demonstration of the framework suggested for an efficient memory design cell. Hence, when designing any memristor-based memory circuit, careful selection of parameters for optimum performance should be targeted.

3.7 CONCLUSION

Memristors are a special kind of electrical components that are used in neuromorphic systems, logic circuits, and memory, among other things. Memristive technologies are appealing because of their nonvolatility, scalability, and compatibility with CMOS. To effectively construct and mimic memory based on memristors, a variety of memristor models have been proposed. The goal of this examination is to make researchers aware of the basic framework that needs to be followed when designing a memristor-based memory cell along with its design parameters. As there isn't a single memristor model that fits all applications, researchers have difficulties when trying to properly simulate and evaluate memristor-based memory circuits. The designed 2T2M memory cell suffers from increased read and write delays as compared to its counterparts. Thus, it is concluded that appropriately selecting a model which will fit your need is a major task. It will take some time before memristor-based circuits are deployed in real-world applications.

REFERENCES

1. D. Blaauw et al., "IoT design space challenges: circuits and systems," In *2014 Symposium on VLSI Technology (VLSI-Technology): Digest of Technical Papers*. IEEE, Hawaii, 2014, pp. 1–2.
2. D. Sepranos and M. Wolf, "Challenges and opportunities in VLSI IoT devices and systems," *IEEE Design and Test*, vol. 36, no. 4, pp. 24–30, 2019. doi: 10.1109/MDAT.2019.2917178.
3. L. Shao et al., "Compact modeling of thin-film transistors for flexible hybrid IoT design," *IEEE Design and Test*, vol. 36, no. 4, pp. 6–14, 2019, doi: 10.1109/MDAT.2019.2899058.
4. C. R. Kumar and I. Aziza, "VLSI design of energy efficient computational centric smart objects for IoT," In *15th Learning and Technology Conference (L&T)*. IEEE, Effat University in Jeddah, Saudi Arabia, 2018, pp. 129–138.
5. J. Singh and B. Raj, "Comparative analysis of memristor models and memories design," *Journal of Semiconductors*, vol. 39, no. 7, 2018, doi: 10.1088/1674-4926/39/7/074006.
6. A. Islam and M. Hasan, "A technique to mitigate impact of process, voltage and temperature variations on design metrics of SRAM Cell," *Microelectronics Reliability*, vol. 52, no. 2, pp. 405–411, 2012, doi: 10.1016/j.microrel.2011.09.034.
7. C. Roy and A. Islam, "TG based 2T2M RRAM using memristor as memory element," *Indian Journal of Science and Technology*, vol. 9, no. 33, 2016, doi: 10.17485/ijst/2016/v9i33/99508.
8. B. Mohammad, D. Homouz, and H. Elgabra, "Robust hybrid memristor-CMOS Memory: modeling and design," *IEEE Transactions on Very Large Scale Integration Systems*, vol. 21, no. 11, pp. 2069–2079, 2013, doi: 10.1109/TVLSI.2012.2227519.
9. S. Pal, V. Gupta, W. H. Ki, and A. Islam, "Design and development of memristor-based RRAM," *IET Circuits, Devices and Systems*, vol. 13, no. 4, pp. 548–557, 2019, doi: 10.1049/iet-cds.2018.5388.
10. H. Said, H. Aziza, and G. C. Sirakoulis., "Memristor based memories: technology, design and test," In *9th IEEE International Conference on Design & Technology of Integrated Systems in Nanoscale Era (DTIS)*. IEEE, Sardinia, Italy, 2014, pp. 1–7.
11. L. Chua, "Memristor-The Missing Circuit Element," *IEEE Transactions on Circuit Theory*, vol. 18, no. 5, pp. 507–519, 1971.
12. D. B. Strukov, G. S. Snider, D. R. Stewart, and R. S. Williams, "The missing memristor found," *Nature*, vol. 453, no. 7191, pp. 80–83, May 2008, doi: 10.1038/nature06932.
13. R. S. Williams, "How we found the missing memristor," *IEEE Spectrum*, vol. 45, no. 12, pp. 28–35, 2008, doi: 10.1109/MSPEC.2008.4687366.
14. A. G. Radwan and M. E. Fouda, "Memristor: models, types, and applications," In A. G. Radwan and M. E. Fouda (Eds.), *Studies in Systems, Decision and Control*, vol. 26. Springer International Publishing, 2015, pp. 13–49. doi: 10.1007/978-3-319-17491-4_2.
15. C. Roy and A. Islam, "TG based 2T2M RRAM using memristor as memory element," *Indian Journal of Science and Technology*, vol. 9, no. 33, 2016, doi: 10.17485/ijst/2016/v9i33/99508.
16. S. P. Adhikari, M. P. Sah, H. Kim, and L. O. Chua, "Three fingerprints of memristor," *IEEE Transactions on Circuits and Systems I: Regular Papers*, vol. 60, no. 11, pp. 3008–3021, 2013, doi: 10.1109/TCSI.2013.2256171.
17. Y. N. Joglekar and S. J. Wolf, "The elusive memristor: properties of basic electrical circuits," *European Journal of Physics*, vol. 30, no. 4, pp. 1069–1078, 2008, https://doi.org/10.1088/0143-0807/30/4/001.

18. Z. Biolek, D. Biolek, Z. Biolek, and V. Biolková, "SPICE model of memristor with nonlinear dopant drift," *Radioengineering*, vol. 18, no. 2, pp. 210–214, 2009. Available: https://www.researchgate.net/publication/26625012.
19. T. Prodromakis, B. P. Peh, C. Papavassiliou, and C. Toumazou, "A versatile memristor model with nonlinear dopant kinetics," *IEEE Transactions on Electron Devices*, vol. 58, no. 9, pp. 3099–3105, 2011, doi: 10.1109/TED.2011.2158004.
20. J. Yu, X. Mu, X. Xi, and S. Wang, "A memristor model with piecewise window function," *Radioengineering*, vol. 22, no. 4, pp. 1107–1113, 2013.
21. J. Zha, H. Huang, and Y. Liu, "A novel window function for memristor model with application in programming analog circuits," *IEEE Transactions on Circuits and Systems II: Express Briefs*, vol. 63, no. 5, pp. 423–427, 2016, doi: 10.1109/TCSII.2015.2505959.
22. T. A. Anusudha and S. R. S. Prabaharan, "A versatile window function for linear ion drift memristor model: a new approach," *AEU–International Journal of Electronics and Communications*, vol. 90, pp. 130–139, 2018, doi: 10.1016/j.aeue.2018.04.020.
23. J. Li, Z. Dong, L. Luo, S. Duan, and L. Wang, "A novel versatile window function for memristor model with application in spiking neural network," *Neurocomputing*, vol. 405, pp. 239–246, 2020, doi: 10.1016/j.neucom.2020.04.111.
24. J. Singh and B. Raj, "An accurate and generic window function for nonlinear memristor models," *Journal of Computational Electronics*, vol. 18, no. 2, pp. 640–647, 2019, doi: 10.1007/s10825-019-01306-6.
25. E. Lehtonen and M. Laiho., "CNN using memristors for neighborhood connections," *In 12th International Workshop on Cellular Nanoscale Networks and their Applications (CNNA 2010)*. IEEE, Berkeley, California, 2010, pp. 1–4.
26. C. Yakopcic, T. M. Taha, G. Subramanyam, R. E. Pino, and S. Rogers, "A memristor device model," *IEEE Electron Device Letters*, vol. 32, no. 10, pp. 1436–1438, 2011, doi: 10.1109/LED.2011.2163292.
27. J. G. Simmons, "Generalized formula for the electric tunnel effect between similar electrodes separated by a thin insulating film," *Journal of Applied Physics*, vol. 34, no. 6, pp. 1793–1803, 1963, doi: 10.1063/1.1702682.
28. S. Kvatinsky, E. G. Friedman, A. Kolodny, and U. C. Weiser, "TEAM: threshold adaptive memristor model," *IEEE Transactions on Circuits and Systems I: Regular Papers*, vol. 60, no. 1, pp. 211–221, 2013, doi: 10.1109/TCSI.2012.2215714.
29. S. Kvatinsky, M. Ramadan, E. G. Friedman, and A. Kolodny, "VTEAM: a general model for voltage-controlled memristors," *IEEE Transactions on Circuits and Systems II: Express Briefs*, vol. 62, no. 8, pp. 786–790, 2015, doi: 10.1109/TCSII.2015.2433536.
30. Z. Deng, C. Wang, H. Lin, and Y. Sun, "A memristive spiking neural network circuit with selective supervised attention algorithm," *IEEE Transactions on Computer-Aided Design of Integrated Circuits and Systems*, vol. 42, no. 8, pp. 2604–2617, 2023, doi: 10.1109/TCAD.2022.3228896.
31. A. S. Elwakil, M. E. Fouda, and A. G. Radwan, "A simple model of double-loop hysteresis behavior in memristive elements," *IEEE Transactions on Circuits and Systems II: Express Briefs*, vol. 60, no. 8, pp. 487–491, 2013, doi: 10.1109/TCSII.2013.2268376.
32. Y. Yang, J. Mathew, R. A. Shafik, and D. K. Pradhan, "Verilog-a based effective complementary resistive switch model for simulations and analysis," *IEEE Embedded Systems Letters*, vol. 6, no. 1, pp. 12–15, 2014, doi: 10.1109/LES.2013.2278740.
33. E. Linn, R. Rosezin, C. Kügeler, and R. Waser, "Complementary resistive switches for passive nanocrossbar memories," *Nature Materials*, vol. 9, no. 5, pp. 403–406, 2010, doi: 10.1038/nmat2748.
34. L. Gao, P. Y. Chen, and S. Yu, "NbOx based oscillation neuron for neuromorphic computing," *Applied Physics Letters*, vol. 111, no. 10, 2017, doi: 10.1063/1.4991917.

35. D. Wheeler et al., "Fabrication and characterization of tungsten-oxide-based memristors for neuromorphic circuits," In *International Workshop on Cellular Nanoscale Networks and their Applications*, IEEE Computer Society, 2014. doi: 10.1109/CNNA.2014.6888611.
36. K. Miller, K. S. Nalwa, A. Bergerud, N. M. Neihart, and S. Chaudhary, "Memristive behavior in thin anodic Titania," *IEEE Electron Device Letters*, vol. 31, no. 7, pp. 737–739, 2010, doi: 10.1109/LED.2010.2049092.
37. S. H. Jo, T. Chang, I. Ebong, B. B. Bhadviya, P. Mazumder, and W. Lu, "Nanoscale memristor device as synapse in neuromorphic systems," *Nano Letters*, vol. 10, no. 4, pp. 1297–1301, 2010, doi: 10.1021/nl904092h.
38. H. Kim, M. P. Sah, C. Yang, S. Cho, and L. O. Chua, "Memristor emulator for memristor circuit applications," *IEEE Transactions on Circuits and Systems I: Regular Papers*, vol. 59, no. 10, pp. 2422–2431, 2012, doi: 10.1109/TCSI.2012.2188957.
39. G. Wang, D. Bai, and X. Wang, " Digital model of TiO_2 memristor for field-programmable gate array," *The Journal of Engineering*, vol. 2014, no. 3, pp. 90–92, 2014, doi: 10.1049/joe.2013.0183.
40. E. S. Yadav, B. Rajesh, C. R. Srinivasan, and P. S. Kalyan, "A study on non-linear behavior of memristor emulator using multisim," *Indonesian Journal of Electrical Engineering and Computer Science*, vol. 16, no. 3, pp. 1213–1220, 2019, doi: 10.11591/ijeecs.v16.i3.pp1213–1220.
41. M. F. Tolba, Y. Halawani, H. Saleh, B. Mohammad, and M. Al-Qutayri, "FPGA-based memristor emulator circuit for binary convolutional neural networks," *IEEE Access*, vol. 8, pp. 117736–117745, 2020, doi: 10.1109/ACCESS.2020.3004535.
42. R. K. Ranjan, N. Rani, R. Pal, S. K. Paul, and G. Kanyal, "Single CCTA based high frequency floating and grounded type of incremental/decremental memristor emulator and its application," *Microelectronics Journal*, vol. 60, pp. 119–128, 2017, doi: 10.1016/j.mejo.2016.12.004.
43. M. F. Tolba et al., "Digital emulation of a versatile memristor with speech encryption application," *IEEE Access*, vol. 7, pp. 174280–174297, 2019, doi: 10.1109/ACCESS.2019.2957300.
44. N. Du, H. Schmidt, and I. Polian, "Low-power emerging memristive designs towards secure hardware systems for applications in internet of things," *Nano Materials Science*, vol. 3, no. 2, pp. 186–204, 2021. doi: 10.1016/j.nanoms.2021.01.001.
45. D. Kim et al., "An overview of processing-in-memory circuits for artificial intelligence and machine learning," *IEEE Journal on Emerging and Selected Topics in Circuits and Systems*, vol. 12, no. 2, pp. 338–353, 2022, doi: 10.1109/JETCAS.2022.3160455.
46. R. Homsi, N. Al-Azzam, B. Mohammad, and A. Alazzam, "Memristive biosensors for cancer biomarkers detection: a review," *IEEE Access*, vol. 11, pp. 19347–19361, 2023. doi: 10.1109/ACCESS.2023.3248683.
47. A. Balaji et al., "Mapping spiking neural networks to neuromorphic hardware," *IEEE Transactions on Very Large Scale Integration Systems*, vol. 28, no. 1, pp. 76–86, 2020, doi: 10.1109/TVLSI.2019.2951493.
48. V. Ntinas, I. Vourkas, A. Abusleme, G. C. Sirakoulis, and A. Rubio, "Experimental study of artificial neural networks using a digital memristor simulator," *IEEE Transactions on Neural Networks and Learning Systems*, vol. 29, no. 10, pp. 5098–5110, 2018, doi: 10.1109/TNNLS.2018.2791458.
49. C. Pehle, L. Blessing, E. Arnold, E. Müller, and J. Schemmel, "Event-based backpropagation for analog neuromorphic hardware," arXiv preprint arXiv:2302.07141, 2023. Available: https://arxiv.org/abs/2302.07141

50. Y. Liu, K. Wang, Y. Qian, X. Liu, and S. Wang, "Neuromorphic oscillators and electronic implementation of One novel MFNN Model," *IEEE Access*, vol. 11, pp. 16076–16084, 2023, doi: 10.1109/ACCESS.2022.3233771.
51. E. Lehtonen, J. H. Poikonen, M. Laiho, and P. Kanerva, "Large-scale memristive associative memories," *IEEE Transactions on Very Large Scale Integration Systems*, vol. 22, no. 3, pp. 562–574, 2014, doi: 10.1109/TVLSI.2013.2250319.
52. M. Zangeneh and A. Joshi, "Design and optimization of nonvolatile multibit 1T1R resistive RAM," *IEEE Transactions on Very Large Scale Integration Systems*, vol. 22, no. 8, pp. 1815–1828, 2014, doi: 10.1109/TVLSI.2013.2277715.
53. S. Pal, S. Bose, W. H. Ki, and A. Islam, "Design of power-and variability-Aware nonvolatile RRAM cell using memristor as a memory element," *IEEE Journal of the Electron Devices Society*, vol. 7, pp. 701–709, 2019, doi: 10.1109/JEDS.2019.2928830.
54. J. Singh and B. Raj, "Design and investigation of 7T2M-NVSRAM With enhanced stability and temperature impact on store/restore Energy," *IEEE Transactions on Very Large Scale Integration Systems*, vol. 27, no. 6, pp. 1322–1328, 2019, doi: 10.1109/TVLSI.2019.2901032.
55. Y. Deng et al., "RRAM crossbar array with cell selection device: a device and circuit interaction study," *IEEE Transactions on Electron Devices*, vol. 60, no. 2, pp. 719–726, 2013, doi: 10.1109/TED.2012.2231683.
56. N. Shaarawy, A. Emara, A. M. El-Naggar, M. E. Elbtity, M. Ghoneima, and A. G. Radwan, "Design and analysis of 2T2M hybrid CMOS-Memristor based RRAM," *Microelectronics Journal*, vol. 73, pp. 75–85, 2018, doi: 10.1016/j.mejo.2018.01.001.
57. M. Uddin, A. S. Shanta, M. Badruddoja Majumder, M. Sakib Hasan, and G. S. Rose, "Memristor crossbar PUF based lightweight hardware security for IoT," In *2019 IEEE International Conference on Consumer Electronics (ICCE)*. IEEE, Las Vegas, Nevada, USA, 2019, pp. 1–4.
58. B. Gao et al., "Concealable physically unclonable function chip with a memristor array," 2022. Available: https://www.science.org
59. M. I. Khan, S. Ali, A. Al-Tamimi, A. Hassan, A. A. Ikram, and A. Bermak, "A robust architecture of physical unclonable function based on Memristor crossbar array," *Microelectronics Journal*, vol. 116, 2021, doi: 10.1016/j.mejo.2021.105238.
60. M. Akbari et al., "True random number generator based on the variability of the high resistance state of RRAMs," *IEEE Access*, vol. 11, pp. 66682–66693, 2023, doi: 10.1109/ACCESS.2023.3290896.
61. A. Yousefi, N. Eslami, and M. H. Moaiyeri, "A Reliable and energy-efficient nonvolatile ternary memory based on hybrid FinFET/RRAM technology," *IEEE Access*, vol. 10, pp. 105040–105051, 2022, doi: 10.1109/ACCESS.2022.3211562.
62. F. Freye, J. Lou, C. Bengel, S. Menzel, S. Wiefels, and T. Gemmeke, "Memristive devices for time domain compute-in-memory," *IEEE Journal on Exploratory Solid-State Computational Devices and Circuits*, vol. 8, no. 2, pp. 119–127, 2022, doi: 10.1109/JXCDC.2022.3217098.
63. Y. Zhang, Y. Shen, X. Wang, and Y. Guo, "A novel design for a memristor-based or gate," *IEEE Transactions on Circuits and Systems II: Express Briefs*, vol. 62, no. 8, pp. 781–785, 2015, doi: 10.1109/TCSII.2015.2435354.
64. K. Cho, S. J. Lee, and K. Eshraghian, "Memristor-CMOS logic and digital computational components," *Microelectronics Journal*, vol. 46, no. 3, pp. 214–220, 2015, doi: 10.1016/j.mejo.2014.12.006.
65. M. Masoumi, "Novel hybrid CMOS/Memristor implementation of the AES algorithm robust against differential power analysis attack," *IEEE Transactions on Circuits and Systems II: Express Briefs*, vol. 67, no. 7, pp. 1314–1318, 2020, doi: 10.1109/TCSII.2019.2932337.

66. L. Yang et al., "Cryptographic key generation and in situ encryption in one-transistor-one-resistor memristors for hardware security," *Advanced Electronic Materials*, vol. 7, no. 5, 2021, doi: 10.1002/aelm.202001182.
67. R P Jain, *Modern Digital Electronics*, vol. 1. Tata McGraw Hill, 2010.
68. N. Shaarawy, M. Ghoneima, and A. G. Radwan., "2T2M memristor-based memory cell for higher stability RRAM modules," In *2015 IEEE International Symposium on Circuits and Systems (ISCAS)*. IEEE, Lisbon, Portugal, 2015, pp. 1418–1421.
69. M. N. Sakib, R. Hassan, S. N. Biswas, and S. R. Das, "Memristor-based high-speed memory cell with stable successive read operation," *IEEE Transactions on Computer-Aided Design of Integrated Circuits and Systems*, vol. 37, no. 5, pp. 1037–1049, 2018, doi: 10.1109/TCAD.2017.2729464.
70. S. Kvatinsky, E. G. Friedman, A. Kolodny, and U. C. Weiser, "The desired memristor for circuit designers," *IEEE Circuits and Systems Magazine*, vol. 13, no. 2, pp. 17–22, 2013, doi: 10.1109/MCAS.2013.2256257.
71. N. Taherinejad and D. Radakovits, "From behavioral design of memristive circuits and systems to physical implementations," *IEEE Circuits and Systems Magazine*, vol. 19, no. 4, pp. 6–18, 2019, doi: 10.1109/MCAS.2019.2945209.
72. M. A. Zidan, H. A. H. Fahmy, M. M. Hussain, and K. N. Salama, "Memristor-based memory: the sneak paths problem and solutions," *Microelectronics Journal*, vol. 44, no. 2, pp. 176–183, 2013, doi: 10.1016/j.mejo.2012.10.001.

4 Innovative Design and Optimization of High-Power Amplifiers

A Comparative Study with GaN HEMT and CMOS Technologies

Manju Yadav, Ramji Gupta, Abhishek Narayan Tripathi, and Alpana Pandey

4.1 INTRODUCTION

Power amplifier (PA) is an electrical device that boosts the strength or amplitude of an electrical signal. It's an important part of communication systems, audio systems, and many other applications where a signal has to be amplified to a greater power level [1–3].

A PA's main job is to take in a low-power electrical signal and boost it up to a level where it can power a load, like a motor, a loudspeaker, or an antenna. It retains the authenticity and quality of the original signal while taking a small input signal and producing a bigger, more powerful output signal [4]. PAs are classified into different types based on the amplification technology used. Some common types include:

1. **Transistor Amplifiers:** Because of their great performance, small size, and efficiency, transistor-based amplifiers are utilized extensively. They can be divided further into field-effect transistor (FET) amplifiers and bipolar junction transistor (BJT) amplifiers [5].
2. **Vacuum Tube Amplifiers:** Due to their distinct sound qualities, vacuum tubes, including triodes, tetrodes, and pentodes, were a common component of early amplifiers and are still preferred by certain audiophiles [6].
3. **Operational Amplifiers, or Op-Amps**, are integrated circuits made expressly to increase the gain and precision of signals. They can be coupled with power stages to form PAs, even though their primary use is in low-power applications.

DOI: 10.1201/9781003483038-4

4. **Class-D Amplifiers:** To get high efficiency, Class-D amplifiers employ pulse-width modulation (PWM) techniques. They are frequently utilized in audio systems, particularly in conjunction with portable speakers and subwoofers [7].

PA design factors include regulating heat dissipation, choosing the right amplification technology, figuring out the necessary power output, assuring stability, and safeguarding against overloads and short circuits [8,9].

In many electronic systems, PAs are essential for providing crystal clear audio [10], dependable wireless connectivity, and effective power transfer. With the development of technology, their design principles and performance characteristics have continued to change, improving power delivery and amplification for a variety of applications [11,12].

The first section of the review covers the fundamental functions and features of PAs, emphasizing important variables, including o/p (output) power, gain, and efficiency. In addition, this chapter examines different approaches which are used in the development of PAs, including GaN HEMT and CMOS technologies. The related work of PAs is discussed in the second section. Further, a comparative analysis is conducted between GaN HEMT and CMOS technologies for wireless communication applications, focusing on the fundamentals of power amplifiers and evaluating the performance differences between these technologies. In the last section, conclusion is discussed.

4.2 RELATED WORK

Tuffy and Neal developed a distributed real frequency technique synthesis algorithm for wideband operation, achieving over 70% efficiency over a 51% bandwidth. The algorithm was linearized using digital predistortion [13]. Using a 2-mm GaN-Si die, Tuffy, Neal, and Lyndon Pattison created a small, hybrid PA, which is class J and its operating frequency range is 1–2.1 GHz. The power output of the PA was 13–18 W, and its drain efficiencies ranged from 60% to 69% [14]. The work by Chen, Peng, Merrick, and Brazil contrasts electromagnetic-based and circuit-based Beamforming Operation (BO) for creating 10 W and 30 W Pas, and the output power is 39.8 dBm. Circuit-based BO attains a drain efficiency of over 60%, whereas Electromagnetic EM-based BO provides an output power of 43.8 dBm and 57% drain efficiency [15]. Pang and Jingzhou discuss the need for broadband and multiband PA design techniques for multiband/multi-mode equipment, using GaN HEMT CGH40010F as an active device and performing load-pull simulations to find optimal impedances [16].

Dong, Yitian, Yi Huang, and Ling Qiu propose a joint optimization of the transmit covariance matrix and active antenna set to maximize energy efficiency in Multi-User Multiple Input Multiple Output (MU–MIMO) systems with nonideal PAs [17]. Chen and Peng, propose a multiobjective Bayesian optimization method for broadband Differential Power Analysis (DPA) design, utilizing EM simulation for more accurate performance and minimizing mismatch between desired and simulated impedance trajectories [8]. You, Fei, and Johannes Benedikt [12] propose a methodology for designing broadband PAs using optimized-load-impedance calculation, I–V curve mining, direct impedance calculation, waveform regulations, and harmonic balance simulations for accurate Zopt solutions. The technique allows for simple, compact multiband matching networks without affecting network architecture or frequency bands. A triple-band amplifier was designed using this method, achieving maximum power-added efficiencies and output powers [18]. Kouhalvandi, Lida, Osman Ceylan, and Serdar Ozoguz talk about an automated process and approach for creating a high-power amplifier (PA) that is optimized. The authors also note that automated RF circuit design tools are gaining popularity among researchers and provide efficient resolutions in analog and radio frequency (RF) system designs [19].

The authors, Li and Chuan, describe an enhanced SA-PSO approach to automate the design of high-efficiency PAs. The choice of starting values is crucial for Electronic Design Automation (EDA) optimization, and because PAs have nonlinear features and a large number of parameters, most initial values end up in the local optimization and cause the Harmonic Balance (HB) simulation to converge slowly [20]. For linearizing RF PAs in wideband communication systems, Le and Duc Han, propose an adaptive Iterative Learning Algorithm (ILA) linearization utilizing the Reduced-Parameter Empirical Model (RPEM) method, which outperforms Least Mean Squares (LMS) or Recursive Least Squares (RLS) and successfully compensates for nonlinear distortion under various circumstances [21]. Zhang, Yumeng, and Bruno Clerckx suggest a channel-adaptive waveform design strategy to maximize transmitter input waveforms while taking energy harvester and high-PA nonlinearities into account. They do this by reformulating the structure of the PA and applying Barrier's method [22].

Zhang, Lei, Kaixue Ma, and Haipeng Fu provide a 60-GHz PA by using 55-nm CMOS for high power. They attain good flatness between the frequency range 57 GHz and 62 GHz [23]. Guixiang Jin discusses the design and implementation of an output transformer for a power amplifier, focusing particularly on its filtering capabilities to reduce harmonic distortion and enhance the transformer's overall efficiency [24]. In order to maximize bias voltages, Rawat and Arvind Singh suggest a 919–923 MHz CMOS PA for LoRa applications [25].

4.3 FUNDAMENTALS OF POWER AMPLIFIER

PAs are electronic devices that boost the power of electrical signals [26], essential in audio, wireless communication, radar applications [27], and RF transmitter applications. Nowadays, PA design is done with the optimization techniques for different applications like Bayesian optimization, optimized-load-impedance calculation, particle swarm optimization, waveform optimization for wireless power transfer with PA, and optimization of cascode structure. [28,29].

Fundamentally, PAs have the following key characteristics:

1. **Power Output:** A PA's power output is its maximum RF output power, determining its ability to deliver desired signal strength or coverage in specific applications [30].

$$P_{out} = I^2 R_{out} \tag{4.1}$$

Measured in watts or decibels relative to a reference power, it is crucial for signal strength. The power output of a PA depends on various factors [31]. PAs' power output is influenced by several factors, including supply voltage, load impedance, linearity, operating frequency, and thermal considerations [32]. Supply voltage increases power output, while load impedance matches the amplifier to the load. Linearity allows for higher output without distortion [33]. Operating frequency affects power output, and thermal management ensures reliable operation at higher power levels.

PA output is the maximum achievable power under ideal conditions, but practical factors like power supply, thermal constraints, and device limitations can lower the actual output [34]. Selecting the right amplifier is crucial.

2. **Power Gain:** PAs increase power levels by amplifying low-power input signals. The gain, the ratio between output and input power, is determined by the amplifier's design and specifications [35]. It can be fixed or adjustable, with some having a manufacturer-determined gain setting or variable gain control for system requirements [36]. There are two ways to express gain:
   I. **Voltage Gain**: The ratio of output voltage [37] to input voltage and is typically expressed by the formula in equation (4.2):

$$A_V = (V_{OUT} / V_{IN}) \tag{4.2}$$

   II. **Power Gain (Ap)**: The relation between the output to input provides gain. It is represented by equation (4.3):

$$A_P = 10 * \log_{10}(P_{OUT} / P_{IN}) \tag{4.3}$$

   The gain of a PA is influenced by input/output impedance matching, load conditions, and design parameters. So, accurate determination requires considering these factors and amplifier specifications [38].

3. **Efficiency:** Efficiency is how much output power is received from the input power [39]. The efficiency is represented by equation (4.4):

$$\text{Eff. } (\eta) = \frac{\text{Output power}}{\text{Input power}} * 100\% \tag{4.4}$$

It is usually expressed as a percentage.

Power-added efficiency (PAE)

Power-added efficiency (PAE) basically converts DC input power into RF output power [40,41]. The PAE formula is given by equation (4.5):

$$\text{PAE} = \frac{\text{RF output power} - \text{RF input power}}{\text{DC input power}} * 100\% \tag{4.5}$$

4.3.1 S-Parameters

The S-parameters are scattering parameters, which are useful for describing PAs. They describe the impedance, the gain, the bandwidth, and the stability [42].

S11 (or S22): In PA, S11 describes about the power reflected back at either the input port or the output port because of an impedance mismatch. The S22 describes the stability factor, which is referred to as the Rollett stability factor; a value of less than 1 indicates stability [43]. S21 (or S12): The higher values of the S21 indicate more power amplification, while the reverse gain is represented by S12.

That's why the S-parameters for PAs are important for PA design prediction, analysis of the parameters, input and output impedance matching, stability, and performance prediction for precise frequency [44,45].

4.4 DISCUSSIONS ON FUTURE ASPECTS AND CHALLENGES OF POWER AMPLIFIERS

For designing the PA, we are using transistors and different CMOS (complementary metal-oxide-semiconductor) to provide high gain and high efficiency for different applications [46,47]. Some general techniques for designing PAs are discussed in this section.

4.4.1 Design of Power Amplifier Circuit with Different Technologies and Details about Parameters with Graphs

4.4.1.1 Using GaN HEMT Power Amplifier

Gallium Nitride (GaN) high-electron-mobility transistor (HEMT) technology provides a wide-bandgap semiconductor substantial with special power handling abilities and also a high breakdown voltage ability [48]. By comparing GaN HEMT [49] PA with conservative technologies such as silicon-based Metal-Oxide-Semiconductor Field-Effect Transistors (MOSFETs) or GaAs-based HEMTs, the GaN HEMT PAs provide advantages of high efficiency, low power, and high gain for high-frequency operations [50,51].

Figure 4.1 shows the block diagram of the PA, in which the first block includes an input matching network with an inductor and a capacitor components. The second block represents the technology used like MOSFET, transistor, and GaN HEMT, and the third block shows the output matching network with a capacitor and an inductor

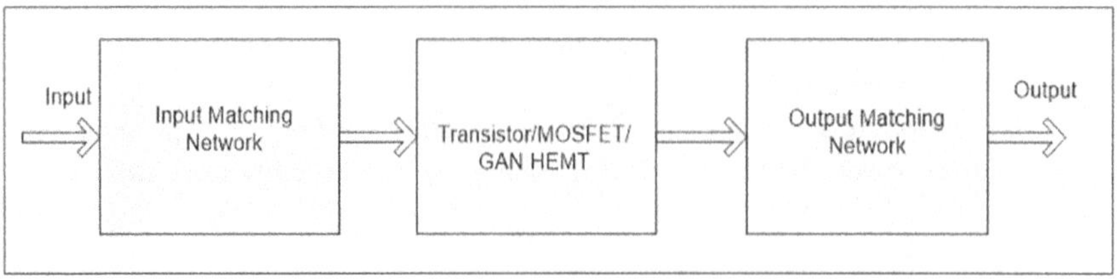

FIGURE 4.1 Block diagram of power amplifier.

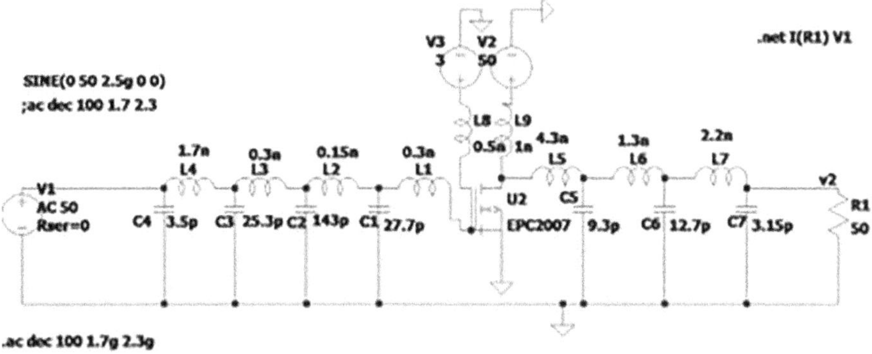

FIGURE 4.2 GaN HEMT power amplifier circuit design [19].

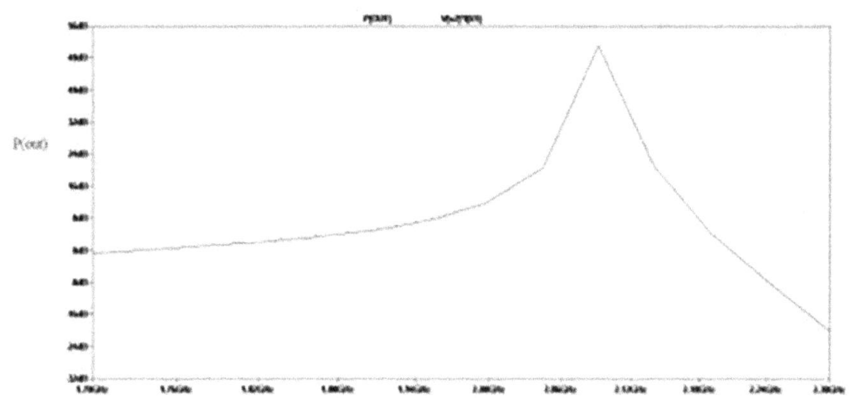

FIGURE 4.3 Output power of GaN HEMT power amplifier.

component. This block diagram represents how input power amplifies the input signal by using an input/output matching network with appropriate component values of *L*, *C*, *R*, V_{gs}, and V_{ds}.

Figure 4.2 represents the GaN HEMT PA circuit, which is implemented on LT Spice software with appropriate values of the *L*, *R*, *C*, V_{ds}, and V_{gs}.

In this simulation, the GaN HEMT EPC 2007 model is used for the verification.

Figure 4.3 represents the output power of the GaN HEMT PA, which shows the output power variation between the frequency range of 1.7–2.5 GHz. The output power starts increasing from a frequency of 1.5 GHz and gets a maximum value of 49 dB at 2.12 GHz. Due to the S11 parameter (return loss), the output power exponentially decreases beyond the frequency of 2.14 GHz.

Because of return loss (S11), the gain of the PA decreases exponentially beyond the 2.12 GHz frequency shown in Figure 4.4, which also starts increasing from the frequency 1.5 GHz and gets a maximum value of 11 dB at 2.12 GHz, after which the frequency gain again decreases up to 2.5 GHz.

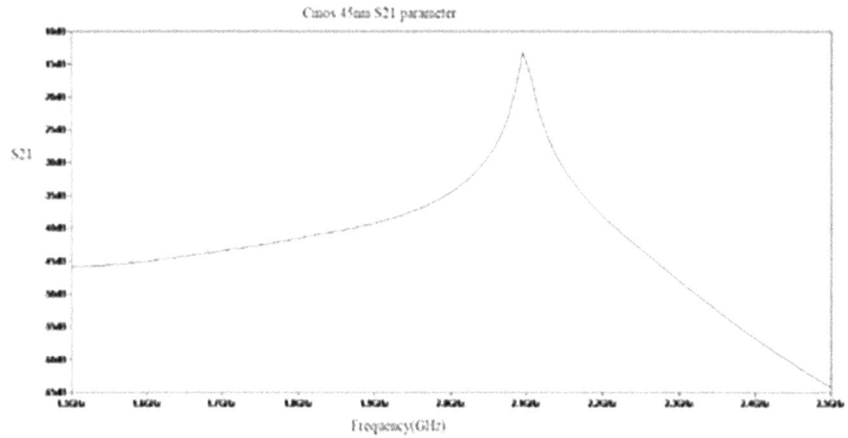

FIGURE 4.4 GaN HEMT power amplifier gain.

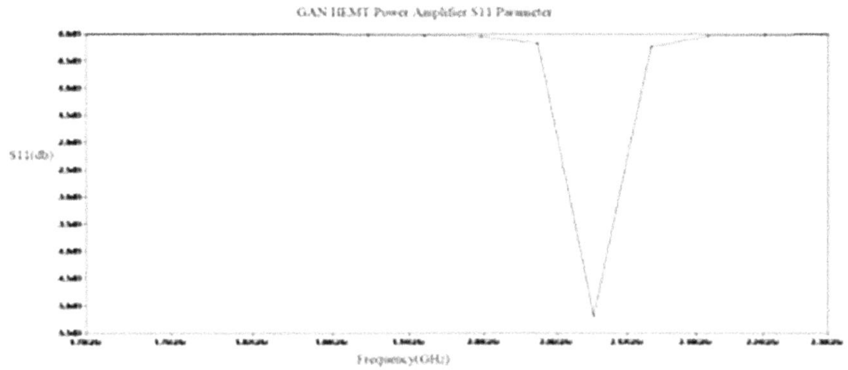

FIGURE 4.5 GaN HEMT power amplifier S11 parameter.

Figure 4.5 represents the S11 (return loss) parameter of the GaN HEMT PA. Its value decreases between the frequency range 2.00–2.22 GHz and a minimum −5dB at frequency 2.14 GHz. Because of this, the high power output is achieved at 2.14 GHz.

Figure 4.6 represents the PAE of GaN HEMT is 49%–86% for the frequency range 1.5–2.5 GHz, and the maximum value is 86% at frequency 2.12 GHz.

Figure 4.7 represents the S21 parameter (transmission coefficient/forward gain), which starts increasing from frequency 1.7–2.12 GHz, and the maximum value obtained −6 dB at 2.12 GHz; so at this frequency, forward gain is high; after this frequency, it starts decreasing exponentially up to 2.3 GHz.

The S22 parameter (output rejection coefficient) is represented in Figure 4.8. It shows that the value of the S22 starts increasing from 1.7 GHz frequency to 2.30 GHz.

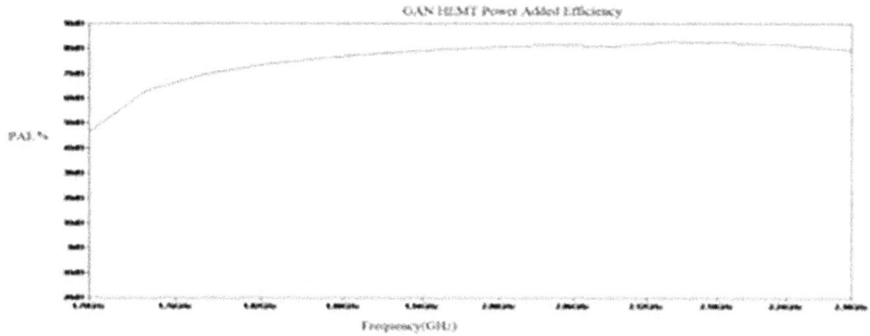

FIGURE 4.6 GaN HEMT power-added efficiency.

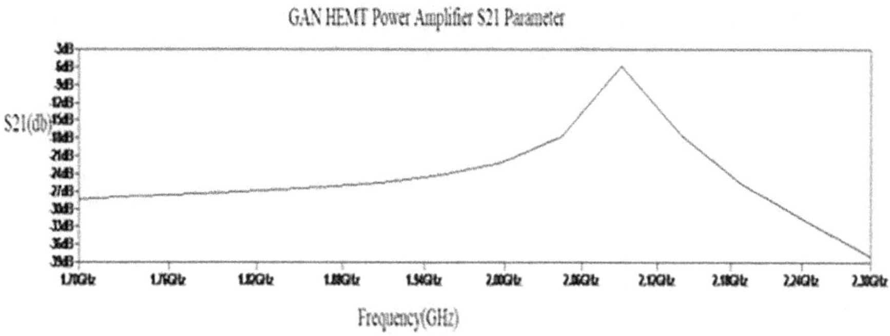

FIGURE 4.7 GaN HEMT power amplifier S21 parameter.

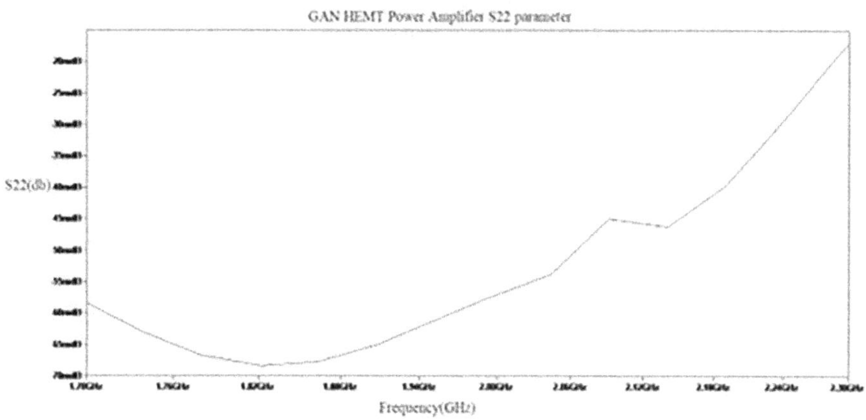

FIGURE 4.8 GaN HEMT power amplifier S22 parameter.

4.4.1.2 CMOS 45 nm Power Amplifier [52]

The CMOS 45 nm PA uses a 45-nm technology to amplify the PA signals [53,54]. There are various applications of the CMOS 45 nm PAs. Basically, CMOS PAs are used for wireless communication systems [55]. They are used to increase the strength of the signal which is transmitted to the antenna. They are also used in various audio devices to increase the strength of the audio signals, and examples are headphones or speakers [56,57]. The below design and graphs of parameters are discussed in detail. Figure 4.9 shows the PA design by using 45 nm CMOS technology. V_{gs}(V3) and V_{ds}(V2) are 4 V and 10 V with L8 = 1 nm, L9 = 1 nm, and R_{out} = 50 ohm.

In Figure 4.10, the output power of the PA using CMOS 45 nm is presented. The pout varies with frequency. From the frequency of 1.5–2.5 GHz, the pout starts increasing up to 2.12 GHz and achieves its maximum value of 47 dB; after that, the pout decreases exponentially up to 2.5 GHz frequency.

Figure 4.11 represents the gain of CMOS 45 nm PA. The gain starts increasing from 1.5 GHz frequency to 2.12 GHz frequency and attains maximum gain at 2.12 GHz, which is 6.7 dB. After this, the frequency gain starts exponentially decreasing up to 2.5 GHz frequency.

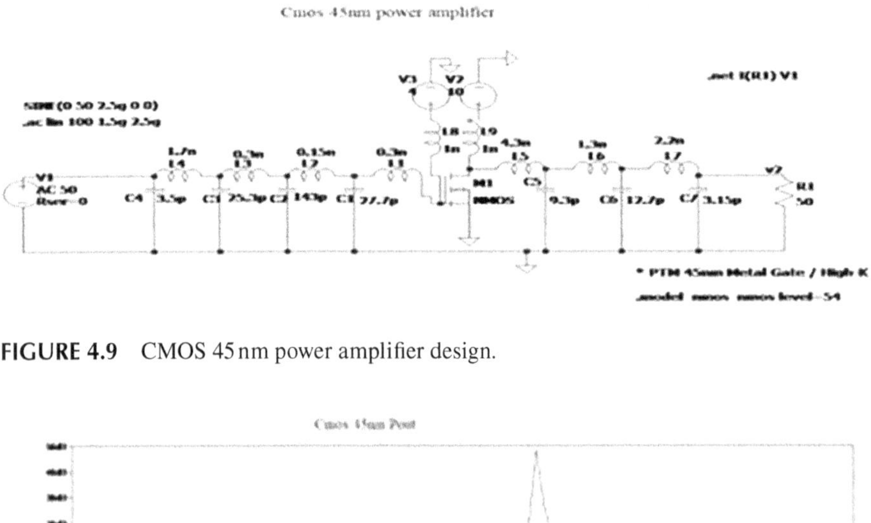

FIGURE 4.9 CMOS 45 nm power amplifier design.

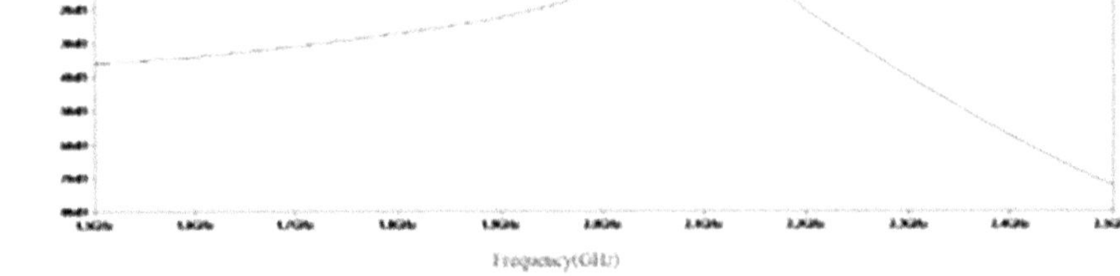

FIGURE 4.10 CMOS 45 nm power output pout.

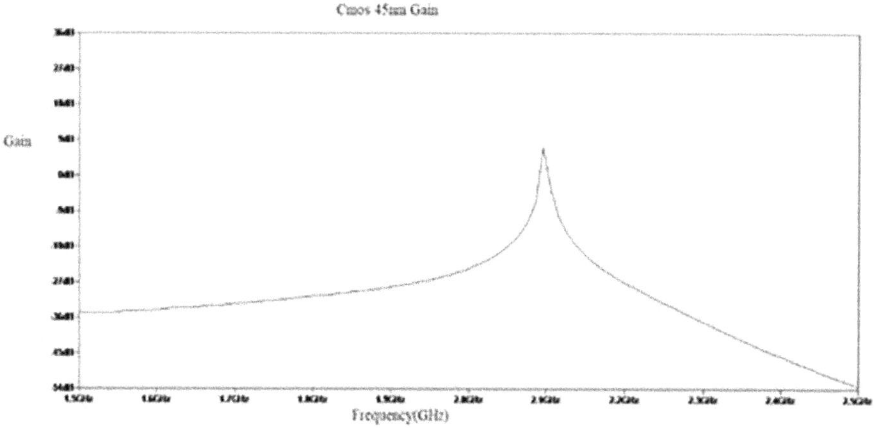

FIGURE 4.11 CMOS 45 nm power amplifier gain.

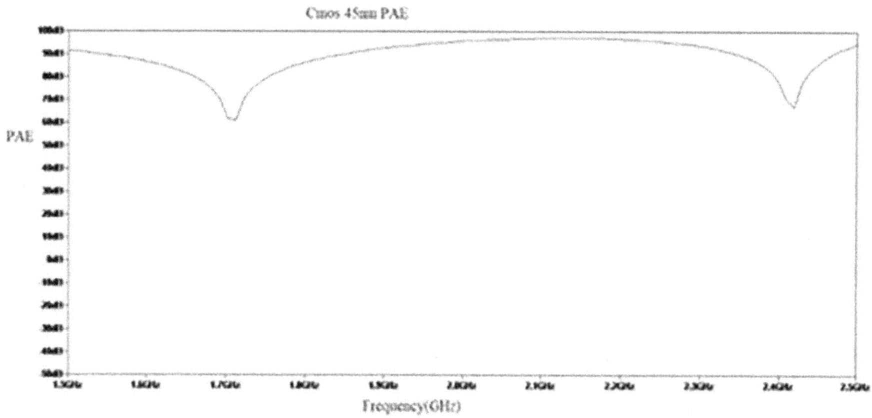

FIGURE 4.12 CMOS 45 nm power amplifier PAE.

Figure 4.12 shows the PAE of the CMOS 45 nm PA. It represents two drops in the PAE value at frequencies 1.7 GHz and 2.2 GHz. The maximum PAE is 83%, and the range of PAE is between 70% and 83%.

The S-parameters S11 of CMOS 45 nm PA are represented by Figure 4.13, which provides the return loss at port 1 with the frequency variation from 1.5 to 2.5 GHz. It also shows the minimum value of −4.4 dB at frequency 2.12 GHz. Because of less return loss at 2.14 GHz frequency, the gain is high at this frequency.

Figure 4.14 shows S21 parameter variation with frequency range from 1.5 to 2.5 GHz; the maximum value is −12 dB at 2.14 GHz.

The S22 parameter (output rejection coefficient) is represented by Figure 4.15. It shows that the value of the S22 starts increasing from 1.5 GHz frequency to 2.50 GHz, and the maximum value of S22 is −1.5 dBm.

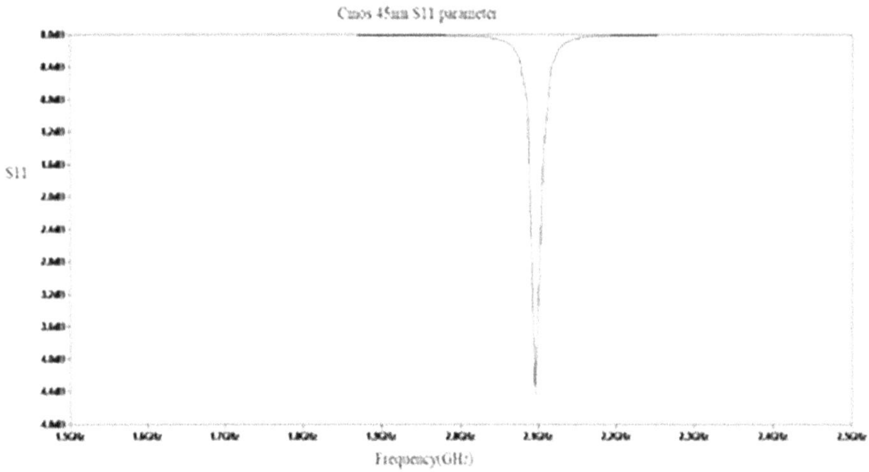

FIGURE 4.13 CMOS 45 nm power amplifier S11 parameter.

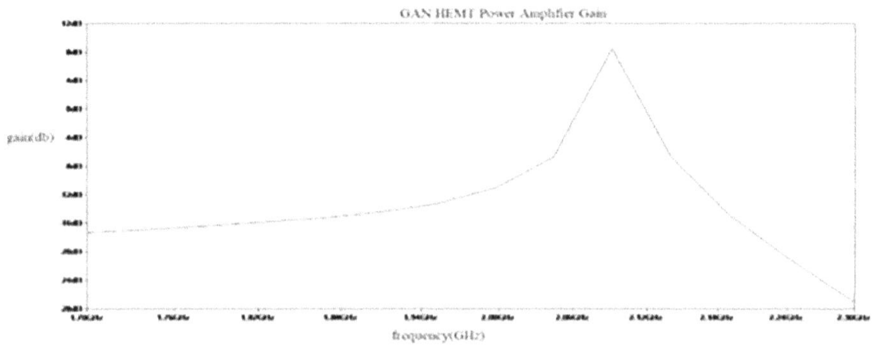

FIGURE 4.14 CMOS 45 nm power amplifier S21 parameter.

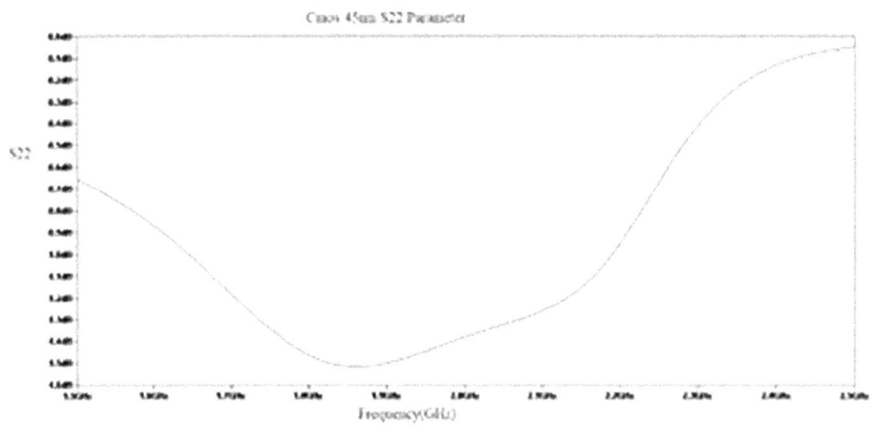

FIGURE 4.15 CMOS 45 nm power amplifier S22 parameter.

4.4.1.3 CMOS 22 nm Power Amplifier Design

One more technology of CMOS PAs is discussed here in detail with the graph analysis. Figure 4.16 represents the CMOS 22 nm PA pout with a frequency range 1.7–2.3 GHz. The P_{out} varies from −21 dB to a maximum value of 41 dB. The maximum value of 41 dB is at 2.12 GHz frequency.

Figure 4.17 represents a CMOS 22 nm PA and a maximum power gain of 4 dB at 2.12 GHz frequency. The graph represents the variation of power gain with a frequency range of 1.72.3 GHz.

Figure 4.18 shows the CMOS 22 nm power amplifier's power-added efficiency with a frequency range of 1.72.3 GHz; the PAE varies from 50% to a maximum value of 81% at a frequency of 12 GHz.

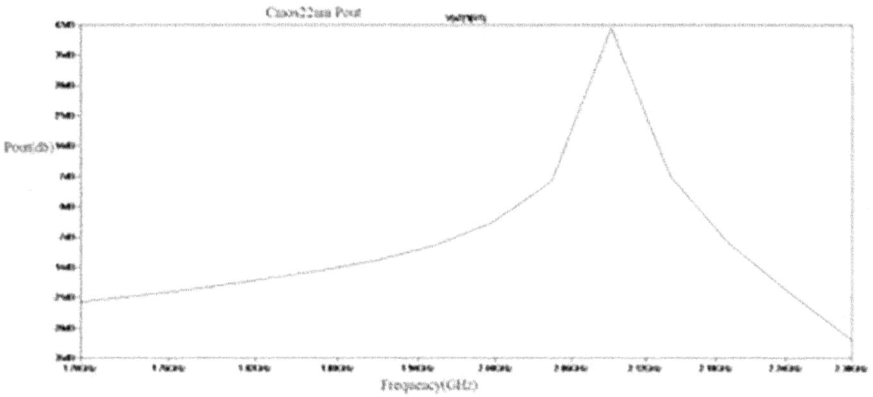

FIGURE 4.16 CMOS 22 nm power amplifier pout.

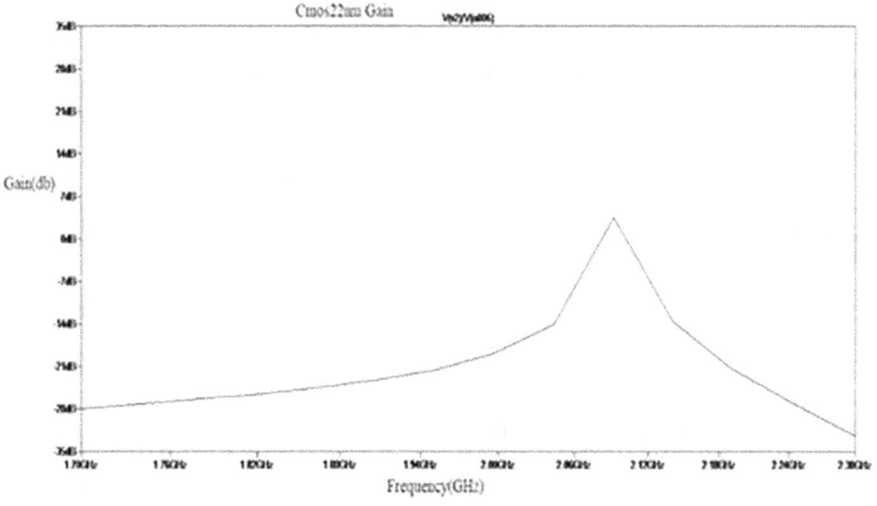

FIGURE 4.17 CMOS 22 nm power amplifier gain.

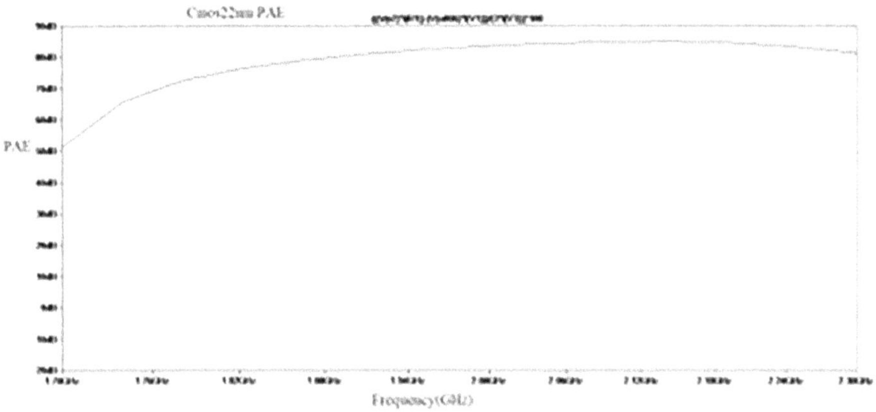

FIGURE 4.18 CMOS 22 nm power amplifier PAE.

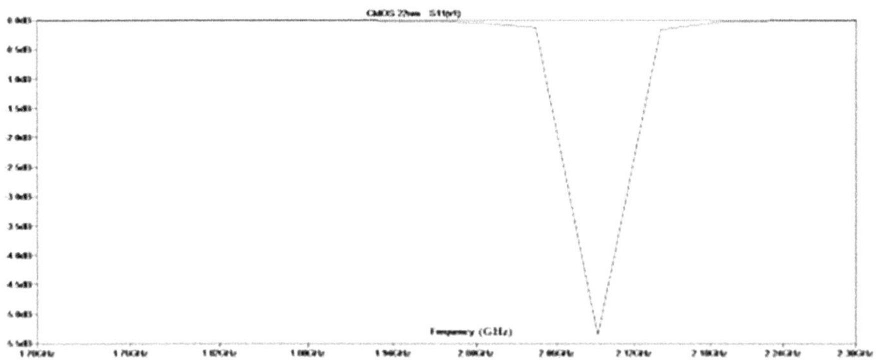

FIGURE 4.19 CMOS 22 nm power amplifier S11 parameter.

The S11 parameter of the CMOS 22 nm PA graph is represented by Figure 4.19. The S11 varies with respect to frequency and its minimum value is −6.5 dB at 2.12 GHz frequency. It represents that at 2.12 GHz frequency, the return loss is minimum so the pout is more.

The S21 (forward gain) of the PA is represented by Figure 4.20. The variation of S21 from frequency 1.70–2.30 GHz is observed in Figure 4.20. The maximum S21 is −11 dB at 2.12 GHz frequency, which shows maximum gain at 2.12 GHz frequency.

Figure 4.21 represents the S22 parameter of the CMOS 22 nm PA. It starts increasing from 1.7 GHz frequency and gets a maximum value of 45 dBm at 2.12 GHz frequency, which shows the power reflected at the PA output.

It's noteworthy that precise circuit formations and designs differ according to the required power output, frequency, and applications [58,59]. The below table provides the details about the input power, output power, gain of PA, and PA's PAE with different technologies.

From Table 4.1, it is easy to understand that the GaN HEMT technology provides best PAE, gain, and pout in comparison with other technologies. We choose one

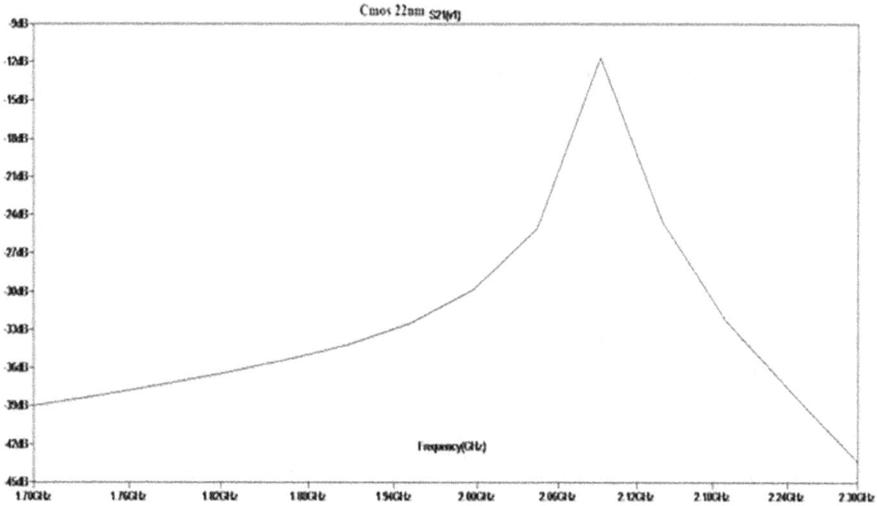

FIGURE 4.20 CMOS 22 nm power amplifier S21 parameter.

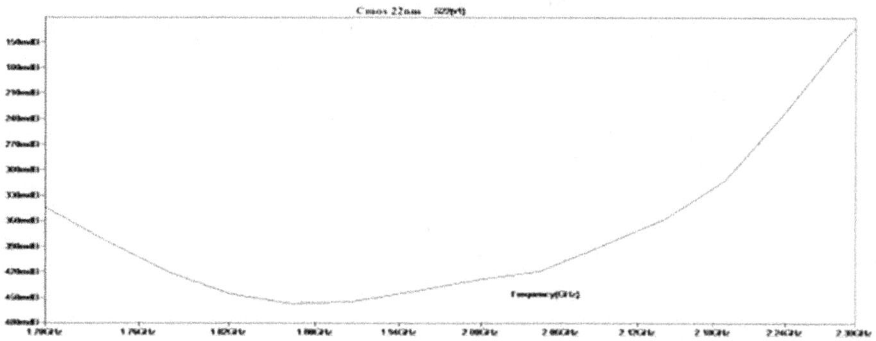

FIGURE 4.21 CMOS 22 nm power amplifier S22 parameter.

TABLE 4.1
Comparison of Different Technologies for Different Parameters

	P(IN) (db)	P(out) (db)	Gain (db)	PAE%
GaN HEMT power amplifiers 180 nm	33	49	11	86
CMOS 45 nm	33	47	6.7	83
CMOS 22 nm	33	41	4	81

circuit design [19] and provide different technologies to get the gain, power output, and PAE and find the best technology GaN HEMT for further research purposes. Also, Table 4.1 provides an analysis of the PAE of different technologies. The paper [19] presents GaN HEMT technology with a maximum PAE of 77%, while the

TABLE 4.2
Comparative Analysis of Power Amplifier's Parameters with Different Technologies

Technology	Power Output(db)	PAE(Power-Added Efficiency)(db)	Gain(db)	References
40nm GaN HEMT	31.1	27	20	[60]
100nm GaN HEMT	20.3	2.6	34.8	[61]
65nm CMOS	19.3	19.2	24.2	[62]
40nm CMOS	20.9	22	18.1	[63]
100nm GaAs pHEMT	28	13	26	[64]
150nm InP HEMT	26	23	10	[65]
90nm SiGe BiCMOS	27.3	12.4	19.3	[66]
70nm GaN HEMT	30	8	16	[67]

simulated results of the technology discussed here show a higher PAE of 86%. This demonstrates that the simulated GaN HEMT technology outperforms the results in paper [19] in terms of power-added efficiency (PAE).

Table 4.2 represents the different technologies of the power amplifier and respective power output, PAE, and gain. The maximum output power is achieved, i.e., 31.1dB, using 40nm GaN HEMT technology with the gain and PAE, i.e., 20 and 27dB, respectively. The maximum gain of 34.8dB was achieved using 100nm GaN HEMT technology where the PAE was 2.6dB and output power delivered was 20.3dB. Here, the comparative table provides a basic analysis that the GaN HEMT technology provides good results for PA designing circuits over other technology, although the trade-off is seen in the parameters while designing the circuit.

4.5 CONCLUSION

The PA is the most important device used in electronics and communication engineering. There are various applications of PAs using CMOS 45nm, CMOS 22nm, and GaN HEMT technologies. The wireless communications and radar systems mostly use GaN HEMT's PA for high power density, high-frequency operation, and to get high efficiency. The various technologies provide different outputs for gain, efficiency, and power output. The best result in terms of PAE is found by using GaN HEMT transistors. The power level, the frequency range, the efficiency, the cost, and the integration are some of the applications that provide the selection of PA technology. The GaN HEMT PA is best suited for high-power and high-frequency applications, whereas the CMOS PA technologies provide various application-specific approaches.

REFERENCES

1. S.T. Cripps. *RF Power Amplifiers for Wireless Communications*, Vol. 250. Artech House (2006).
2. R. Tong. *Efficient Solid-State Power Amplifiers for RF Power Source Applications*. Doctoral dissertation, Acta Universitatis Upsaliensis (2022).

3. P. Chen, B.M. Merrick, and T.J. Brazil. "Bayesian optimization for broadband high-efficiency power amplifier designs." *IEEE Transactions on Microwave Theory and Techniques* 63(12) (2015): 4263–4272.
4. S. Hu, F. Wang, and H. Wang. "A 28-/37-/39-GHz linear doherty power amplifier in silicon for 5G applications." *IEEE Journal of Solid-State Circuits* 54(6) (2019): 1586–1599.
5. Y. Dong, Y. Huang, and L. Qiu. "Energy-efficient sparse beamforming for multiuser MIMO systems with nonideal power amplifiers." *IEEE Transactions on Vehicular Technology* 66(1) (2016): 134–145.
6. C. Li, F. You, S. He, X. Tang, W. Shi, and J. Wang, "High-efficiency power amplifier employing minimum-power harmonic active load modulator." *IEEE Transactions on Circuits and Systems II: Express Briefs*, 66(8) (2019): 1371–1375.
7. J. Hur, H. Kim, O. Lee, K. Kim, F. Bien, K. Lim, C. Lee, and J. Laskar, "A multilevel class-D CMOS power amplifier for an outphasing transmitter with a nonisolated power combiner." *IEEE Transactions on Circuits and Systems II: Express Briefs*, 63(7) (2016): 618–622.
8. P. Chen, et al. "Multiobjective bayesian optimization for active load modulation in a broadband 20-W GaN Doherty power amplifier design." *IEEE Transactions on Microwave Theory and Techniques* 65(3) (2016): 860–871.
9. T.C. Yu, and C.-L. Yang. "Design and analysis of dual-frequency power amplifier for wireless power and data transfer application." *2017 IEEE Wireless Power Transfer Conference (WPTC)*. IEEE, Dresden, Germany (2017).
10. J. Pang, et al. "Analysis and design of highly efficient wideband RF-input sequential load modulated balanced power amplifier." *IEEE Transactions on Microwave Theory and Techniques* 68(5) (2020): 1741–1753.
11. A. Vasjanov, and V. Barzdenas. "A review of advanced CMOS RF power amplifier architecture trends for low power 5G wireless networks." *Electronics* 7(11) (2018): 271.
12. F. You, and J. Benedikt. "An optimized-load-impedance calculation and mining method based on I–V curves: using broadband class-E power amplifier as example." *IEEE Transactions on Industrial Electronics* 66(7) (2018): 5254–5263.
13. N. Tuffy, et al. "A simplified broadband design methodology for linearized high-efficiency continuous class-F power amplifiers." *IEEE Transactions on Microwave Theory and Techniques* 60(6) (2012): 1952–1963.
14. N. Tuffy, and L. Pattison. "A compact high efficiency GaN-Si PA implemented in a low cost DFN package with 71% fractional bandwidth." *2014 IEEE MTT-S International Microwave Symposium (IMS2014)*. IEEE, Tampa, Florida, USA (2014).
15. P. Chen, BM. Merrick, and T.J. Brazil. "Bayesian optimization for broadband high-efficiency power amplifier designs." *IEEE Transactions on Microwave Theory and Techniques* 63(12) (2015): 4263–4272.
16. J. Pang, et al. "A novel design of concurrent dual-band high efficiency power amplifiers with harmonic control circuits." *IEEE Microwave and Wireless Components Letters* 26(2) (2016): 137–139.
17. Y. Dong, Y. Huang, and L. Qiu. "Energy-efficient sparse beamforming for multiuser MIMO systems with nonideal power amplifiers." *IEEE Transactions on Vehicular Technology* 66(1) (2016): 134–145.
18. A. Munir, A. Faroqi, A.M. Ridwan, and A.F. Pratiwi, "Power transfer efficiency of planar inverted-F wireless charging radiator." *Proceedings of International Symposium on Electronics and Smart Devices (ISESD)*, pp. 1–4. IEEE, Bandung, Indonesia (2018).
19. L. Kouhalvandi, O. Ceylan, and S. Ozoguz. "Automated matching network modeling and optimization for power amplifier designs." *2019 11th International Conference on Electrical and Electronics Engineering (ELECO)*. IEEE, Bursa, Turkey (2019).
20. C. Li, et al. "Simulated annealing particle swarm optimization for high-efficiency power amplifier design." *IEEE Transactions on Microwave Theory and Techniques* 69(5) (2021): 2494–2505.

21. D.H. Le, et al. "Linearization of RF power amplifiers in wideband communication systems by adaptive indirect learning using RPEM algorithm." *Mobile Networks and Applications* 25 (2020): 1988–1997.
22. Y. Zhang, and B. Clerckx. "Waveform optimization for wireless power transfer with power amplifier and energy harvester non-linearities." *ICASSP* 2022-2022 *IEEE International Conference on Acoustics, Speech and Signal Processing (ICASSP)*. IEEE, Singapore (2022).
23. L. Zhang, K. Ma, and H. Fu. "A 60-GHz 32-way hybrid power combination power amplifier in 55-nm bulk CMOS." *IEEE Transactions on Microwave Theory and Techniques* 71(2) (2022): 613–627.
24. G. Jin, et al. "A linear CMOS power amplifier with efficiency-optimized transformer matching." *IEEE Microwave and Wireless Components Letters* 32(9) (2022): 1059–1062.
25. A.S. Rawat, et al. "A 919 MHz—923 MHz, 21 dBm CMOS power amplifier with bias modulation linearization technique achieving PAE of 29% for LoRa application." *IEEE Access* 10 (2022): 79365–79378.
26. V. Camarchia, et al. "The Doherty power amplifier: review of recent solutions and trends." *IEEE Transactions on Microwave Theory and Techniques* 63(2) (2015): 559–571.
27. J. Li, YZ. Xiong, Y. Li et al. Analysis and optimization of cascode structure in power amplifier for X-Band phase array radar application. *Circuits, Systems, and Signal Processing* 34 (2015): 1–20.
28. A. Papadimitriou, and M. Bucher Multi-objective low-noise amplifier optimization using analytical model and genetic computation. *Circuits, Systems, and Signal Processing* 36 (2017): 4963–4993.
29. C. Talarico, G. Agrawal, J. Wang-Roveda et al. Design optimization of a transimpedance amplifier for a fiber optic receiver. *Circuits, Systems, and Signal Processing* 34 (2015): 2785–2800.
30. N. Tuffy, et al. "A simplified broadband design methodology for linearized high-efficiency continuous class-F power amplifiers." *IEEE Transactions on Microwave Theory and Techniques* 60(6) (2012): 1952–1963.
31. D. Zhang, B. Zhong, Z. Jin et al. High-voltage amplifier with high dynamic response for stick–slip driving. *Circuits, Systems, and Signal Processing* 39 (2020): 2759–2775.
32. K. Ding, D.M.W. Leenaerts, and H. Gao. "A 28/38 GHz dual-band power amplifier for 5G communication." *IEEE Transactions on Microwave Theory and Techniques* 70(9) (2022): 4177–4186.
33. R. Schumacher, E.G. Lima & G.H.C. Oliveira RF power amplifier behavioral modeling based on Takenaka–Malmquist–Volterra series. *Circuits, Systems, and Signal Processing* 35 (2016): 2298–2316.
34. D. Joshi, S. Dash, S., Reddy et al. Multi-objective hybrid particle swarm optimization and its application to analog and RF circuit optimization. *Circuits, Systems, and Signal Processing* 42 (2023): 4443–4469.
35. S.C. Cripps, et al. "On the continuity of high efficiency modes in linear RF power amplifiers." *IEEE Microwave and Wireless Components Letters* 19(10) (2009): 665–667.
36. Z. Li, B. Liu, Y. Duan et al. "Flat-high-gain design and noise optimization in SiGe low-noise amplifier for S–K band applications". *Circuits, Systems, and Signal Processing* 40 (2021) 2720–2740.
37. L. Ma, Z. Wang, J. Xu et al. "Noise analysis and optimization of programmable gain amplifier with DC offset cancelation". *Circuits, Systems, and Signal Processing* 35 (2016): 753–770.
38. T.S. Delwar, A. Siddique, U. Aras et al. "A Design of adaptive genetic algorithm-based optimized power amplifier for 5G applications". *Circuits, Systems, and Signal Processing* 43 (2024): 2–21.

39. É.A. Mindubaev, and N.O. Savel'ev. "Methods for improving the efficiency of transcutaneous wireless power transmission systems using a class E power amplifier." *Biomedical Engineering* 57 (2023) 45–47.
40. R. Raja, R. Theegala, and B. Venkataramani. "A class-E power amplifier with high efficiency and high power-gain for wireless sensor network." *Microsystem Technologies* 23 (2017): 4179–4193.
41. M. Li, et al. "A 17–26.5 GHz 42.5 dBm broadband and highly efficient gallium nitride power amplifier design." *Frontiers of Information Technology & Electronic Engineering* 23(2) (2022): 346–350.
42. A. Eroglu. *Introduction to RF Power Amplifier Design and Simulation.* CRC press (2018).
43. N. Tuffy, et al. "A simplified broadband design methodology for linearized high-efficiency continuous class-F power amplifiers." *IEEE Transactions on Microwave Theory and Techniques* 60(6) (2012): 1952–1963.
44. B. Park, D. Kim, S. Kim, Y. Cho, J. Kim, D. Kang, S. Jim, K. Moon, and B. Kim, "High-performance CMOS power amplifier with improved envelope tracking supply modulator." *IEEE Transactions on Microwave Theory and Techniques* 64(3) (2016): 798–809.
45. S. Manjula, and D. Selvathi. "Optimal design of low power CMOS power amplifier using particle swarm optimization technique." *Wireless Personal Communications* 82(4) (2015): 2275–2289.
46. A.V. Rambabu,·K. Kumar, D. Haripriya, and·A.K. Tamilarasan. *Design of Low Power RF CMOS Power Amplifer Structure with an Optimal Linear Gain Controller for Future Wireless Communication.* Springer Nature (2021).
47. Y. Yin, B. Chi, Z. Xia, and Z. Wang, "A reconfigurable dual-mode CMOS power amplifier with integrated T/R switch for 0.1–1.5-GHz multistandard applications." *IEEE Transactions on Circuits and Systems II: Express Briefs*, 61(7) (2014): 471–475.
48. K. Chen, D. Peroulis, "Design of adaptive highly efficient GaN power amplifier for octave-bandwidth application and dynamic load modulation." *IEEE Transactions on Microwave Theory and Techniques*, 60(6) (2012): 1829–1839.
49. N.T.A. Nia, and V. Nayyeri. "A 0.85–5.4 GHz 25-W GaN power amplifier." *IEEE Microwave and Wireless Components Letters* 28(3) (2018): 251–253.
50. J. Pang, S. He, Z. Dai, C. Huang, J. Peng, and F. You. "Design of continuous-mode GaN power amplifier with compact fundamental impedance solutions on package plane." *IET Microwaves, Antennas & Propagation* 10(10) (2016): 1056–1064.
51. J.J.M. Rubio, et al. "A 0.6–3.8 GHz GaN power amplifier designed through a simple strategy." *IEEE Microwave and Wireless Components Letters* 26(6) (2016): 446–448.
52. N. Rostomyan, M. Ozen, and P. Asbeck. "Comparison of pMOS and nMOS 28 GHz high efficiency linear power amplifiers in 45 nm CMOS SOI." *2018 IEEE Topical Conference on RF/Microwave Power Amplifiers for Radio and Wireless Applications (PAWR).* IEEE, Anaheim, California, USA (2018).
53. J.-H Chen, et al. "A broadband stacked power amplifier in 45-nm CMOS SOI technology." *IEEE Journal of Solid-State Circuits* 48(11) (2013): 2775–2784.
54. A. Agahr, et al. "Multi-drive stacked-FET power amplifiers at 90 GHz in 45 nm SOI CMOS." *IEEE Journal of Solid-State Circuits* 49(5) (2014): 1148–1157.
55. S. Du, and X. Zhu. "A low-power CMOS class-E Chireix RF outphasing power amplifier for WLAN applications." *Wireless Personal Communications* 90 (2016): 1547–1561.
56. J. Feng, and M. Onabajo. "Wide dynamic range CMOS amplifier design for RF signal power detection via electro-thermal coupling." *Journal of Electronic Testing* 30 (2014): 101–109.
57. S.B. Hamidi, and D. Dawn. "Fully integrated CMOS tunable power amplifier using reconfigurable input/interstage/output matching networks." *Analog Integrated Circuits and Signal Processing* 107 (2021): 73–82.

58. B. Francois and P. Reynaert, "A fully integrated watt-level linear 900-MHz CMOS RF power amplifier for LTE-applications." *IEEE Transactions on Microwave Theory and Techniques* 60(6) (2012): 1878–1885.
59. S. J. Wilk, W. Lepkowski, and T. J. Thornton, "32 dBm power amplifier on 45 nm SOI CMOS." *IEEE Microwave and Wireless Components Letters* 23(3) (2013): 161–163.
60. A. Margomenos et al., "GaN technology for E, w and G-band applications." In *Proceedings IEEE Compound Semiconductor Integrated Circuit Symposium* (CSICS), San Diego, California, USA (2014): pp. 1–4.
61. R. Weber et al., "A beyond 110 GHz GaN cascode low-noise amplifier with 20.3 dBm output power." In *IEEE MTT-S International Microwave Symposium Digest* , Philadelphia, Pennsylvania, USA (2018): pp. 1499–1502.
62. K.-Y. Wang, T.-Y. Chang, and C.-K. Wang, "A 1 V 19.3dBm 79GHz power amplifier in 65nm CMOS." In *IEEE International Solid-State Circuits Conference Digest of Technical Papers*, San Francisco, California, USA (2012): pp. 260–262.
63. D. Zhao and P. Reynaert, "14.1 a 0.9 V 20.9dBm 22.3%-PAE Eband power amplifier with broadband parallel-series power combiner in 40nm CMOS." In *IEEE International Solid-State Circuits Conference Digest of Technical Papers*, San Francisco, California, USA (2014): pp. 248–249.
64. J.T. Bessemoulin, M. Rodriguez, M. G. McCulloch, A. E. Parker, and S. J. Mahon, "Reduced-size E-band GaAs power amplifier MMIC" In *Proceedings 10th European Microwave Integrated Circuits Conference (EuMIC)*, Paris, France (2015): pp. 25–28.
65. Z. Griffith, M. Urteaga, P. Rowell, and R. Pierson, "340–440 mW broadband, high-efficiency E-Band PA's in InP HBT." In *Proceedings IEEE Compound Semiconductor Integrated Circuit Symposium* (CSICS), Monterey, California, USA (2015): pp. 1–4.
66. H.-C. Lin and G. M. Rebeiz, "A 70–80-GHz SiGe amplifier with peak output power of 27.3 dBm." *IEEE Transactions on Microwave Theory and Techniques* 64(7) (2016): 2039–2049.
67. D. Schwantuschke, B.-J. Godejohann, P. Bruckner, A. Tessmann, and R. Quay, "Mm-wave operation of AlN/GaN-devices and MMICs at V- & W-band." In *22nd International Microwave and Radar Conference* (MIKON) (2018): pp. 238–241.

5 Exploring FPGA Architecture Designs for Matrix Multiplication in Machine Learning

Jyoti Kori

5.1 INTRODUCTION

Matrix multiplication is a fundamental operation in machine learning algorithms, forming the backbone of many computational tasks such as neural network training and inference. As the demand for efficient and high-performance computing systems in machine learning continues to rise, exploring field-programmable gate array (FPGA) architecture designs for accelerating matrix multiplication has garnered significant interest. This introduction provides an overview of the importance of matrix multiplication in machine learning, the challenges associated with traditional computing architectures, and the potential benefits of leveraging FPGA-based solutions [1].

Machine learning algorithms often involve large-scale matrix operations, including matrix-vector products, matrix-matrix products, and convolutions. Matrix multiplication is particularly prevalent in deep learning models, where layers of neurons are interconnected through weighted connections represented as matrices. Efficient matrix multiplication is crucial for accelerating training and inference tasks, enabling faster model convergence and improved performance [2,3].

There can be a number of difficulties with implementing machine learning processes on FPGA boards which are listed below.

1. **Restricted Resources**: In comparison to conventional Central Processing Units (CPUs) or Graphics Processing Units (GPUs), FPGAs have fewer resources. In order to fit the model and its operations within the available hardware resources, resource utilization must be carefully considered while designing effective implementations of machine learning algorithms on FPGAs.
2. **High Complexity of Development**: There is a high learning curve involved in creating FPGA-based machine learning solutions, and it calls for proficiency in hardware design languages (HDLs), like Verilog or VHSIC Hardware Description Language (VHSIC stands for Very High-Speed

DOI: 10.1201/9781003483038-5

Integrated Circuit) (VHDL), in addition to understanding machine learning algorithms and optimization strategies.

3. **Algorithm Mapping**: To map machine learning algorithms onto FPGA hardware, the algorithms must be divided into operations that are compatible with the hardware and effectively implemented using the FPGA's capabilities. Trade-offs between accuracy, speed, and resource use are frequently present in this process.
4. **Limited Precision**: Floating-point arithmetic, which is frequently utilized in machine learning methods, is generally not supported very well by FPGAs. Reduced precision methods and fixed-point arithmetic can cause accuracy to be lost, necessitating careful optimization and quantization strategies.
5. **Programming and Optimization**: Hardware-level tweaking and optimization must be done by hand in order to optimize machine learning algorithms for FPGA implementation. It takes time and requires knowledge of FPGA design and optimization methods to complete this operation.
6. **Data Movement**: To achieve high performance in machine learning applications, there must be efficient data movement between the FPGA and external memory. In FPGA-based systems, minimizing data transfer latency and optimizing memory bandwidth are critical factors to take into account.
7. **Power Efficiency**: Depending on the tasks being carried out, FPGAs dynamically use power. Achieving power-efficient machine learning implementations on FPGAs requires careful power management techniques and optimization of hardware resources.
8. **Development Tools and Ecosystem**: Compared to traditional CPU or GPU-based implementations, the quality and availability of development tools, libraries, and frameworks for FPGA-based machine learning are not as advanced. It could be difficult for developers to locate materials and tools that are appropriate for their projects.
9. **Testing and Debugging**: FPGA designs are hardware-centric, and testing and debugging FPGA-based machine learning implementations can be difficult. To find and address design flaws, methods like hardware debugging tools, simulation, and emulation are crucial.

Traditional CPU and GPU architectures are not fully optimized for matrix multiplication tasks, leading to suboptimal performance and energy inefficiencies. CPUs are designed for general-purpose computing tasks and may not exploit parallelism effectively for matrix operations. GPUs offer parallel processing capabilities but may suffer from memory bandwidth limitations and high power consumption when scaling to large matrix sizes.

5.2 INTRODUCTION TO FPGA ARCHITECTURE

FPGAs are reconfigurable hardware devices that offer customizable logic blocks interconnected through programmable routing resources. FPGAs provide flexibility in designing custom hardware accelerators tailored to specific computational tasks, including matrix multiplication. Unlike fixed-function ASICs, FPGAs can be reprogrammed to adapt to evolving machine learning algorithms and application requirements [3,4].

FPGA-based accelerators have demonstrated promising results in accelerating matrix multiplication tasks, offering high throughput, low latency, and energy efficiency. By customizing FPGA architectures for matrix multiplication, it is possible to exploit parallelism at various levels (data, task, and pipeline) to achieve significant speedups compared to CPU and GPU implementations. Additionally, FPGAs enable co-design opportunities, where hardware and software components can be optimized in tandem to maximize performance and resource utilization [5].

The primary objective of this study is to explore FPGA architecture designs tailored for accelerating matrix multiplication operations in machine learning applications. We aim to investigate various optimization techniques, including parallelization strategies, memory hierarchy designs, and precision formats, to maximize the efficiency and scalability of FPGA-based accelerators. Furthermore, we seek to evaluate the performance, resource utilization, and energy efficiency of FPGA implementations compared to traditional CPU and GPU counterparts [6].

5.3 FPGA OVERVIEW

FPGAs have undergone significant evolution since their inception, expanding their capabilities and applications across various domains, including digital signal processing (DSP), image processing, video processing, and machine learning. Here's a detailed overview of how FPGAs have evolved and their usage in these applications.

Early Development: FPGAs emerged in the 1980s as a more flexible alternative to application-specific integrated circuits (ASICs). Unlike ASICs, which are custom-designed for specific applications, FPGAs are reprogrammable devices that allow users to configure the logic gates and interconnections within the chip to implement various digital circuits.

Increased Density and Performance: Over the years, FPGA manufacturers have continually increased the density and performance of their devices. Modern FPGAs contain millions of logic gates and embedded resources such as digital signal processing blocks (DSPs), memory blocks, and high-speed transceivers. This increased density enables the implementation of complex algorithms and processing tasks directly on the FPGA [7].

Integration of Hard IPs: FPGA vendors have started integrating specialized hard intellectual property (IP) cores directly into their chips. These hard IPs include DSP blocks, dedicated memory blocks, and high-speed interfaces. Integrating these IPs into the FPGA fabric improves performance and efficiency for applications such as DSP, image processing, and video processing.

High-Speed interfaces: FPGAs often include high-speed serial transceivers capable of transmitting data at multi-gigabit per second rates. These interfaces enable FPGAs to interface with high-speed external devices such as sensors, cameras, and storage devices, making them suitable for real-time video processing applications.

Parallel Processing: FPGAs excel at parallel processing due to their highly parallel architecture. This makes them well-suited for tasks that involve massive parallelism, such as image and video processing algorithms that can be parallelized across multiple processing elements within the FPGA.

Customizable Architectures: One of the key advantages of FPGAs is their customizable architecture. Users can design and implement custom hardware accelerators tailored to specific algorithms and applications. This flexibility allows for highly optimized and efficient implementations of DSP, image processing, and video processing tasks [8].

Hardware Acceleration for Machine Learning: With the rise of machine learning and artificial intelligence, FPGAs have gained traction as hardware accelerators for training and inference tasks. FPGAs can be used to accelerate neural network inference by implementing custom hardware architectures optimized for specific neural network models. This hardware acceleration can significantly improve the performance and energy efficiency of machine learning applications.

Development Tools and Libraries: FPGA vendors provide comprehensive development tools and libraries specifically designed for DSP, image processing, video processing, and machine learning applications. These tools simplify the design process and enable developers to leverage the full capabilities of FPGAs for their applications. Overall, FPGAs have evolved into highly versatile and powerful devices capable of performing a wide range of signal processing and computational tasks. Their reconfigurable nature, parallel processing capabilities, and integration of specialized IPs make them invaluable tools for DSP, image processing, video processing, and machine learning applications [9,10].

5.4 COMPARISON OF FPGAS WITH ASICS, GPPS, AND DSPS

ASICs are custom-designed integrated circuits optimized for a specific application or product. They are created to perform a particular set of functions with maximum efficiency and minimal power consumption. ASICs offer the highest level of performance and lowest power consumption among the four types of devices due to their custom design. However, ASICs require significant upfront investment in design and fabrication, making them suitable mainly for high-volume production where the cost can be amortized over many units. ASICs typically have longer development cycles compared to other devices due to the complexity of the design process and the need for extensive testing and validation.

General-Purpose Processors (GPPs) are standard microprocessors designed to execute a wide range of tasks, making them versatile and suitable for general-purpose computing. They are characterized by their ease of programming, with developers using high-level languages like C or C++ to write software applications. GPPs offer a balance between performance, versatility, and cost-effectiveness, making them suitable for a variety of applications ranging from desktop computers to embedded systems. However, GPPs may not provide optimal performance for tasks requiring high parallelism or specialized hardware acceleration.

DSPs are specialized microprocessors optimized for digital signal processing tasks such as audio and video processing, communications, and control systems. They excel at executing algorithms that involve processing and analyzing signals in real-time, thanks to their specialized instruction sets and hardware architecture. DSPs offer high performance and efficiency for signal processing tasks, making them suitable for applications where real-time processing is critical. However, DSPs may lack the flexibility and versatility of GPPs, as they are optimized for specific types of computations and may not perform as well for general-purpose computing tasks.

FPGAs are reconfigurable hardware devices that can be programmed after manufacturing to implement custom logic functions. They offer flexibility and parallelism, allowing designers to create custom hardware accelerators tailored to specific applications. FPGAs are suitable for prototyping, testing, and deploying custom hardware designs, offering a balance between performance, flexibility, and cost-effectiveness. However, FPGA design requires specialized knowledge of HDLs and hardware design concepts, and the resources available on an FPGA are limited compared to ASICs.

In terms of operational performance, DSPs face limitations due to their reliance on clock speed and sequential operation, which hinders their full parallelization potential. Conversely, FPGAs exhibit the capability to attain high speeds through well-designed parallel architectures. When it comes to re-configurability, DSPs necessitate alterations to the program's memory content, while FPGAs achieve re-configurability through the download of reconfiguration data into RAM. Power consumption in DSPs is influenced by the quantity of memory elements utilized, whereas in FPGAs, power consumption is dictated by the specific circuit design. FPGAs offer notable advantages, especially in the implementation of parallel algorithms where various components operate concurrently to execute system functions. As a consequence, the execution speed in FPGAs remains unaffected by the number of modules, in contrast to DSP systems where execution speed decreases as the number of functionalities increases. Typically, FPGAs deliver performance levels approximately an order of magnitude higher than DSPs [11,12].

5.5 DESIGN METHODOLOGY

The approach to designing hardware for implementing computation-intensive algorithms involves a collaborative endeavor utilizing electronic design automation (EDA) tools, methodologies, and FPGA technology. This collaboration facilitates the creation of optimized circuits tailored to specific applications. By leveraging FPGA hardware, meticulously crafted IP cores, and EDA tools, the efficiency of the design process is significantly enhanced. This design methodology outlines a step-by-step process for FPGA design, providing a guiding framework for realizing algorithm hardware.

Even though different FPGA vendors have slightly different design flows, most of them follow a similar work sequence. These are essential steps in contemporary FPGA design techniques, and they are present in all FPGA EDA tools. To create optimized digital circuits with FPGAs, EDA tools like Mentor Graphics' FPGA Advantage, Altera's Quartus II, and Xilinx Integrated Software Environment (ISE)

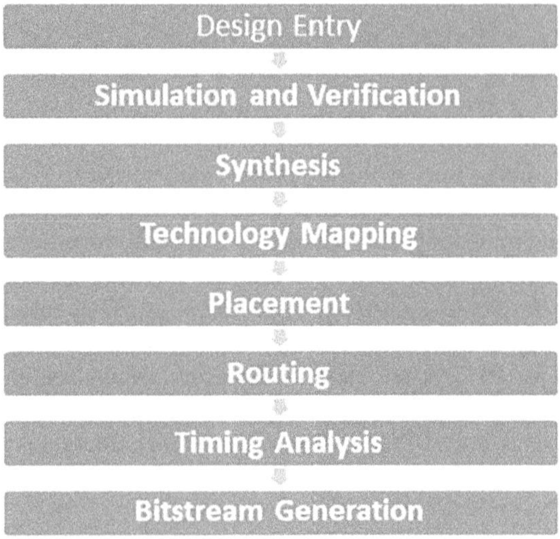

FIGURE 5.1 FPGA design process.

are essential. This work follows the standard FPGA design process, as shown in Figure 5.1. The algorithm or circuit meant for implementation on the FPGA device is described in this flow during the design entry stage. The two common approaches to design definition are schematic-based and HDL-based, and they vary based on the FPGA design's complexity. Nonetheless, FPGA designers typically use HDL-based (VHDL or Verilog) design entries for complex and computationally intensive algorithms. Once the design has been specified using schematics or HDLs, the designer needs to confirm that the design makes sense logically. Functional or behavioral simulation approaches are usually used for this validation process.

This step is usually completed by designers right after they finish the logic synthesis and design entry stages. A critical step in the design process is logic synthesis, which converts schematic-based or HDL design entries into a netlist of the actual gates and blocks found in FPGA devices. This is a crucial phase in the entire design process. Technology mapping enters the picture as a crucial step in the normal FPGA design flow after logic synthesis. Here, a netlist of technology-neutral logic gates is converted into one that contains logic cells and Input/Output Blocks (IOBs) unique to the target FPGA architectures by means of the EDA tool. The quality of the implemented circuits is greatly influenced by mapping. Technology mapping is followed by placement, which determines the best location for each block in the circuit. Since placement directly affects the ability to route and performance on the FPGA, it is critical for FPGA designs. Timing-driven and routability-driven algorithms are the two main categories for FPGA placement algorithms.

In the FPGA design pipeline, routing comes after placement and is the final stage before creating the bitstream that programs the FPGA. FPGA routing is a laborious process that uses routing resources such as wire segments, programmable switches, and multiplexers. For contemporary designs with a large number of gates and intricate algorithms, analysis is essential. The first steps involve analysis tools such as Quartus II, ISE simulator, and ModelSim. Depending on how serious the problem is, the design could need to go back to earlier phases or even the beginning in the case of an error. Timing simulation is essential for verifying that the design is logically accurate while taking the delays of FPGA devices into account. The last phase in the FPGA design flow is bitstream creation and uploading the created bit file into the FPGA [13,14].

5.6 SIMULINK MODEL TO HDL CODE GENERATION

Simulink's model-based design is particularly flexible since model modifications can be made quickly. Simulink makes effective use of the right libraries to implement the block-based functionalities. The algorithm design process is automated by Simulink HDL Coder, from modeling through FPGA implementation. It generates bit-true, cycle-accurate, synthesizable Verilog, and VHDL code from Simulink models, MATLAB code, and Stateflow diagrams [15]. The resultant HDL code can be synthesized and simulated with common tools before being implemented on an FPGA. Simulink HDL Coder can create reports on the use of hardware resources and control HDL implementation and architecture. For speedy verification, Simulink HDL Coder generates test benches, EDA co-simulation models, and code traceability. The algorithm can be modeled using Stateflow, MATLAB, or Simulink before HDL code is written.

In order to facilitate quick verification of the generated HDL code, Verilog and VHDL test benches are created using Simulink HDL Coder. Because there are no design misunderstandings and the design time between the implementation engineer and design developer is substantially reduced, an automatic approach of translating an algorithm into a hardware design has shown to be viable. Based on Simulink models, the Simulink HDL Coder program generates an HDL code. After the model is created, the HDL code is generated via a command line interface or graphical user interface. To generate HDL code, the model needs to be constructed with building blocks that support HDL code production. Before the code is generated, the model parameters for HDL code creation must be configured. The hdlsetup command uses the set param function to quickly and reliably build up models for the production of HDL code [9]. Hdlsetup configures the solver options that the HDL Coder recommends or needs. Any changes to the parameters need to be checked for compatibility with the model. HDL code is created with the makehdl command.

5.7 DESIGN AND IMPLEMENTATION OF MATRIX-VECTOR MULTIPLICATION

In this case, we describe the design process and present the outcomes of applying matrix-vector multiplication, an expensive computing operation needing substantial resources.

Multiple multiply and add units must be used in order to perform this operation. The ability of digital signal processors to execute adds and multiplications simultaneously sets a performance constraint. DSPs typically require multiple clock cycles in order to complete all necessary multiply-add operations.

On the other hand, modern FPGAs have a multitude of hardware resources built right into the structure, such as multipliers, DSP48 blocks, Block RAMs, and more. Because of this natural resource availability, FPGAs can process data at higher and more effective rates that are appropriate for these types of demanding applications—that is, if the algorithm is skillfully written to take advantage of these embedded resources. When the algorithm is skillfully written to take advantage of these integrated resources, FPGAs can reach higher and more efficient processing speeds appropriate for such demanding applications, thanks to their natural availability of resources.

5.8 HARDWARE ARCHITECTURE

Using matrix-vector multiplication, a computationally costly operation needing substantial resources, we describe here the design process and present the results.

The essential operation is matrix-vector multiplication, as shown in equation (5.1). We choose integer arithmetic to guarantee effective implementation and maximize speed-up. This decision is motivated by the fact that, in comparison to integer arithmetic, floating-point arithmetic units function more slowly and require more FPGA silicon real estate [13].

The goal of the design is to compute $\mathbf{g}=\mathbf{A} \times \mathbf{C}$, where **A** is a matrix and **c** and **g** are vectors, as shown in Table 5.1. We want to figure out the vector **g.** We use the broadcast technique for matrix-vector multiplication to accomplish this. The broadcasting of matrix **A** rows and their multiplication by matching column elements of vector **c** constitute this methodology. The subsequent actions take place as follows.

1. Extracting column elements of vector **c** and row elements of matrix **A**.
2. Keeping these components in internal buffers in rows and columns.
3. Carrying out the row and column elements' multiplication.
4. Logging the outcomes in the output buffers.

The FPGA is used to implement input and output buffers. Multiply and accumulate operations are involved in matrix-vector multiplications, forming a multiply-accumulate unit furnished with an adder and a multiplier. The multiplier accepts row and column items as inputs, while the adder receives the output directly. Reusing the adder's first output as the second input allows for iterative accumulation.

TABLE 5.1
Matrix and Vector Dimensions

Symbol	Matrix Dimension
A	1024 × 28
c	28 × 1
g	1024 × 1

5.9 MATHEMATICAL REPRESENTATION

We represent the vector c as $(c_1, c_2, c_3, \ldots, c_m)^T$ and vector $\boldsymbol{g}$ represents the image data. Suppose we want to multiply matrix A with vector **c** represented by equation (5.1):

$$g = A \times c \tag{5.1}$$

The dimension of the given matrices depends on the application. One example is considered as shown in Table 5.1. The multiply-accumulate unit performs iterative multiplication and addition on each member of vector **c** and matrix **A**, processing each element of row **A** and column **c** in turn. Up until every row in matrix **A** is processed, this sequential process is carried out. Upon the activation of the reset signal, the contents of registers A and $\boldsymbol{c}$ are cleared, and then an implementation-resulted delay occurs [7]. The initial element of vector **g** is then accessible at the serial output and is subsequently kept in on-chip memory. Until all rows of matrix, **A** are analyzed, this iterative process continues, producing the output vector **g**, whose elements are all stored in memory locations.

Figure 5.2 shows a simplified block diagram of the matrix-vector multiplication procedure. We used the Xilinx ISE tool to translate the algorithm into VHDL and run it on the Xilinx Artix-7, AES-A7MB-7A35T-G, 410-319 speed grade: -1 family [16]. Synthesis was performed on the design with an emphasis on Artix-7 FPGA speed optimization. Table 5.2 offers comprehensive information on how hardware resources are used.

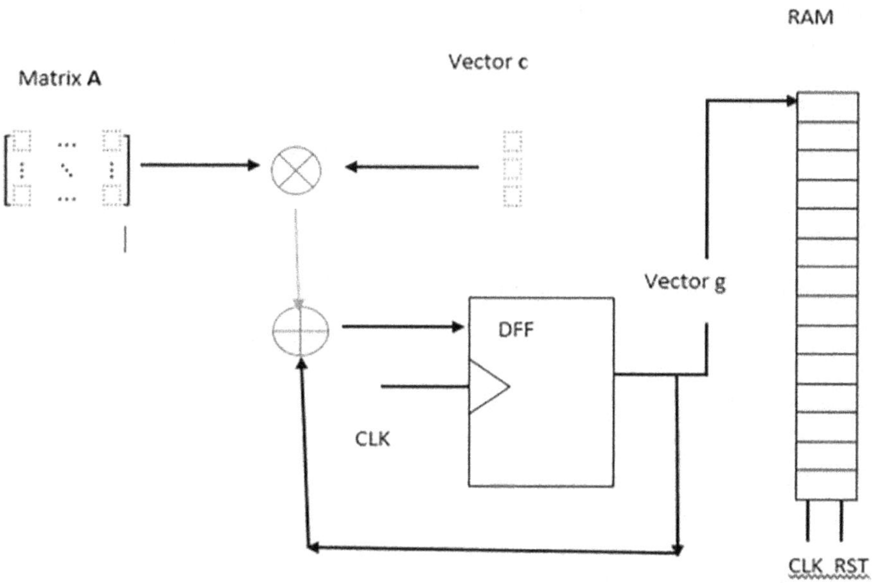

FIGURE 5.2 Matrix-vector multiplication.

TABLE 5.2
FPGA Resources Utilization

Resources	Used	Available	% Utilization
Slices	13000	33000	39
LUTs	9500	50000	19
DSPs	55	90	61
Frequency		17.376 MHz	

Table 5.2 data shows that about 39% of the slices and 61% of the DSP48s are in use, meaning there is a good chance that more parallel processors can be implemented on the same FPGA chip. The Xilinx ISE tool, set up for speed optimization, produced the outcomes shown in Table 5.2. With the Artix-7 FPGA, the total processing time is 58 μs, which translates to a throughput of 17241 frames per second. These results highlight the usefulness of using FPGA for matrix-vector multiplication in real-time, high-speed image processing applications.

5.10 MATRIX INVERSE

Matrix **X** is said to be invertible if there exists an equal-sized matrix **Y** with the formula **XY** = **YX** = **I**, where **I** is the identity matrix and **Y** is the name of the **X** matrix's inverse. An inverse-free matrix is called a singular matrix. A square matrix is unique only when its determinant is exactly zero. The steps involved in matrix inverse computation are as follows: calculate the cofactor matrix, transpose the cofactor matrix in Step 2, and then multiply the reciprocal of the input matrix's determinant by the cofactor matrix's transposed form.

Consider a matrix **A** where

$$\mathbf{A} = \begin{matrix} 4 & 12 & -16 \\ 12 & 37 & -43 \\ -16 & -43 & 98 \end{matrix}$$

The matrix of minors will be used to compute the cofactors of "**A**"

$$\text{cofactor}(\mathbf{A}) = \begin{matrix} 1777 & 488 & 76 \\ 488 & 136 & 20 \\ 76 & 20 & 4 \end{matrix}$$

The cofactor matrix's transposition will be

$$\left(\text{cofactor}(\mathbf{A})\right)' = \begin{matrix} 1777 & 488 & 76 \\ 488 & 136 & 20 \\ 76 & 20 & 4 \end{matrix}$$

Transpose the cofactor matrix and multiply the reciprocal of the determinant of "**A**"

$$\text{Inverse}(\mathbf{A}) = (1/\text{determinant}(\mathbf{A})) * (\text{cofactor}(\mathbf{A}))'$$

$$\text{Inverse}(\mathbf{A}) = \begin{matrix} 49.3611 & -13.5556 & 2.1111 \\ -13.5556 & 3.7778 & -0.5556 \\ 2.1111 & -0.5556 & 0.1111 \end{matrix}$$

5.11 ELIMINATION OF THE MATRIX BY THE GAUSS-JORDAN METHOD

The inverse of a matrix can be found by Gauss-Jordan elimination. Finding the most basic row operations that convert a matrix into an identity matrix (**I**) is the first step in doing this. Next, apply those same operations on the identity matrix (**I**) to obtain the inverse(**A**) [17].

5.12 MATRIX INVERSE COMPUTATION USING GAUSS-JORDAN ELIMINATION

Let's begin by writing down matrix *A* and placing an identity matrix next to it: [**A** | **I**].

By using row transformations and the right-side matrix, which **I** also contributed to, the objective is to reduce **A** to an identity matrix or inverse(**A**).

The stages involved in computing inverse(**A**) are as follows:

1. Swap the rows.
2. Reverse the diagonal elements.
3. Set the components that are not diagonal to 0.

Consider a matrix **A** and identity matrix **I**:

$$\begin{matrix} & (\mathbf{A}) & & (\mathbf{I}) \\ [\mathbf{A} \mid \mathbf{I}] = & \begin{matrix} 1 & 2 & 3 \\ 2 & 5 & 3 \\ 1 & 0 & 8 \end{matrix} & & \begin{matrix} 1 & 0 & 0 \\ 0 & 1 & 0 \\ 0 & 0 & 1 \end{matrix} \end{matrix}$$

Locate the element in the first column with the highest value and then swap the current row with that row's highest element value. R_1 and R_2 rows should be switched since R_2 has the highest values

$$\begin{matrix} (\mathbf{A}) & & (\mathbf{I}) \\ \begin{matrix} 2 & 5 & 3 \\ 1 & 2 & 3 \\ 1 & 0 & 8 \end{matrix} & & \begin{matrix} 0 & 1 & 0 \\ 1 & 0 & 0 \\ 0 & 0 & 1 \end{matrix} \end{matrix}$$

Make the first column's diagonal element "1"

$$R_1 \rightarrow R_{1|2} = \begin{matrix} 1 & 2.5 & 1.5 & & 0 & 0.5 & 0 \\ 1 & 2 & 3 & & 1 & 0 & 0 \\ 1 & 0 & 8 & & 0 & 0 & 1 \end{matrix}$$

Make the first column's non-diagonal elements "0"

$$\text{New } R_2 \rightarrow R_2 - R_1$$

$$\text{New } R_3 \rightarrow R_3 - R_1$$

$$= \begin{matrix} 1 & 2.5 & 1.5 & 0 & 0.5 & 0 \\ 0 & - -0.5 & 1.5 & 1 & -0.5 & 0 \\ 0 & -2.5 & 6.5 & 0 & -0.5 & 1 \end{matrix}$$

Now, the diagonal elements in column 1 are "1," and the other components are "0." The other columns are subjected to the same process, resulting in the reduction of matrix **A** to identity matrix and identity matrix to inverse(**A**) [15,18]

$$\begin{matrix} 1 & 0 & 0 \\ 0 & 1 & 0 \\ 0 & 0 & 1 \end{matrix} \qquad \begin{matrix} -40 & 16 & 9 \\ 13 & -5 & -3 \\ 5 & -2 & -1 \end{matrix}$$

$$(\mathbf{I}) \qquad\qquad \text{Inverse}(\mathbf{A})$$

5.13 SUBSYSTEM FOR MATRIX INVERSE

As shown in Figure 5.3, three subsystems make up this sample model: the input subsystem, matrix inverse, and output subsystem. The input subsystem serializes the matrix input to the processing module when the ready signal is activated. A processing module called matrix inverse performs the matrix inverse operation. Upon

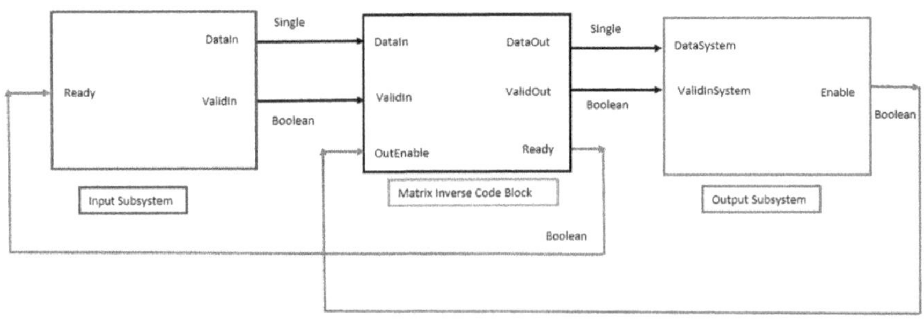

FIGURE 5.3 Matrix inverse subsystem.

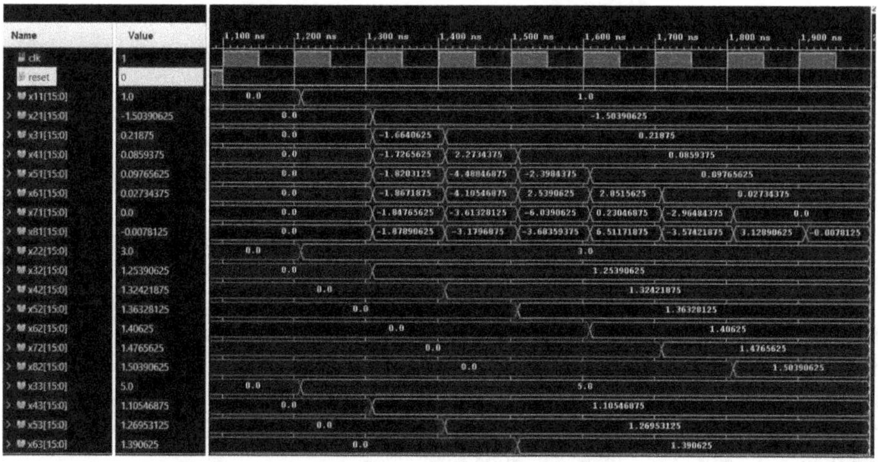

FIGURE 5.4 Testbench output of matrix inverse.

TABLE 5.3
Resource Utilization for Matrix Inverse

Size	DSPs	LUTs	RAMs	Latency
4×4	12	11386	45	958
8×8	24	20415	76	2803
16×16	48	37884	138	9122

arrival of the enable signal, the output subsystem deserializes the matrix inverse result and outputs it as a matrix [19]. The test bench output is shown in Figure 5.4.

5.14 SYNTHESIS STATISTICS FOR MATRIX INVERSE

Resource utilization for matrix inverse is shown in Table 5.3. The synthesis tool used is Altera Quartos.

The number of DSPs used is 12, 24, and 48 for 4 × 4, 8 × 8, and 16 × 16 matrix size, respectively. The number of Look-Up Tables (LUTs) used is 11386, 20415, and 37884 for 4 × 4, 8 × 8, and 16 × 16 matrix sizes, respectively. Furthermore, the RAM used is 4576138 bytes for 4 × 4, 8 × 8, and 16 × 16 matrix sizes. Latency observed for 4 × 4, 8 × 8, and 16 × 16 matrix sizes is 958, 2803, and 9122 clocks, respectively.

5.15 CONCLUSION

Matrix multiplication stands as a fundamental operation in numerous machine learning tasks, posing significant computational demands that traditional computing architectures often struggle to meet. This introduction underscores the pivotal role of matrix multiplication in machine learning while shedding light on the limitations inherent in CPU and GPU architectures. Consequently, it introduces FPGAs as a promising solution, owing to their flexibility and potential for hardware acceleration.

FPGAs have undergone significant evolution, boasting increased density, integrated hard IPs, high-speed interfaces, and customizable architectures. Their proficiency in parallel processing tasks renders them apt for a wide array of applications, including DSP, image processing, video processing, and machine learning. FPGAs strike a balance between performance, flexibility, and cost-effectiveness, thus presenting opportunities for hardware acceleration in machine learning algorithms.

Comparatively, FPGAs offer flexibility and parallel processing capabilities that outshine ASICs, GPPs, and DSPs. While ASICs deliver high performance and low power consumption, they entail substantial upfront investment. GPPs, though versatile, may not optimize performance for parallel tasks, while DSPs excel in signal processing but lack flexibility. FPGAs amalgamate flexibility with parallelism, making them well-suited for diverse applications.

The design methodology for FPGA-based hardware implementation encompasses several stages, including design entry, logic synthesis, technology mapping, placement, routing, and analysis. Utilizing tools like Simulink HDL Coder simplifies this process by generating HDL code from Simulink models, facilitating swift verification and implementation on FPGAs.

Matrix-vector multiplication emerges as a computationally intensive task demanding substantial hardware resources. The design entails broadcasting rows of matrix **A** and multiplying them with corresponding column elements of vector **c**. Leveraging FPGA-based implementation, which employs multiply-accumulate units and input/output buffers, enables efficient computation. Synthesis results illustrate resource utilization and processing time, underscoring the FPGA's suitability for real-time image processing applications.

The computation of matrix inverse involves various methods, including cofactor matrix calculation and Gauss-Jordan elimination. Discussing the design and implementation of a matrix inverse subsystem on an FPGA, along with synthesis statistics for different matrix sizes, provides insight into this process.

This work showcases the potential of FPGAs in accelerating matrix operations in machine learning applications. By optimizing FPGA architectures and harnessing parallel processing, we can enhance the efficiency and scalability of matrix operations, thus contributing to the advancement of machine learning algorithms.

REFERENCES

1. S. M. Qasim, A. A. Telba, and A. AlMazroo, "FPGA design and implementation of matrix multiplier architectures for image and signal processing applications", *International Journal of Computer Science and Network Security*, Vol. 10, pp. 168–176, 2010.
2. S. Ogrenci, A. K. Katsaggelos, and M. Sarrafzadeh, "Analysis and FPGA implementation of image restoration under resource constraint," *IEEE Transactions on Computers*, Vol. 52, No. 3, pp. 390–399, 2003.

3. G. T. Tchendjou, E. Simeu, and F. Lebowsky, "FPGA implementation of machine learning based image quality assessment," *2017 29th International Conference on Microelectronics (ICM),* Beirut, Lebanon, 2017, pp. 1–4, doi: 10.1109/ICM.2017.8268848.
4. A. G. Ye and D. M. Lewis, "Procedural texture mapping on FPGAs," In *Proceedings of ACM/SIGDA 7th International Symposium on Field Programmable Gate Arrays*, Monterey, California, pp. 112–120, 1999.
5. S. Knapp, "Using Programmable Logic to Accelerate DSP Functions". Available: https://www.xilinx.com/appnotes/dspintro.pdf.
6. J. Ma, "Signal and image processing via reconfigurable computing," In *Procedings of the First Workshop on Information and Systems Technology*, 2003.
7. F. Otto and Z. Pavel, *"Hardware accelerated imaging algorithms,"* In *Proceedings of AUTOS'2002 Automatizace systému*, Prague, Czech Republic, pp. 165–171, 2002.
8. G. R. Goslin, "A guide to using field programmable gate arrays for application-specific digital signal processing performance," *Microelectronics Journal*, Vol. 28, Issue 4, pp. 24–35, 1997.
9. H. R. Wu, W. Lin, and L. J. Karam, "An overview of perceptual processing for digital pictures", *2012 IEEE International Conference on Multimedia and Expo Workshops (ICMEW)*, Melbourne, Australia, pp. 113–120, 2012.
10. D. Johnson, K. Gribbon, D. Bailey, and S. Demidenko, "Implementing digital signal processing algorithm's in FPGA's: Digital spectral warping," In *Proceedings of 9th Electronics New Zealand Conference*, Auckland, New Zealand, pp. 72–77, 2002.
11. K. Compton and S. Hauck, "Reconfigurable computing: A survey of systems and software," *ACM Computing Surveys*, Vol. 34, No. 2, pp. 171–210, 2002.
12. R. Tessier and W. Burleson, "Reconfigurable computing for digital signal processing: A survey," *Journal of VLSI Signal Processing*, Vol. 28, No. 3, pp. 7–27, 2001.
13. T. J. Todman, G. A. Constantinides, S. J. E. Wilton, O. Mencer, W. Luk, and P. Y. K. Cheung, "Reconfigurable computing: Architectures and design methods," *IEE Proceedings of Computer Digital Techniques*, Vol. 152, No. 2, pp. 193–207, 2005.
14. I. Grout, *Digital Systems Design with FPGAs*. Elsevier Ltd., Oxford, 2008.
15. MathWorks, "Simulink HDL Coder". Available: https://www.mathworks.com/products/slhdlcoder/.
16. Xilinx, "ISE Design Suite". Available: https://www.xilinx.com/support/documentation/index.html.
17. O. Gustafsson, E. Bertilsson, J. Klasson, and C. Ingemarsson, "Approximate neumann series or exact matrix inversion for massive MIMO?" *Proceedings Symposium on Computer Arithmetic,* pp. 1063–6889, 2017, https://doi.org/10.1109/ARITH.2017.11.
18. HDL Coder manual, MathWorks, *HDL Coder manual.* Available: https://www.mathworks.com/help/hdlcoder/
19. A. Krishnamoorthy and D. Menon, "Matrix inversion using Cholesky decomposition," *2013 Signal Processing: Algorithms, Architectures, Arrangements, and Applications (SPA)*, Poznan, Poland, pp. 70–72, 2013.

6 Silicon Chip Design and Testing

Shasanka Sekhar Rout and Jagana Bihari Padhy

6.1 INTRODUCTION

VLSI plays a crucial role in silicon chip design by enabling high levels of integration, handling complexity, improving performance, facilitating miniaturization, enabling system integration, enhancing design productivity, and serving as an enabler for innovation in the semiconductor industry. Its impact is pervasive across various application domains, ranging from consumer electronics and telecommunications to automotive and healthcare. VLSI also plays a critical role in testing by enabling the design of testable chips, implementing efficient testing methodologies, analyzing test coverage, developing test programs, and conducting comprehensive testing to ensure the quality and reliability of semiconductor devices [1]. Silicon chip design and testing lie at the heart of the semiconductor industry, driving innovation and enabling the technological advancements that shape our modern world [2]. From the microprocessors powering our computers to the integrated circuits in our smartphones, silicon chips play a pivotal role in countless electronic devices. This chapter delves into the intricacies of silicon chip design and the rigorous testing processes that ensure their functionality and reliability. From the inception of an idea to the final integration into functional systems, this chapter provides an in-depth exploration of the stages involved in designing and testing silicon chips, which is shown in Figure 6.1.

Absolutely, the chapter on semiconductor chip design and testing provides a comprehensive examination of various testing methodologies employed throughout the semiconductor manufacturing process. From the initial stages of wafer testing to the final quality assurance measures, each step is crucial in ensuring the functionality, reliability, and performance of semiconductor chips. Furthermore, the chapter underscores the tangible impact of chip design and testing on shaping the landscape of modern technology. Semiconductor chips are the backbone of countless electronic devices, and their quality and reliability are paramount in fostering innovation and enhancing the efficiency of various industries.

6.2 SPECIFICATION AND ARCHITECTURE DESIGN

At the outset of a silicon chip project, engineers embark on a journey to define the chip's specifications and architectural design. This phase involves close collaboration between design teams and stakeholders to articulate the desired functionalities and performance metrics of the chip. Through feasibility studies and architectural

 DOI: 10.1201/9781003483038-6

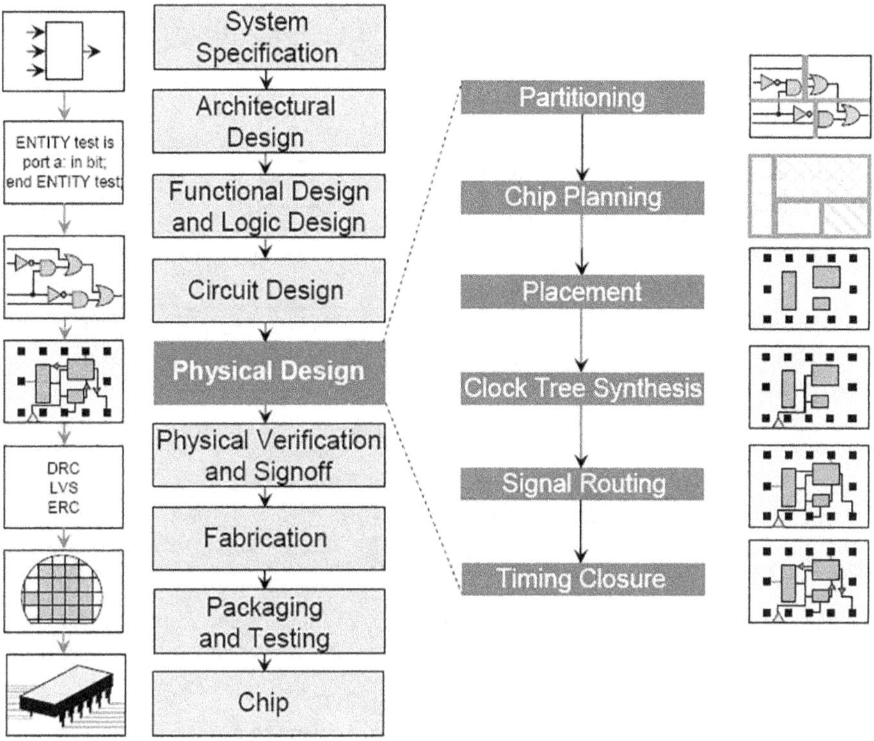

FIGURE 6.1 VLSI design flow.

exploration, engineers establish the groundwork for subsequent design stages. Specification and architecture design is a critical phase in the development of silicon chips, laying the groundwork for subsequent stages of the design process. This phase involves defining the requirements and high-level structure of the chip to ensure that it meets the desired functionality, performance, and power targets [3,4].

Example: Consider the development of a microcontroller for Internet of Things (IoT) devices. Engineers collaborate with stakeholders to define requirements such as processing power, memory, and connectivity. Architectural decisions are made regarding the type of processor core, peripherals, and communication protocols to be integrated into the chip.

The block diagram shown in Figure 6.2 provides a high-level overview of the major functional blocks and their interconnections within the silicon chip, illustrating the architecture and design considerations during the specification and architecture design phase [5].

Input/Output Interfaces: These interfaces are for communication with external devices, such as sensors, actuators, and communication modules.

Processing Unit: This block represents the central processing unit (CPU) or digital signal processor (DSP) responsible for executing instructions and performing computational tasks.

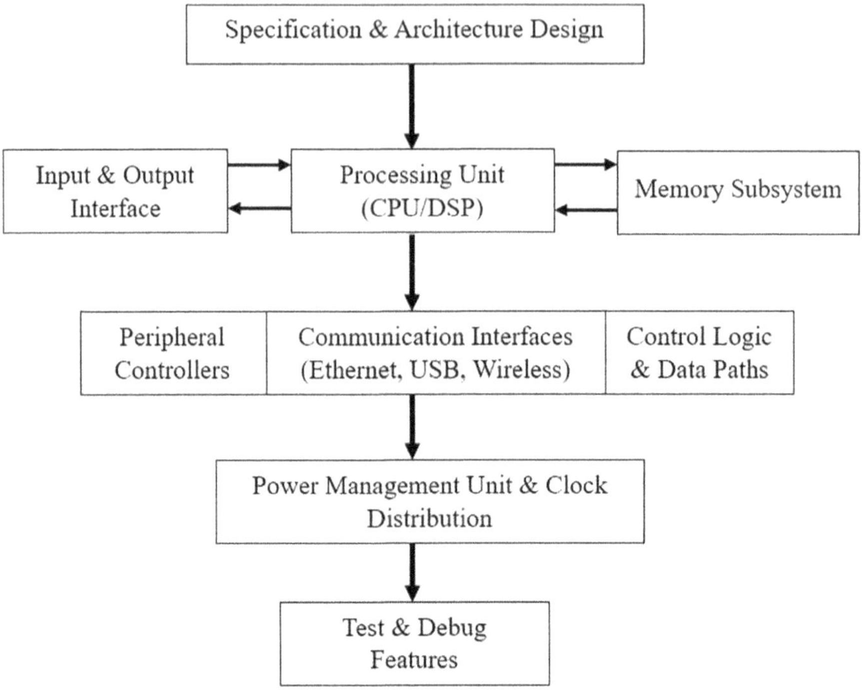

FIGURE 6.2 Specification and architecture design of a silicon chip.

Memory Subsystem: This includes various types of memory components, such as cache memory, RAM, and Read-Only Memory (ROM), for storing data and instructions.

Peripheral Controllers: These controllers are for managing peripheral devices, such as timers, DMA controllers, and other input/output peripherals.

Communication Interfaces: These interfaces are for communication with other devices or systems, including Universal Asynchronous Receiver-Transmitter (UART), Ethernet, USB, and wireless communication protocols.

Control Logic and Data Paths: These logic circuits are responsible for controlling the operation of the chip and managing data flow between different functional blocks.

Power Management Unit and Clock Distribution: This controls power distribution and manages clock signals within the chip to optimize energy efficiency and ensure proper timing.

Test and Debug Features: This includes features designed for testing, debugging, and validation purposes, such as test access ports (TAPs), built-in self-test (BIST) circuits, debug interfaces, and security mechanisms.

6.3 LOGIC DESIGN AND SYNTHESIS

With specifications in hand, engineers delve into the logic design phase, where they conceptualize the digital circuitry that will implement the desired functionalities. Leveraging hardware description languages (HDLs), such as Verilog or VHSIC

Hardware Description Language (VHSIC stands for Very High-Speed Integrated Circuit) (VHDL), designers craft intricate logic structures, employing a mix of combinatorial and sequential elements. Once the logic design is complete, synthesis tools map the design onto the target silicon technology, optimizing for factors like area, power consumption, and timing [6].

Logic design and synthesis are fundamental stages in silicon chip development, where the conceptual design is translated into a detailed digital circuit implementation. This phase involves converting high-level descriptions of desired functionalities into logic-level representations, optimizing for performance, area, and power consumption. The fundamental stages involved in this phase are as follows.

a. **Hardware Description Languages (HDLs):** Engineers use hardware description languages, such as Verilog or VHDL, to describe the behavior and structure of digital circuits. HDLs allow designers to specify logical operations, timing constraints, and interconnections between different components.
b. **Logic Design:** Logic design involves conceptualizing the digital circuitry required to implement the desired functionalities of the chip. Designers create logic diagrams, state diagrams, and truth tables to represent the logical relationships and operations within the circuit.
c. **RTL (Register Transfer Level) Design:** RTL design focuses on specifying the behavior of the digital circuit at the register transfer level, describing how data is transferred between registers and how operations are performed. Designers define the behavior of individual logic elements, registers, and interconnections using RTL constructs [7].
d. **Component Selection:** Designers select appropriate digital components, such as logic gates, flip-flops, multiplexers, and arithmetic units, to implement the desired functionality. Component selection is based on factors such as performance requirements, area constraints, and power consumption considerations.
e. **Optimization Techniques:** Various optimization techniques are applied to improve the efficiency and performance of the digital circuit. Techniques include logic optimization, timing optimization, area optimization, and power optimization.
f. **Synthesis:** Synthesis is the process of transforming the RTL description of the digital circuit into a gate-level netlist. Synthesis tools analyze the RTL code and map it onto the target technology library, selecting appropriate logic gates and optimizing the circuit for area, timing, and power.
g. **Technology Mapping:** Technology mapping involves mapping the synthesized logic onto specific technology libraries, which contain information about the available logic cells, flip-flops, and interconnect resources. Designers select the optimal mapping based on factors such as performance requirements, area constraints, and power targets.
h. **Timing Analysis:** Timing analysis ensures that the synthesized circuit meets the required timing constraints, such as setup time, hold time, and clock frequency. Designers use timing analysis tools to simulate the behavior of the circuit and identify any timing violations that need to be addressed.

i. **Verification:** Verification is performed to ensure that the synthesized circuit behaves correctly according to the specified requirements. Designers use simulation, formal verification, and other techniques to verify the functional correctness of the synthesized logic.
j. **Iterative Refinement:** The logic design and synthesis process often involve iterative refinement, where designers make incremental improvements to the design based on simulation results, synthesis reports, and feedback from verification.

Example: Using Verilog HDL, designers create the logic circuits necessary to implement the desired functionalities of the microcontroller. This includes designing arithmetic logic units (ALUs), memory controllers, and input/output interfaces. Synthesis tools then map this logical design onto the chosen silicon technology, optimizing for factors like power consumption and speed.

6.4 PHYSICAL DESIGN

The transition from logical abstraction to physical realization marks the onset of the physical design phase. Engineers grapple with the intricacies of floor planning, placement, and routing, striving to optimize the layout for manufacturability and performance. Advanced algorithms and tools aid in the placement of millions of components on the silicon die, while routing algorithms navigate the complex maze of interconnections, mindful of signal integrity and power distribution. The physical design flow is a sequence of steps involved in transforming a logical circuit representation into a physical layout that can be manufactured on silicon [8,9]. The overview of physical design flow is summarized in Figure 6.3. The Place and Route (PnR) flow mainly follows these stages: Design planning/Import/Partitioning => Floorplan & Powerplan => Placement => CTS => Routing => Sign Off.

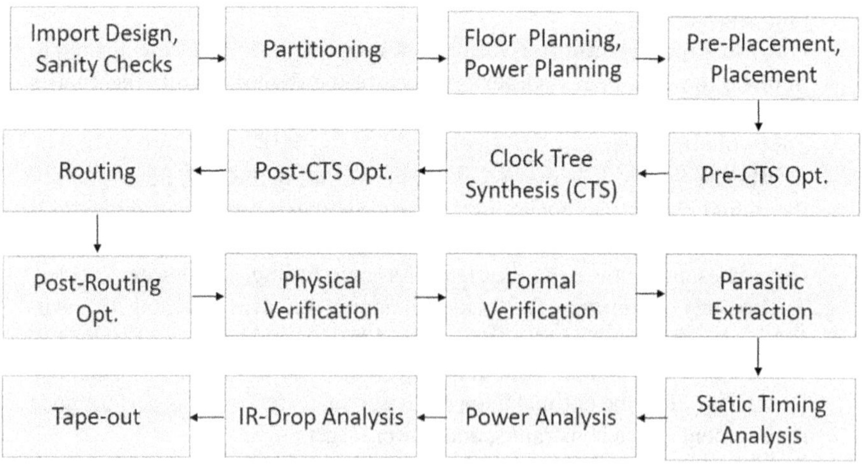

FIGURE 6.3 Physical design flow.

a. **Import Design:**

   The following input file information is loaded into the PnR tool.

   i. Netlist (.v/.vhd/.edif)
   ii. Physical Libraries (.lef)
   iii. Timing Libraries (.lib)
   iv. Technology File
   v. Constraints (.sdc)
   vi. IO Info. File
   vii. Power Spec. File
   viii. Optimization Directives
   ix. Clock Tree Spec. File
   x. DEF/FP

b. **Floorplanning:** The process begins with floorplanning, where designers allocate space on the silicon die for different functional blocks and components. Constraints such as chip size, power grid, signal routing, and thermal considerations are taken into account during floorplanning.

c. **Placement:** Placement involves positioning the individual components (cells) within the allocated chip area. Algorithms place cells based on optimization objectives such as minimizing wire length, meeting timing constraints, and avoiding congestion.

d. **Clock Tree Synthesis (CTS):** Clock tree synthesis generates a hierarchical network of clock distribution lines to ensure synchronized timing signals across the chip. CTS aims to minimize clock skew and ensure consistent clock arrival times at all clocked elements.

e. **Routing:** Routing connects the placed components with wires to establish functional connections according to the logical design. Global routing determines the paths of major interconnects, while detailed routing focuses on individual wires and signal paths.

f. **Power Distribution Network (PDN) Design:** The PDN design involves creating a network of power supply lines and distribution capacitors to ensure stable power delivery to all components. PDN design aims to minimize voltage drop and power distribution noise to maintain reliable operation of the chip.

g. **Timing Closure:** Timing closure ensures that the physical design meets the required timing constraints specified during synthesis. Techniques such as buffering, resizing, and optimization are applied to achieve timing closure and meet the desired performance targets.

h. **Physical Verification:** Physical verification checks the layout against design rules, ensuring compliance with manufacturing constraints and standards. DRC (design rule check), LVS (layout vs. schematic), and ERC (electrical rule check) are performed to identify and resolve layout errors.

i. **Design for Manufacturability (DFM):** DFM techniques are applied to enhance manufacturability and yield of the chip. Techniques include lithography-aware layout optimization, metal density control, and spacing rules adjustment to address manufacturing challenges [10].

j. **Signoff:** After completing physical design and verification, the layout undergoes signoff to ensure readiness for manufacturing. Designers review the layout, perform final checks, and generate design files for fabrication.

Example: Physical design engineers tackle challenges such as floor planning, placement, and routing to ensure efficient use of silicon real estate. For our microcontroller example, considerations include the proximity of memory to the processor core for faster access and the placement of input/output pins for optimal connectivity.

6.5 VERIFICATION

Verification is a cornerstone of silicon chip design, ensuring that the implemented design behaves as intended. Engineers employ a plethora of verification techniques, ranging from simulation at various levels of abstraction to formal verification methods. Through exhaustive testing, designers uncover and rectify design flaws, striving for correctness and robustness before proceeding to manufacturing [11].

Verification is a critical stage in the silicon chip design process, aimed at ensuring that the chip functions correctly and meets the specified requirements before it is manufactured. The overview of the verification process is as follows.

a. **Functional Verification:** Functional verification verifies that the chip behaves as intended and performs the desired functions correctly. Techniques include simulation, emulation, and formal verification to validate the behavior of the chip against the functional specifications.
b. **Simulation-Based Verification:** Simulation involves running tests on a software model of the chip (RTL simulation) or on an actual prototype (gate-level simulation) to verify its functionality. Testbenches, which are sets of stimuli and expected responses, are used to simulate various scenarios and verify the chip's behavior.
c. **Emulation:** Emulation uses specialized hardware platforms called emulators to run the chip design at near-real-time speeds. Emulation is used for testing large and complex designs that cannot be efficiently simulated on traditional simulators.
d. **Formal Verification:** Formal verification uses mathematical techniques to formally prove the correctness of the chip design. Model checking, theorem proving, and equivalence checking are common formal verification methods used to analyze the chip's behavior against its specifications.
e. **Code Coverage Analysis:** Code coverage analysis measures the extent to which the chip's design has been exercised by the verification tests. Coverage metrics such as statement coverage, branch coverage, and condition coverage help assess the completeness of the verification effort.
f. **Assertion-Based Verification:** Assertions are statements embedded in the RTL code that specify expected behavior or constraints. Assertion-based verification checks whether these assertions hold true during simulation, helping to detect violations of design intent.

g. **Testbench Automation:** Testbenches are automated to generate stimuli, execute tests, and analyze results efficiently. Verification environments such as UVM (universal verification methodology) provide a standardized framework for testbench development and management.
h. **Gate-Level Simulation:** Gate-level simulation verifies the functionality of the chip's synthesized netlist at the gate level. Timing information, delays, and signal propagation are considered in gate-level simulation to assess the chip's timing behavior.
i. **Performance Verification:** Performance verification ensures that the chip meets timing, power, and area targets specified during the design phase. Timing analysis, power analysis, and physical verification checks are performed to validate performance metrics.
j. **Integration Testing:** Integration testing verifies the interaction and compatibility of individual chip components when integrated into larger systems. Interface protocols, communication protocols, and system-level functionality are tested to ensure seamless integration.

Example: Simulation is crucial at this stage to verify that the design meets specifications. Engineers create test benches and use simulations to verify functionality under various conditions. For our microcontroller, this involves testing different instructions, memory access patterns, and communication protocols to ensure correct operation.

6.6 MANUFACTURING

The manufacturing phase heralds the transformation of design blueprints into tangible silicon wafers. Foundries employ cutting-edge lithography techniques to etch intricate patterns onto silicon substrates, laying the foundation for the chip's physical realization. Doping, etching, deposition, and metallization processes sculpt the silicon into functional circuits, culminating in the birth of countless chips on a single wafer, which is shown in Figure 6.4 [12,13]. The manufacturing process can be represented as follows:

a. **Wafer Preparation:** Silicon wafers are prepared by slicing a cylindrical ingot of semiconductor-grade silicon into thin, circular wafers. These wafers undergo polishing and cleaning processes to remove surface imperfections and contaminants.
b. **Wafer Cleaning:** Wafer surfaces are thoroughly cleaned using chemical and mechanical processes to remove any particles, residues, or contaminants that could affect the fabrication process.
c. **Thin Film Deposition:** Thin films of various materials, such as silicon dioxide (SiO_2), silicon nitride (Si_3N_4), and metal layers, are deposited onto the wafer surface using techniques like chemical vapor deposition (CVD) or physical vapor deposition (PVD).
d. **Photolithography:** Photolithography is a key process for defining the intricate patterns of the chip's circuitry on the wafer surface. A photosensitive

material called photoresist is applied to the wafer, exposed to UV light through a photomask containing the desired circuit patterns, and developed to transfer the patterns onto the wafer.

e. **Etching:** Etching processes selectively remove material from the wafer surface based on the pattern defined by the photoresist. Wet etching and dry etching techniques are used to etch away unwanted material layers, leaving behind the desired circuit features.
f. **Doping:** Doping introduces impurities into specific regions of the silicon wafer to modify its electrical properties and create semiconductor junctions. Dopants such as boron, phosphorus, and arsenic are diffused or implanted into the wafer to create p-type and n-type regions required for transistor formation.
g. **Ion Implantation:** Ion implantation is a precise doping technique where dopant ions are accelerated and implanted into the wafer surface at controlled depths using high-energy beams. Ion implantation allows for precise control over dopant concentration and distribution within the semiconductor substrate.
h. **Annealing:** Annealing processes heat the wafer to high temperatures to activate dopants, repair crystal lattice defects, and relieve stress induced during fabrication processes. Rapid thermal annealing (RTA) and furnace annealing are commonly used to achieve the desired electrical and structural properties of the semiconductor material.
i. **Chemical Mechanical Polishing (CMP):** CMP is used to planarize the wafer surface and remove any uneven topography resulting from previous fabrication steps. Abrasive slurries and mechanical polishing pads are employed to achieve a smooth and flat wafer surface suitable for subsequent processing.
j. **Metallization and Interconnect Formation:** Metal layers, typically aluminum or copper, are deposited onto the wafer surface to form interconnects that connect individual components of the chip's circuitry. Dielectric layers such as silicon dioxide are deposited and patterned to insulate the metal interconnects and provide electrical isolation between different circuit layers.
k. **Back-End-of-Line (BEOL) Processing:** BEOL processing involves the formation of metal interconnects, vias, and bond pads that connect the chip's active components to external connections such as package pins. Insulating layers and passivation coatings are applied to protect the chip's surface and improve reliability.
l. **Testing and Quality Assurance:** The manufactured chips undergo rigorous testing to ensure they meet the specified electrical, functional, and reliability requirements. Wafer-level testing and final package testing are performed to screen for defects, verify functionality, and ensure that high-quality chips are shipped to customers [14].
m. **Packaging:** Packaged chips are encapsulated in protective materials such as ceramic or plastic packages to shield them from environmental hazards and provide mechanical support. Packaging also includes wire bonding or flip-chip bonding to connect the chip to the package leads or external connectors.

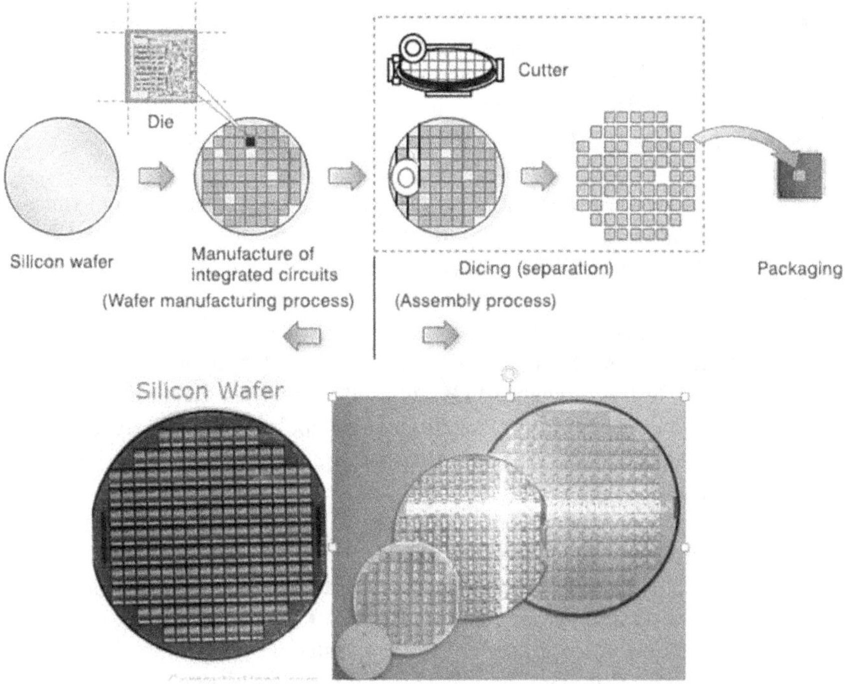

FIGURE 6.4 Manufacturing process of basic silicon wafers [13].

n. **Final Testing and Burn-In:** Packaged chips undergo final testing and burn-in to verify their functionality and reliability under operating conditions. Burn-in involves subjecting the chips to elevated temperatures and electrical stresses to accelerate aging and detect early failures.
o. **Quality Control and Packaging:** After passing final testing, the chips are sorted, labeled, and packaged into trays or tape-and-reel formats for shipment to customers. Quality control measures ensure that only chips meeting the specified criteria are packaged and shipped, maintaining high standards of reliability and performance.

Example: Once the design is verified, it moves to the manufacturing phase. The microcontroller design is translated into a series of photolithographic masks used to pattern the silicon wafer. Semiconductor fabrication facilities (fabs) employ advanced manufacturing processes to etch, deposit, and layer materials onto the wafer, creating the intricate circuitry of the microcontroller.

6.7 TESTING

Testing is a critical juncture in the lifecycle of a silicon chip, encompassing a spectrum of tests to ensure functionality, quality, and reliability. Wafer-level testing scrutinizes chips, identifying defects and anomalies before packaging. Subsequent package-level testing subjects individual chips to rigorous examination, encompassing functional, electrical, and reliability tests. Through exhaustive testing regimes, manufacturers safeguard the integrity and performance of their silicon chips [15,16]. Absolutely, wafer-level testing is a critical step in the semiconductor manufacturing process, which includes the following stages.

a. **Purpose:** Wafer-level testing is conducted to assess the quality and functionality of individual chips on a silicon wafer before they are separated and packaged. By testing chips in bulk, manufacturers can efficiently identify defects and anomalies early in the production process, thereby reducing waste and improving yield.
b. **Testing Equipment:** Specialized testing equipment, such as wafer probers or automatic test equipment (ATE), is used to perform electrical tests on the entire wafer. These tests typically involve applying various signals and stimuli to the chips and measuring their responses to assess functionality and performance [14].
c. **Test Patterns:** Test patterns are programmed into the testing equipment to evaluate different aspects of chip functionality, including logic operations, memory integrity, and input/output functionality. Manufacturers may use pre-defined test patterns or develop custom patterns tailored to the specific design and requirements of the chips being tested.
d. **Defect Identification:** During wafer-level testing, defective chips are identified based on deviations from expected test results. Common defects include shorts, opens, excessive leakage, and performance variations. Advanced testing techniques, such as BIST and scan testing, help detect and diagnose these defects accurately.
e. **Data Analysis:** Test data collected during wafer-level testing is analyzed to assess overall wafer yield and identify trends or patterns that may indicate underlying process issues. Statistical analysis tools are often employed to extract valuable insights from large volumes of test data, enabling manufacturers to optimize production processes and improve quality control.
f. **Yield Enhancement:** The primary goal of wafer-level testing is to maximize yield by identifying and segregating defective chips from the good ones. Manufacturers may employ techniques like binning, where chips are sorted into different quality categories based on test results to optimize resource allocation and ensure that only high-quality chips proceed to the next stage of the manufacturing process.
g. **Feedback Loop:** Insights gained from wafer-level testing contribute to continuous improvement efforts in semiconductor manufacturing. By analyzing test data and identifying root causes of defects, manufacturers can implement corrective actions to enhance process reliability, reduce defect rates, and ultimately improve product quality.

Testing is significant and multifaceted, encompassing various aspects of semiconductor testing processes, which can be explained as follows [17,18].

i. **Testability Features in Design:** VLSI designers incorporate testability features into chip designs to facilitate testing at later stages. These features include BIST structures, scan chains, boundary scan cells, and other design-for-testability (DFT) techniques. BIST structures enable on-chip test pattern generation and result verification, while scan chains facilitate serial access to internal chip nodes for testing purposes.
ii. **Scan-Based Testing:** VLSI chips often employ scan-based testing methodologies, where test patterns are serially shifted into and out of the chip through scan chains. This allows for efficient testing of sequential and combinational logic elements within the chip. Scan-based testing is widely used for manufacturing testing, as well as for post-manufacturing in-system testing (IST) and field testing.
iii. **Functional Testing:** VLSI chips undergo functional testing to verify correct operation under normal operating conditions. Functional tests are designed to validate the chip's functionality against its intended specifications, ensuring that it performs the required operations accurately and reliably. Functional testing may involve the execution of test vectors on the chip's inputs and monitoring its outputs for correct responses.
iv. **Structural Testing:** VLSI chips are subjected to structural testing to detect and diagnose manufacturing defects and faults within the chip's internal components. Structural testing techniques, such as stuck-at fault testing, transition fault testing, and delay fault testing, aim to identify faulty transistors, interconnects, and other circuit elements that may impact chip performance or reliability.
v. **Fault Coverage Analysis:** VLSI testing involves analyzing the effectiveness of test patterns in detecting faults and defects within the chip. Fault coverage analysis assesses the percentage of faults that are detected by the test patterns, providing insights into the quality of the test set and the overall test strategy. VLSI designers and test engineers use fault coverage metrics to optimize test patterns and improve test effectiveness.
vi. **Automated Test Equipment (ATE):** VLSI chips are tested using specialized automated test equipment (ATE), such as semiconductor testers or ATE systems. ATE systems generate test patterns, apply them to the chip under test, capture and analyze the chip's responses, and report test results. VLSI engineers develop test programs and test vectors for ATE systems to conduct various tests, including functional tests, scan tests, and structural tests.
vii. **Post-Silicon Validation:** VLSI chips undergo post-silicon validation to verify their performance and functionality in real-world operating conditions. Post-silicon validation involves testing chips in actual system environments, such as evaluation boards or prototype systems, to ensure compatibility, reliability, and performance. VLSI engineers use a combination of hardware and software tools to conduct post-silicon validation tests and debug chip issues [12].

Example: After manufacturing, the silicon wafers undergo testing to identify defective chips. This involves wafer probing, where each chip's functionality is tested while still on the wafer. For our microcontroller, tests include checking processor functionality, memory integrity, and input/output operations.

6.8 PACKAGING AND QUALITY ASSURANCE

Packaging encapsulates chips in protective casings, providing electrical connections and safeguarding against environmental stressors. Quality assurance measures permeate every facet of packaging and testing, ensuring adherence to stringent standards and regulatory requirements. Through meticulous quality control, manufacturers uphold the reliability and consistency of their chip offerings.

Example: Functional chips are then packaged into protective enclosures. Quality assurance measures ensure that each packaged chip meets performance and reliability standards. For the microcontroller, packaging considerations include thermal management, ESD protection, and ensuring proper electrical connections.

6.9 INTEGRATION AND SYSTEM TESTING

The culmination of the silicon chip journey occurs with integration into larger systems, where chips synergize to deliver complex functionalities. System-level testing validates the interoperability and performance of integrated systems, encompassing functional, performance, and compatibility tests. Through seamless integration and comprehensive testing, silicon chips seamlessly integrate into the fabric of modern technology [19].

Example: The packaged microcontrollers are integrated into IoT devices, undergoing system-level testing to validate functionality. This involves testing interactions with sensors, actuators, and communication networks. For instance, the microcontroller might be tested in a smart thermostat to ensure accurate temperature sensing, efficient control algorithms, and reliable communication with a central server.

6.10 CONCLUSION

Silicon chip design and testing epitomize the integration of creativity and engineering precision, underpinning the technological marvels that define our era. From the inception of design concepts to the seamless integration into complex systems, the journey of a silicon chip is a testament to human ingenuity and perseverance. As the semiconductor industry continues to evolve, the quest for smaller, faster, and more efficient chips perpetuates, driving innovation and shaping the course of technological progress. The design phase involves meticulous planning, where engineers translate specifications into architectural blueprints, leveraging advanced techniques such as VLSI design and logic synthesis to optimize performance, power consumption, and area utilization. Wafer-level testing enables early detection of defects, setting the stage for subsequent testing phases. The comprehensive exploration of testing methodologies in semiconductor chip design underscores the importance of rigorous testing processes in ensuring the functionality, reliability, and performance of semiconductor chips. Testing methodologies play a pivotal role in validating chip functionality, from functional and structural testing to post-silicon validation, ensuring compliance with stringent quality standards. The integration of testability features and automated test equipment streamlines the testing process, enhancing efficiency and accuracy. This chapter highlights how chip design and testing contribute to shaping the technological landscape and driving innovation across various industries. Moreover, the tangible impact of silicon chip design and testing reverberates across diverse industries, fueling innovation, driving technological advancements, and revolutionizing the way we live, work, and communicate. From smartphones and computers to automotive electronics and IoT devices, silicon chips underpin a myriad of applications, shaping the landscape of modern technology.

REFERENCES

1. Reis, R., Lubaszewski, M. S., and Jess, J. A. (Eds.). *Design of Systems on a Chip: Design and Test*, Springer Science & Business Media, 2007.
2. Pecht, M., Radojcic, R., and Rao, G. (Eds.). *Guidebook for Managing Silicon Chip Reliability*. CRC Press, 2019.
3. Beanato, G., Giovannini, P., Cevrero, A., Athanasopoulos, P., Zervas, M., Temiz, Y., and Leblebici, Y. Design and testing strategies for modular 3-D-multiprocessor systems using die-level through silicon via technology, *IEEE Journal on Emerging and Selected Topics in Circuits and Systems*, 2(2), 295–306, 2012.
4. Huang, C. M., Lee, K. J., Yang, C. C., Hu, W. H., Wang, S. S., Chen, J. B., and Jou, J. Y. *Multi-Project System-on-Chip (MP-SoC): A novel test vehicle for SoC silicon prototyping*, In *IEEE International SOC Conference*, San Diego, California, USA, 137–140, 2006.
5. Zhang, B., Pan, W., Zheng, Y., Shi, Z. and Yan, X. *A Fully Automated Large-Scale Addressable Test Chip Design with High Reliability*, In *20th IEEE European Conference on Circuit Theory and Design (ECCTD)*, Linköping, Sweden, 61–64, 2011.
6. Arden, W. M. The international technology roadmap for semiconductors-perspectives and challenges for the next 15 years, *Current Opinion in Solid State and Materials Science*, 6(5), 371–377, 2002.
7. Rout, S. S., Mahapatro, S., Jayaswal. G. and Hooda, M. Low-power integrated circuit smart device design, In *Intelligent Green Technologies for Sustainable Smart Cities*, Wiley, 227–246, 2022.
8. M. Sarrafzadeh and C. K. Wong. *An Introduction to VLSI Physical Design*, McGraw-Hill, 1996.
9. N. A. Sherwqni. *Algorithms for VLSI Physical Design Automation*, Springer, 1993.
10. Abramovici, M. In-system silicon validation and debug, *IEEE Design & Test of Computers*, 25(3), 216–223, 2008.
11. Knechtel, J., Sinanoglu, O., Elfadel, I. A. M., Lienig, J. and Sze, C. C. Large-scale 3D chips: Challenges and solutions for design automation, testing, and trustworthy integration, *IPSJ Transactions on System and LSI Design Methodology*, 10, 45–62, 2017.
12. Mitra, S., Seshia, S. A. and Nicolici, N. June. *Post-silicon validation opportunities, challenges and recent advances*, In *Proceedings of the 47th Design Automation Conference*, Anaheim, California, USA, 12–17, 2010.
13. Nazir, H. Z., Hussain, T., Akhtar, N., Abid, M. and Riaz, M. Robust adaptive exponentially weighted moving average control charts with applications of manufacturing processes, *The International Journal of Advanced Manufacturing Technology*, 105, 733–748, 2019.
14. Mourad, S. and Zorian, Y. *Principles of Testing Electronic Systems*, John Wiley and Sons, 2000.
15. Mir, S. and Charlot, B. On the integration of design and test for chips embedding MEMS, *IEEE Design & Test of Computers*, 16(04), 28–38, 1999.

16. Bhaskar, U., Passi, V., Houri, S., Escobedo-Cousin, E., Olsen, S.H., Pardoen, T. and Raskin, J. P. On-chip tensile testing of nanoscale silicon free-standing beams, *Journal of Materials Research*, 27(3), 571–579, 2012.
17. Wang, L. T., Stroud, C. E. and Touba, N. A., *System-on-Chip Test Architectures: Nanometer Design for Testability*, Morgan Kaufmann, 2010.
18. Sowmya, N., Rout, S. S. and Patjoshi, R. K. Implementation of ultra low-power electronics for biomedical applications, electronic devices, In *Circuits, and Systems for Biomedical Applications*, 153–176, 2021.
19. Larsson, E. *Introduction to Advanced System-on-Chip Test Design and Optimization*, Springer Science & Business Media, 2006.

7 A Novel Deep Learning Approach for Early Brain Tumor Detection

Ankit Singh, Ajay Singh Raghuvanshi, B. Ganesh, and K.V.S. Harsha Vardhan

7.1 INTRODUCTION

A brain tumor is an abnormal proliferation of cells within or adjoining the brain, which can emerge from intrinsic neural tissue or extraneous sites through metastasis. These growths can be classified as either primary, originating from native brain structures, or secondary, resulting from dissemination from other organs. Additionally, brain tumors can be characterized as benign or malignant, contingent upon whether they are non-cancerous or cancerous. Brain tumors can influence numerous aspects of brain functionality, such as thought processes, communication, locomotory capabilities, and perception. The manifestations of brain tumors depend on their type, size, and position within the brain.

Common symptoms related to brain tumors include headache, nausea, vomiting, visual difficulties, muscle weakness, seizures, and changes in character traits and mental capacities. The severity and presentation of symptoms are influenced by the tumor's size, location, and aggressiveness. Early detection and appropriate intervention are crucial for optimizing outcomes and improving quality of life for individuals affected by brain tumors.

Over 12,000 people a year, including 500 children and young adults, are affected by primary brain tumors, i.e., 33 people per day [1]. Every year, brain tumors claim the lives of more than 5,300 individuals. A tumor that originates in the brain or spinal cord is referred to as a primary brain or spinal cord tumor. A projected 24,810 persons in the US (14,280 men and 10,530 women) will have an initial malignant tumor diagnosis in their brain or spinal cord in 2023 [2]. In 2020, an estimated 308,102 cases of primary brain or spinal cord tumors were identified worldwide [3]. In the United States, 5,230 children under the age of 20 are predicted to receive a CNS tumor diagnosis in 2023 [2]. Of all Americans, 23–24 have a brain tumor (malignant or non-malignant) for every 100,000 [1]. Glioblastoma represents around 14.5% of all brain tumors and is the most prevalent type of malignant brain tumor [3]. Meningioma represents approximately 38.3% of all brain tumor cases and is the most often reported benign (or non-malignant) brain tumor.

Magnetic resonance imaging (MRI) and computed tomography (CT) can help us identify the presence and location of a brain tumor, as well as its size and shape.

DOI: 10.1201/9781003483038-7

MRI and CT can also help differentiate between primary brain tumors (those that originate from the brain) and secondary brain tumors (those that spread from other parts of the body). Classification: MRI and CT can help determine the type and grade of a brain tumor, based on its appearance, enhancement, perfusion, and other features [4].

Brain tumors demonstrate remarkable diversity in terms of morphology, size, location, and appearance. These neoplasms can be further stratified according to their histologic subtypes, grades, and stages, influencing their clinical management and prognostication [5]. For instance, brain tumors can be classified as primary or secondary, benign or malignant, and can present with distinctive histological features, such as gliomas, meningiomas, and pituitary tumors.

Moreover, brain tumors can possess heterogeneity within themselves, featuring distinct regions with differing biological properties [6]. This heterogeneity complicates the assessment of tumor margins during surgical procedures and affects the efficacy of therapy.

Labeled datasets, however, may suffer from noise, incompleteness, or inconsistency due to factors such as human errors or variations in annotation standards. Despite these challenges, ongoing efforts in machine learning and computational modeling aim to improve the accuracy and reliability of brain tumor diagnostics and prognostications [7].

In summary, brain tumors exhibit considerable variation in their morphology, location, and composition, necessitating careful consideration of these attributes when developing diagnostic and treatment protocols. Labeled datasets, although valuable, must be approached cautiously due to the possibility of noise, incompleteness, or inconsistency. Nonetheless, technological advances continue to enhance our capacity to understand and manage brain tumors effectively.

Within the realm of computer vision, Residual Network 50 (ResNet-50) serves as a prominent classification model, demonstrating exceptional precision in distinguishing photographs as tumorous or non-tumorous. Compared to traditional architectures, ResNet-50 offers notable advantages, such as:

1. High accuracy rates, often exceeding those of alternative models like AlexNet, VGG16, GoogleNet, and ResNet18.
2. The utilization of a "bottleneck" design for building blocks, enabling faster training by reducing parameter counts and matrix multiplication operations [8].
3. Implementation of three layers per block instead of two, thereby increasing expressive power without significantly augmenting computational requirements.
4. Superior performance across various domains, including medical imaging, despite potential differences in content compared to general-purpose images [9].
5. Application of transfer learning strategies which facilitate knowledge transfer from pre-existing models to expedite training and boost performance.

Despite its strengths, ResNet-50 demands substantial computational resources and memory due to its deep architecture, and the training process can be time-consuming and complex. Nevertheless, its proven effectiveness makes it a preferred choice among both researchers and practitioners [10].

7.2 LITERATURE REVIEW

Because brain tumors are varied, complex, and frequently have overlapping traits, classifying them is a difficult undertaking. Additionally, several imaging techniques, including MRI, CT, positron emission tomography (PET), and single-photon emission computed tomography (SPECT), may yield distinct information about the tumor that needs to be integrated and analyzed. Therefore, advanced methodologies and procedures that can manage these difficulties and achieve high accuracy and resilience are needed for brain tumor classification.

Autonomous brain tumor categorization from MRI scans is non-invasive, removing the need for biopsies and improving the precision of the diagnosis process. The research community has spent a lot of time and energy creating automated methods for classifying brain tumors from the beginning of these millennia until the late 1990s. The categorization of brain tumors into their respective histologic categories is another job that has been accomplished with outstanding performance.

Brain tumor classification using deep learning. This chapter presents a method for brain tumor classification using a convolutional neural network (CNN) that extracts features from brain MRI images and then classifies them using fully connected networks. This chapter also compares different CNN architectures, such as basic CNN and VGG-16, and reports an accuracy of 95.71% on an online data set [11,12].

Studies propose machine learning-based algorithms for brain architecture analysis and classification of pictures. In-depth research on brain MRI scan diagnosis techniques was done by Abd-Ellah et al. [13], who compared and contrasted the benefits and drawbacks of deep learning and conventional machine learning techniques. The authors also introduced a novel semi-automatic segmentation method for brain tumor images [6]. The T1W configuration was used by this model for three-dimensional MRI segmentation. To classify breast cancer images, a CNN-based architecture was introduced. The system's architectural design, which retrieved information from fitting scales, allowed for the highest level of accuracy in tumor segmentation and localization. In addition, a CNN-based model for brain tumor diagnosis employed AlexNET, DenseNet201, GoogLeNet, and Inception-v3 [14]. The outcomes showed that the suggested approach could accurately identify and classify tumors in MRI. The collection of work demonstrates a significant progress in 3D visualization and the segmentation and classification of brain cancers from MRI data. To improve the effectiveness of extracting features, tumor classification, localization, and novel approaches are still needed.

Somasundaram and Gobinath [15] suggested a number of techniques for using MR images to identify brain malignancies. The efficiency of segmentation enhancement utilizing 3D CNNs, multi-class Support Vector Machines (SVMs), and SVMs were investigated by the researchers. Deep learning techniques and CNNs demonstrated exceptional performance in a range of medical image processing applications, including brain tumor identification. The classification and segmentation of brain tumors using deep learning techniques produced better results than conventional machine learning classifiers.

A different study used a deep learning neural model to extract characteristics from MRI [16]. These features were then fed into machine learning classifiers, including multilayer perceptrons (MLPs), SVMs, and naive Bayes. The suggested approach achieved a 96% classification accuracy when SVMs were used. In the research by Kumar et al. [17], several machine learning and deep learning methods for brain tumor identification and segmentation were investigated. These methods included support vector machines, K-nearest neighbors (KNN), MLPs, naive Bayes, and random forest algorithms. Remarkably, at 92.4%, the classification accuracy of the traditional SVMs was the greatest. The researchers also demonstrated a custom CNN design with five layers for MRI brain tumor identification, which achieved an astounding 97.2% accuracy.

Khan et al. [18] used the VGG19 CNN architecture with K-means clustering to identify and segment brain tumors in MRI data. The proposed approach initially divided the input MR modality into slices before pre-processing the intensities with a statistical normalization procedure. The total accuracy of their approach was 94%.

Machine learning-based brain tumor identification and categorization: an extensive overview. This study offers a thorough assessment of the literature on machine learning methods, such as deep learning, for the identification and classification of brain tumors. The study discusses deep learning approaches for brain tumor analysis, segmentation, feature extraction, classification, augmentation techniques, publicly accessible data sets, and the anatomy of brain tumors. The benefits, drawbacks, advancements, and potential future directions in this field are also covered in the study.

An innovative approach called VGG-SCnet, which combines the VGGNet stacked upon a classifier, was presented by Majib et al. [19]. For automated brain tumor diagnosis from MRI images, the scientists improved a pre-trained VGG-16 model with extra layers suggested by their methodology to enable quicker and more efficient training. During the data preparation stage, the region of interest was identified using the most prominent contours. Using augmentation techniques, the class imbalance in the dataset was corrected. Because it had less features, the sixth layer was chosen for feature extraction. Ultimately, a multilayer classifier was used to detect tumors in imaging.

Deep transfer learning methods for brain tumor detection performance analysis. In order to automatically predict brain tumor cells, this research compares the performance of CNN-pretrained VGG-16, ResNet-50, and Inception-v3 models that use transfer learning. The BRATS 2015 and BRATS 2017 data sets are used in the paper's evaluation of the models. The article reports an accuracy of 97.5% on BRATS 2017 and on BRATS 2015 [10].

Deep Learning-Based Brain Tumor Detection and Classification. Using the state-of-the-art object detection framework YOLO (You Only Look Once) and the deep learning library fastai, respectively, this study offers two deep learning-based techniques for brain tumor detection and classification. Using a portion of the BRATS 2018 data set, which included 1992 brain MRI images, the research assesses the methodologies. The study indicates YOLO and fastai accuracy of 96.8% and 97.6%, respectively.

7.3 PROPOSED METHODOLOGY

ResNet-50 (Residual Network 50) is a CNN design that has excelled at picture classification and other computer vision applications. It is a deep neural network with 50 layers, and its fundamental innovation is the use of residual connections, which allows it to train much deeper networks without having to deal with the vanishing gradient problem. This improves the performance of deep neural networks by adding additional neural layers while reducing error percentage. In other words, skip connections combine the outputs of prior layers with the outputs of stacked layers, allowing far deeper networks to be trained than previously possible.

CNNs are a class of deep learning models primarily employed for processing grid-like data structures, such as images. They consist of multiple interconnected layers, namely convolutional, pooling, non-linear activation, and fully connected layers, arranged sequentially to extract high-level abstractions from low-level pixel inputs. The cornerstone of CNNs revolves around local connectivity and weight-sharing principles, enabling spatial feature hierarchy formation through progressive aggregation of neighboring pixels. As a result, CNNs exhibit superior representation capabilities over traditional handcrafted features, rendering them highly effective for various computer vision applications.

Residual Networks (ResNets), exemplified by the flagship model ResNet-50, introduce innovative design elements aimed at combating deleterious effects induced by increased depth. Specifically, ResNets incorporate residual blocks composed of multiple convolutional layers equipped with shortcut connections, also termed skip connections. These links allow direct transmission of activations from earlier layers to later stages, effectively forming parallel signal paths within the overall architecture. Notably, residual blocks empower ResNets to learn residual mappings instead of direct feature representations, thereby simplifying optimization landscapes and mitigating vanishing gradient problems plaguing deeper architectures.

To elucidate further, let us examine the internal composition of a typical residual block found in ResNet-50. It generally comprises two or three consecutive convolutional layers sandwiched between batch normalization and rectified linear unit (ReLU) operations. A key differentiator resides in the addition of a shortcut connection that bypasses intermediate layers, performing element-wise summation with activated outputs derived from preceding layers. During backpropagation, these skip connections facilitate smoother gradient flow, reducing optimization difficulties encountered in standard feedforward networks devoid of such provisions.

By stacking multiple residual blocks together, ResNet-50 constructs a hierarchical framework that captures increasingly complex patterns embedded within input data. Additionally, ResNet-50 adopts bottleneck designs for computational efficiency, wherein each residual block consists of three nested sub-blocks featuring dimensionality reduction, transformation, and restoration operations. Overall, ResNet-50 epitomizes the power of residual learning paradigms, demonstrating marked improvements in performance relative to its contemporaries while maintaining comparable complexity levels.

In summary, both CNNs and ResNet-50 belong to the family of deep learning models specializing in grid-like data analysis. While CNNs lay the foundation for modern computer vision architectures, innovations introduced by ResNet-50 address critical limitations imposed by escalating depth, culminating in enhanced expressiveness and stability. Through a judicious combination of convolutional layers, batch normalization, activation functions, and residual connections, ResNet-50 emerges as a potent tool for tackling diverse real-world challenges encompassing medical diagnostics, autonomous driving, and surveillance systems.

At the core of ResNet-50 lies a sophisticated deep architecture, meticulously constructed to harness the potential of residual connections and deliver outstanding performance on challenging computer vision tasks. Composed of 50 concatenated residual blocks, this pyramidal edifice systematically expands and contracts channel dimensions while retaining spatial resolution, thus striking an optimal equilibrium between representational capacity and computational feasibility. Each residual block incorporates a triad of convolutional layers: a 1×1 convolution for dimension reduction, followed by a 3×3 convolution tasked with mining contextual dependencies and concluding with another 1×1 convolution for restoring initial dimensionality. This elegant configuration not only streamlines information flow but also curtails parameter proliferation, contributing to ResNet-50's lauded parsimony.

A salient characteristic of ResNet-50's deep architecture is the presence of skip connections, which traverse residual blocks and establish direct communication channels between distant layers. These conduits ensure that authentic input signals percolate through the network, reinforcing feature hierarchies and tempering the detrimental impacts of amplified noise or distortion. Furthermore, they imbue ResNet-50 with immunity against vanishing gradient adversities, sustaining robust learning dynamics even when confronted with daunting depths.

Collectively, ResNet-50's deep architecture weaves an intricate tapestry of learned representations, progressively disentangling latent factors governing visual concepts and unlocking novel opportunities for transferring knowledge across related endeavors. Whether deployed as standalone entities or integrated within larger ensemble frameworks, these deeply rooted models continue to serve as linchpins propelling advances in artificial intelligence research and industrial application development.

Skipping Connections: Because they enable gradients to "skip" over layers, these residual connections are often referred to as skip connections, which enable the training of extremely deep networks.

Bottleneck Architectures: Using the same number of filters in each layer of deeper networks is computationally costly. With its "bottleneck" design, ResNet-50 has three layers per residual block: 1×1, 3×3, and 1×1 convolutions. It is computationally more efficient due to the 1×1 convolutions, which first lower and then expand the dimensions.

Global Average Pooling: Instead of completely linked layers, global average pooling is utilized at the network's conclusion. This method reduces overfitting and the amount of parameters in the network.

Pre-Trained Models: ResNet-50, whose weights have previously been learned on a huge dataset (such as ImageNet), is frequently utilized as a pre-trained model. With smaller datasets, these pre-trained weights can be adjusted for certain tasks.

The comparison of different models is shown in Table 7.1, highlighting the variations in performance metrics.

TABLE 7.1
Comparison with Various Models

Model	Training Parameters	Validation Accuracy	Validation Loss	Test Accuracy	Test Loss
Mobilenet	3206976	0.9648	0.1321	0.9531	0.3220
Inception-v3	21768352	0.9688	0.2712	0.9621	0.3091
DenseNet-121	8068994	0.9812	0.1708	0.9745	0.0723
ResNet-50 (proposed)	25698306	0.9910	0.0429	0.9949	0.0177

7.4 RESULTS

During the training process, our deep learning model is classified into two distinct categories. The initial category, labeled "yes," comprises a collection of images displaying evident signs of brain tumors. Conversely, the second category, labeled "no," consists of images depicting brains that appear healthy or non-tumorous.

Upon completion of the prediction phase, our model renders verdicts regarding each input image, designating them as either belonging to the "yes" or "no" class. When an image falls under the "yes" label, it signifies the presence of a brain tumor within the depicted brain tissue. Alternatively, if an image receives a "no" classification, it implies that the analyzed brain image does not exhibit any discernible indicators of a tumor.

In summary, our deep learning model serves as a reliable tool for distinguishing between brain images showcasing tumors and those appearing healthy. By accurately categorizing inputs, it enables medical professionals to make informed decisions about further diagnostic procedures and potential treatment plans.

Our extensive dataset encompasses a grand total of 21672 individual MRI images, carefully curated for the purpose of training and validating our deep learning model. Among these images, there exist 3066 instances that represent healthy or non-tumorous brains. On the contrary, a substantial number of 18606 images reveal the presence of varying types of brain tumors (Figure 7.1).

Figure 7.1a demonstrates examples of upper-side MRI images, while Figure 7.1b highlights side-angle MRI images included in our dataset. These diverse representations enable our model to develop a more comprehensive understanding of the intricacies involved in brain tumor detection across multiple viewpoints.

To ensure a balanced and representative distribution of data during the training and validation processes, we meticulously divided our dataset into two portions. An impressive 80% of the entire collection was allocated for intensive training, allowing our model to learn effectively from a vast array of samples. Meanwhile, the remaining 20% served as valuable test data, enabling us to rigorously evaluate the model's performance and fine-tune its parameters accordingly.

Throughout the course of our research, we remain committed to exploring innovative approaches and refining our deep learning model to achieve unparalleled precision and accuracy in the realm of brain tumor detection.

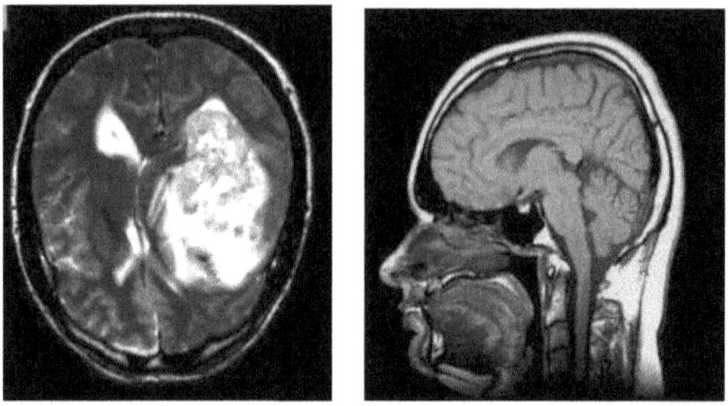

FIGURE 7.1 Image with Brain Tumor.

Our deep learning model, trained on a comprehensive dataset of brain tumor images, demonstrated commendable performance in classifying tumors. The model achieved an accuracy of 99%, sensitivity of 99.36%, and specificity of 98.78%. These metrics indicate the robustness of the model in distinguishing between tumorous and non-tumorous brain images.

$$\text{Sensitivity} = \frac{\text{True positives}}{\text{True positives} + \text{false negatives}}$$

$$\text{Specificity} = \frac{\text{True negatives}}{\text{True negatives} + \text{false positives}}$$

7.4.1 Confusion Matrix

True Positives (TP): 3737

In these instances, your model accurately inferred that the examples are members of the positive class (tumorous).

True Negatives (TN): 566

These are the instances when your model accurately predicted that the given examples are from the non-tumorous negative class.

False Positives (FP): 7

These are the cases where your model predicted tumorous, but the actual class was non-tumorous.

False Negatives (FN): 24

These are the cases where your model predicted non-tumorous, but the actual class was tumorous (Figure 7.2).

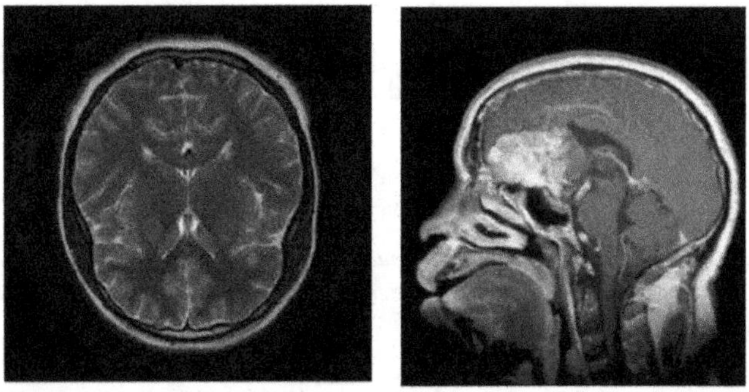

FIGURE 7.2 Images without Brain Tumor.

7.4.2 Accuracy

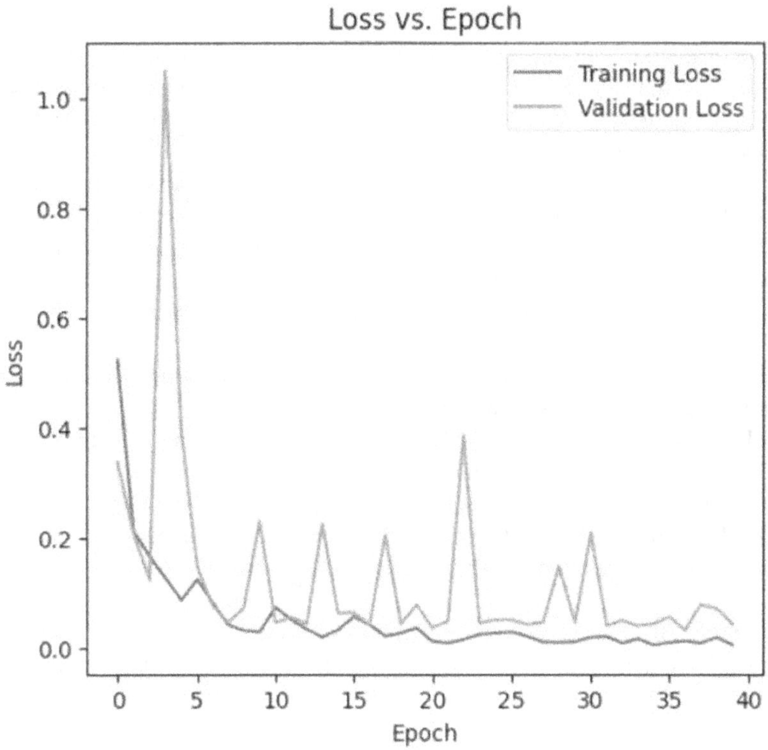

7.5 DISCUSSION

In this comprehensive exploration, we present a cutting-edge deep learning solution devised for detecting brain tumors at their earliest stages using MRI. Our groundbreaking approach relies upon the powerful capabilities of ResNet-50 as its underlying structure. To validate the efficacy of our proposed methodology, we conduct an

TABLE 7.2
Comparative Analysis of the Proposed Model

Paper	Source	Method	Accuracy
Brain tumor classification using deep learning	Springer	CNN with fully connected networks	95.71%
Deep transfer learning approaches in performance analysis of brain tumor detection	*International Journal of Advanced Computer Science and Applications*	Transfer learning-based CNN-pre-trained VGG-16, ResNet-50, and Inception-v3	97.5% on BRATS 2017
Deep learning-based brain tumor detection and classification	*International Journal of Innovative Technology and Exploring Engineering*	YOLO and fastai	96.8% for YOLO and 97.6% for fastai
Proposed research	-	ResNet-50	99.1%

extensive evaluation comparing it to prevailing tactics employed for classification. Moreover, semantic segmentation, a critical component encompassing tumor localization, is meticulously considered in our analysis. Simultaneously, we investigate the influence of transfer learning using ResNet-50.

A comparative analysis of the proposed model with existing approaches is provided in Table 7.2. All investigations were executed using the widely recognized BRATS dataset. However, our advanced model incorporates distinct datasets featuring both top and side angles of the brain.

Key disparities between the presented models and ours include:

1. **Data Composition**: Previous models utilize solely top-angle MRIs, whereas our model integrates side-angle data.
2. **Model Architecture**: Our model builds upon the robustness of ResNet-50.
3. **Transfer Learning Integration**: We explore the implications of implementing transfer learning using ResNet-50.
4. **Evaluation Metrics**: A thorough assessment of these factors contributes to a comprehensive understanding of the models' performances.

By presenting this detailed account, we aim to shed light on the significance of considering diverse perspectives and expanding the scope of data usage in developing sophisticated deep learning solutions for brain tumor detection.

7.6 CONCLUSION

Our research demonstrates the effectiveness of deep learning, specifically the ResNet-50 model, in identifying and categorizing brain tumors from MRI scans. By employing a multi-stage network design that integrates both feature extraction and classification, we achieved high levels of accuracy. Our model has shown exceptional performance, achieving a noteworthy accuracy rate of 99.1% in differentiating between brain images with tumors and those without.

In addition, our methodology utilizes an extensive dataset consisting of MRI images taken from multiple angles, including both the top and side views. This diverse dataset enhances the comprehensiveness of brain tumor detection, as it considers various perspectives. The combination of this extensive dataset with the robustness of the ResNet-50 architecture and the implementation of transfer learning techniques significantly improves the performance of our proposed approach.

Through rigorous testing and analysis, we have proven the reliability and accuracy of our deep learning model in identifying and categorizing brain tumors. The model's impressive accuracy, sensitivity, and specificity underscore its significance as a valuable resource for healthcare professionals in diagnosing brain tumors at an early stage. Early detection, facilitated by our model, can lead to timely intervention and, ultimately, better patient outcomes.

Looking ahead, we plan to fine-tune and enhance our deep learning model, as well as investigate additional datasets and imaging techniques, to expand its versatility and utility. Furthermore, ongoing research and innovation in deep learning applications for medical imaging hold great potential for enhancing disease diagnosis and treatment, ultimately leading to enhanced healthcare outcomes for patients on a global scale.

REFERENCES

1. L. M. DeAngelis, Brain tumors, *New England Journal of Medicine*, vol. 344, no. 2, pp. 114–123, 2001.
2. J. L. Fisher, J. A. Schwartzbaum, M. Wrensch and J. L. Wiemels, Epidemiology of brain tumors, *Neurologic Clinics*, vol. 25, no. 4, pp. 867–890, 2007.
3. T. M. Mitchell, *Machine Learning*, New York: McGraw-Hill, vol. 1, 2007.
4. A. C. Heron, A. Madabhushi, N. M. Rajpoot and B. Yener, Histopathological image analysis: a review, *IEEE Reviews in Biomedical Engineering*, vol. 2, pp. 147–171, 2009.
5. E. I. Zacharaki et al., Classification of brain tumor type and grade using MRI texture and shape in a machine learning scheme, *Magnetic Resonance in Medicine*, vol. 62, no. 6, pp. 1609–1618, 2009.
6. P. A. Yushkevich, Y. Gao, and G. Gerig, ITK-SNAP: an interactive tool for semi-automatic segmentation of multi-modality biomedical images. In *Proceedings of the 2016 38th Annual International Conference of the IEEE Engineering in Medicine and Biology Society (EMBC)*, Orlando, FL, USA, 16–20 August 2016, pp. 3342–3345.
7. T. Araújo, G. Aresta, E. Castro, J. Rouco, P. Aguiar, C. Eloy, A. Polonia and A. Campilho, Classification of breast cancer histology images using convolutional neural networks. *PLoS One,* vol. 12, p. e0177544, 2017.
8. O. Sampetrean and H. Saya, Modeling phenotypes of malignant gliomas, *Cancer Science*, vol. 109, no. 1, pp. 6–14, 2018.
9. D. Bzdok, M. Krzywinski and N. Altman, Machine learning: supervised methods, *Nature Methods*, vol. 15, no. 1, p. 5, 2018.
10. M. D. Z. Alom, et al. The history began from AlexNet: a comprehensive survey on deep learning approaches. *ArXiv* abs/1803.01164, 2018.
11. M. W. Nadeem, M. A. A. Ghamdi, M. Hussain, M. A. Khan, K. M. Khan, S. H. Almotiri and S. A. Butt, Brain tumor analysis empowered with deep learning: a review, taxonomy, and future challenges, *Brain Sciences,* vol. 10, p. 118, 2020. https://doi.org/10.3390/brainsci10020118

12. S. A. Abdelaziz Ismael, A. Mohammed and H. Hefny, An enhanced deep learning approach for brain cancerMRI images classification using residual networks, *Artificial Intelligence in Medicine*, vol. 102, p. 101779, 2020.
13. M. K. Abd-Ellah, A. I. Awad, A. A. M. Khalaf and H. F. A. Hamed, A review on brain tumor diagnosis from MRI images: practical implications, key achievements, and lessons learned. *Magnetic Resonance Imaging,* vol. 61, pp. 300–318, 2019.
14. A. Çinarand M. Yildirim, Detection of tumors on brain MRI images using the hybrid convolutional neural network architecture, *Medical Hypotheses,* vol. 139, p. 109684, 2020.
15. S. Somasundaram and R. Gobinath, Current trends on deep learning models for brain tumor segmentation and detection: a review. In *Proceedings of the 2019 International Conference on Machine Learning, Big Data, Cloud and Parallel Computing (COMITCon 2019)*, Faridabad, India, 14–16 February 2019, pp. 217–221.
16. G. Latif, G. B. Brahim, D. N. F. A. Iskandar, A. Bashar and J. Alghazo, Glioma tumors' classification using deep-neural-network-based features with SVM classifier, *Diagnostics*, vol. 12, p. 1018, 2022.
17. G. Kumar, P. Kumar and D. Kumar, Brain tumor detection using convolutional neural network. In *Proceedings of the International Conference on Mobile Networks and Wireless Communications (ICMNWC),* Tumkur, Karnataka, India, 3–4 December 2021, pp. 1–6.
18. A. R. Khan, S. Khan, M. Harouni, R. Abbasi, S. Iqbal and Z Mehmood, Brain tumor segmentation using K-means clustering and deep learning with synthetic data augmentation for classification. *Microscopy Research and Technique*, vol. 84, pp. 1389–1399, 2021.
19. M. S. Majib, M. M. Rahman, T. M. S. Sazzad, N. I. Khan and S. K. Dey, VGG-SCNet: a VGG net-based deep learning framework for brain tumor detection on MRI images, *IEEE Access*, vol. 9, pp. 116942–116952, 2021.

8 TCAD-Augmented Machine Learning for the Prediction of Device Behaviour and Failure Analysis

Sandeep Moparthi and Gopi Krishna Saramekala

8.1 INTRODUCTION

In order to keep up with the increasing momentum in industrial automation and address time-to-market concerns, the integrated circuit (IC) industry is integrating machine learning (ML) at several levels of its process hierarchy. ML-assisted TCAD simulation of devices has emerged as a feasible solution for the performance analysis and reverse engineering of emerging devices. The idea here is to predict the electrical characteristics of a device to carry out the device-level performance analysis and optimization with the help of a well-trained ML algorithm in very little time. The TCAD-aided ML simulation of emerging devices offers reverse engineering and faster decision-making in R&D. A TCAD-ML model is presented in this chapter with the help of a cylindrical silicon nanotube. The critical device parameters such as channel length (L) and core radius (t_c) are varied to create the necessary dataset.

For an efficient ML design, an accurate modelling of the input and output parameter dependency is needed. This can be achieved by a simple polynomial equation. However, the complexity of the ML model is directly proportional to the degree of polynomial which makes the model challenging [1,2]. It is quite difficult and impractical to model a non-linear problem of a higher degree with polynomials. The use of the polynomial model in turn limits the number of input variables in the dataset. Besides this, decision tree (DT) and neural network (NN)-based algorithms offer flexibility to model the problems with a higher degree of non-linearity. The literature reveals that the majority of TCAD-assisted ML efforts have employed the linear regressor (LR), NN, and random forest (RF) techniques to model the problem. Despite its simple design, the LR method restricts the prediction of non-linear dependencies. On the other hand, when it comes to non-linear regression, NN results in increased processing time and design complexity. However, because of its straightforward DT-based design, the RF regressor offers fast computing performance with minimal complexity for issues with non-linearity [3].

DOI: 10.1201/9781003483038-8

Furthermore, it is observed that the literature extensively addressed linear models with a single variable. It is seen that none of the newly developed gate-all-around devices used TCAD-assisted ML for performance prediction as far as we are aware. As a result, a ML model is developed in this chapter to forecast the electrical characteristics by altering important device parameters like t_c and L on the Si NTFET's performance. We adopt and evaluate the performance of two popular DT-based algorithms: the gradient-boosting (GB) regressor and the RF regressor. The performance of algorithms that are being trained is evaluated using metrics such as maximum error (ME) and mean square error (MSE). During the testing phase, the method that performs better during training is chosen based on performance metrics to forecast the threshold voltage V_{TH}, SS, I_{ON}, and I_{OFF} of Si NTFET. The type and size of the dataset as well as the quantity of input parameters have a major impact on the appropriateness of the RF and GB algorithms. Thus, it is crucial to compare the algorithms' performances in order to determine which one is most appropriate for a given dataset. The chapter's remaining sections are arranged as follows. The creation of the dataset and the ML methods is explained in detail in the following section. The training and testing results are shown in the results section along with a discussion of the algorithms under consideration.

8.2 DATA GENERATION AND MACHINE LEARNING METHODOLOGY

An ML method supported by TCAD is presented to forecast the Si NTFET's output characteristics. Figure 8.1 and Table 8.1 depict the cross-sectional view of Si NTFET and device parameters taken into account for the Si NTFET simulation, respectively. Shockley-Read-Hall (SRH) recombination model and Carrier Velocity-Temperature

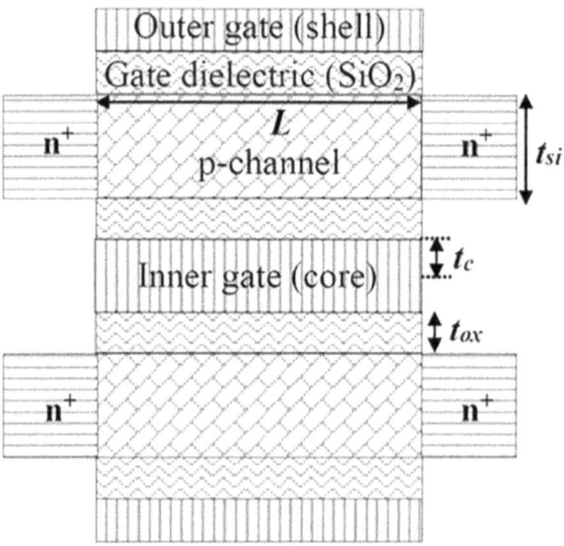

FIGURE 8.1 Cross-section of Si NTFET along the channel.

TABLE 8.1
Device Parameters

Parameters	Values
Core radius (t_c)	2–7 nm
Channel length (L)	10–100 nm
Silicon channel thickness (t_{si})	10 nm
Gate dielectric thickness (t_{ox})	1.5 nm
Gate work function (WF)	4.71 eV
Channel doping (N_A)	10^{16}cm^{-3}
Source/drain doping (N_D)	10^{20}cm^{-3}
Drain voltage (V_d)	0.5 V
Temperature (T)	300 K

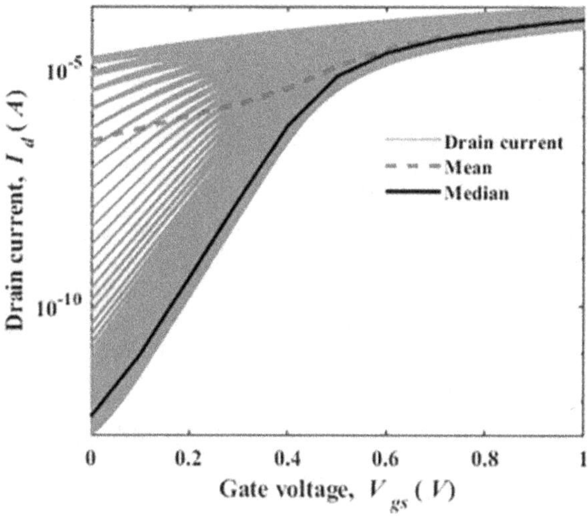

FIGURE 8.2 TCAD simulation results of Si NTFET.

(CVT) (Lombardi) mobility model are used in a 3D TCAD tool Atlas from Silvaco to simulate the device structures. By adjusting the two most important parameters, t_c and L, within the given range while holding all other parameters constant, as indicated in Table 8.1, a dataset of 4641 samples is produced. With a step size of 0.1 nm, the critical parameter t_c is changed from 2 nm to 7 nm. The parameter L is adjusted with a step size of 1 nm, ranging from 10 nm to 100 nm, in order to capture the short-channel and long-channel trend. Thirty percent of the 4641 extracted data samples are used for testing, and 70% are used for training. It is important to remember that the amount of data samples mostly determines the split ratio of the dataset [3]. The transfer characteristics of Si NTFET structures for each data sample, as determined by TCAD simulation, are displayed in Figure 8.2. Figure 8.2 also displays the simulated dataset's mean and median values. The transfer characteristics of each device are used to obtain the output parameters V_{TH}, SS, I_{ON}, and I_{OFF}. The gate voltage (V_{gs}) at which the parameters I_{ON} and I_{OFF} are measured is 1 V and 0 V, respectively.

The non-linear dependency between the input parameter variables and all of the output parameters can be seen from Figure 8.3. Furthermore, as the input parameters vary, all of the output parameters also change continuously. Regression is an ML technique that is used for the training and prediction of continuous values (outputs), whereas classification is meant for the prediction of discrete values. Owing to the previously described factors, the output parameters for the supplied input parameters are predicted using a multi-variable non-linear regression. The two popular DT-based ML algorithms are GB and RF regressors [3,4]. The Python scikit-learn library [4] is where the RF and GB algorithms are imported from. When a trained algorithm is applied to new data in ML, data overfitting reduces its accuracy. To limit the data overfit issue which is more serious in a single DT, GB and RF regressors are constructed using a group of DTs. The characteristics of the DT have

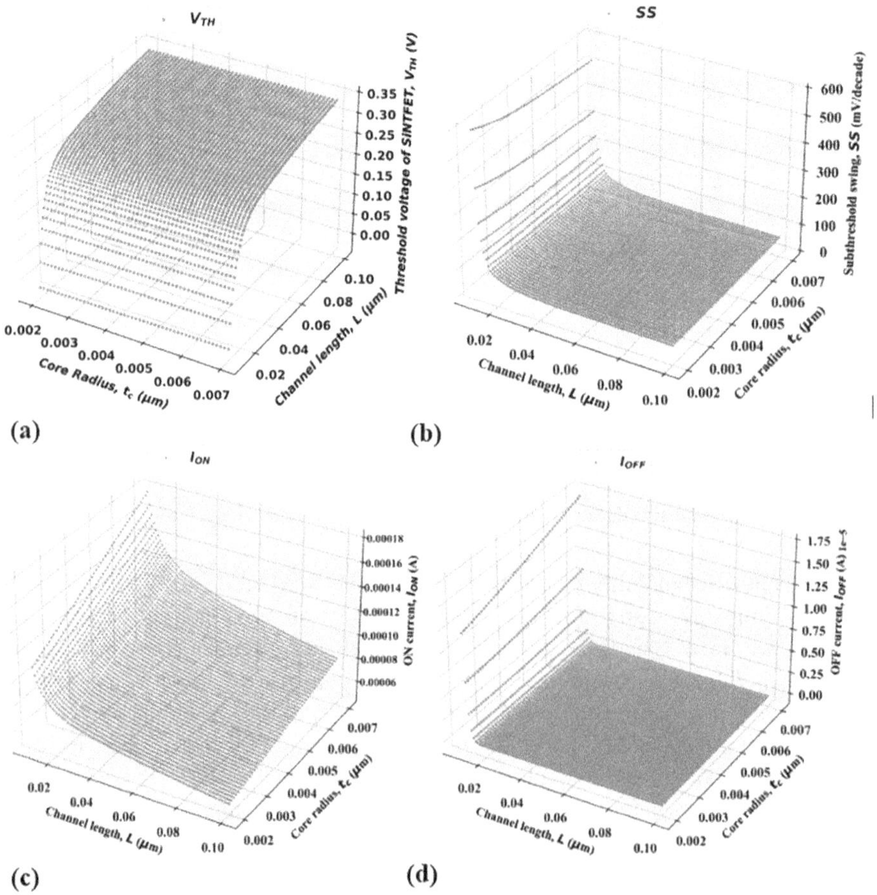

FIGURE 8.3 Variation of (a) V_{TH}, (b) SS, (c) I_{ON}, and (d) I_{OFF} with t_c and L.

TABLE 8.2
Key Hyperparameters Considered for GB and RF Regressors

	V_{TH} (V)	SS mV/decade	I_{ON} (A)	I_{OFF} (A)
Number of estimators	100	100	100	100
Maximum tree depth	18	20	16	34
Minimum samples per split	2	2	2	2
Minimum samples per leaf	1	1	1	1

an impact on how well the RF and GB regressor algorithms function. To improve the efficiency of these algorithms, these characteristics known as hyperparameters must be tuned.

In this study, the number of estimators, maximum tree depth, minimum samples per split, and minimum samples per leaf are taken into account as the main hyperparameters. The number of estimators represents the highest number of DTs that may be used to successively minimize the MSE in RF and GB. The maximum length of a DT is determined by its maximum tree depth. The smallest quantity of data samples needed to split an internal node in a DT is known as the minimum samples per split [3,4]. The bare minimum of samples needed for a leaf (end) node is specified by the minimum number of samples per leaf. For both RF and GB regressors, the default parameters for the number of estimators, minimum samples per split, and minimum samples per leaf are 100, 2, and 1, respectively [4]. However, for RF and GB regressors, the default value for maximum tree depth is NONE and 3, respectively [4]. When it comes to RF regressors, the default value of NONE for maximum tree depth denotes that the programmer has not provided a predefined value. As a result, the RF algorithm is free to extend the tree depth to its optimal value, which is when all leaves fall below the minimum samples per split value. The optimum values of maximum tree depth for the provided dataset are found to be 18, 20, 16, and 34 for the associated output parameters V_{TH}, SS, I_{ON}, and I_{OFF}, respectively, according to the RF regressor. In addition, the GB regressor's default maximum tree depth of 3 is far less than the ideal values that the RF regressor was able to achieve. This is the reason why, when compared to their default hyperparameter values, the GB regressor outperforms the RF regressor. To provide a level playing field in comparison, this study takes the maximum tree depth values that were produced from the RF regressor and applies them to the GB regressor. Table 8.2 displays the hyperparameter values that were utilized to compare the RF and GB regressor methods.

8.3 TRAINING AND TESTING RESULTS

8.3.1 Section 1 RF and GB Algorithm Results

The outcomes of the training and testing of the examined algorithms are shown in this section. Figure 8.4 and Table 8.3 display the absolute MSE and ME values for all of the outputs for the RF and GB algorithms that are undergoing training. When the

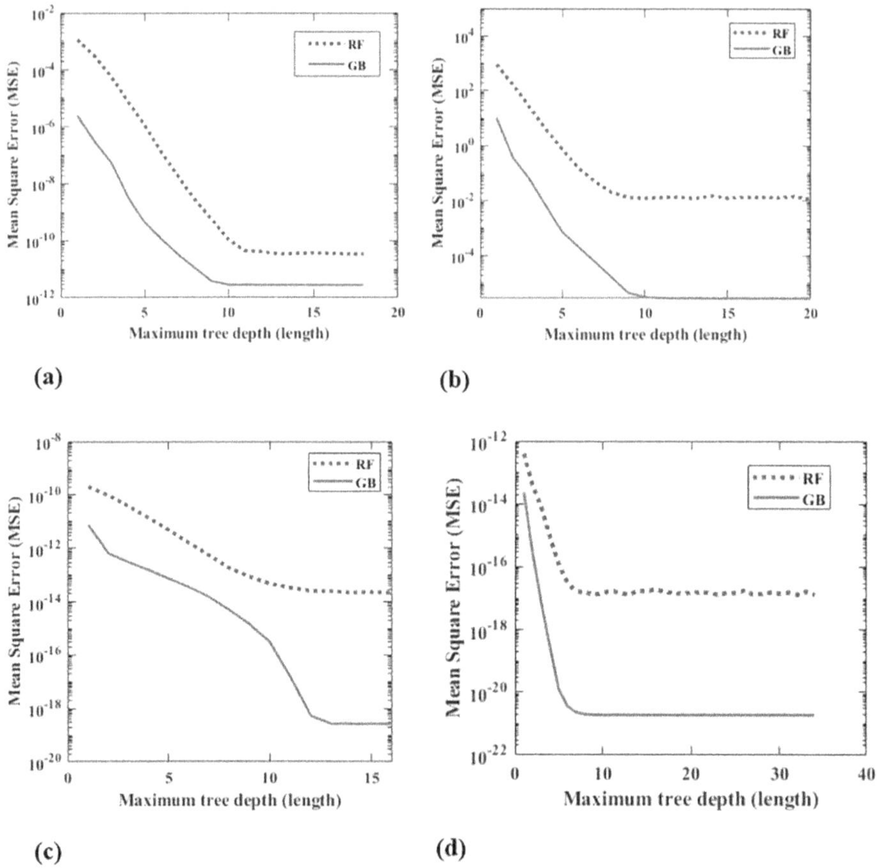

FIGURE 8.4 MSE versus maximum tree depth for RF and GB regressors (a) V_{TH}, (b) SS, (c) I_{ON}, and (d) I_{OFF}.

TABLE 8.3
Performance Metrics Used for Training

	Training	V_{TH} (V)	SS mV/decade	I_{ON} (A)	I_{OFF} (A)
MSE	RF regression	3.25E-11	1.16E-02	2.23E-14	1.31E-17
	GB regression	2.85E-12	2.79E-06	2.66E-19	1.88E-21
ME	RF regression	9.72E-05	3.21	1.91E-06	7.12E-08
	GB regression	8.11E-06	0.01369	2.355E-09	4.33E-10

identical hyperparameter values are taken into account for both of these methods, it is observed that the GB regressor performs better than the RF regressor. The MSE for the GB regressor attained its optimal value of 2.86×10^{-12}, 2.81×10^{-6}, 2.68×10^{-19}, and 1.89×10^{-21} for V_{TH}, SS, I_{ON}, and I_{OFF}, respectively, as can be shown in

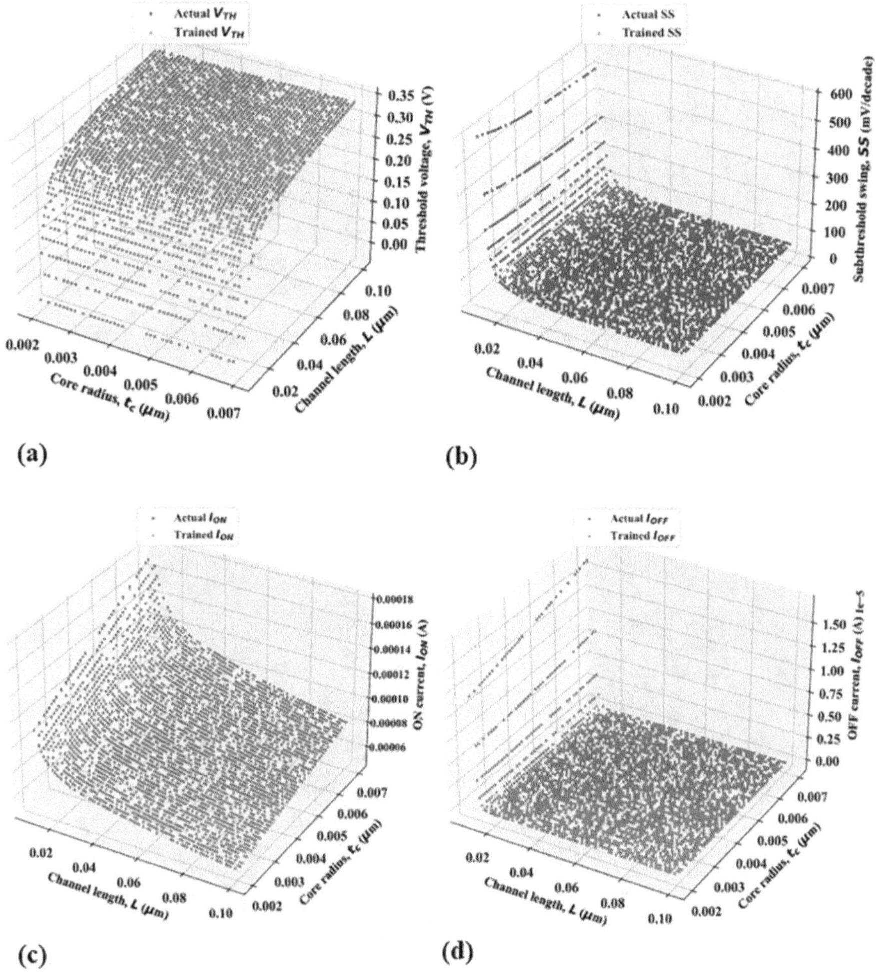

FIGURE 8.5 Training results: (a) V_{TH}, (b) SS, (c) I_{ON}, and (d) I_{OFF} with RF regression.

Figure 8.4 and Table 8.3. The training of V_{TH}, SS, I_{ON}, and I_{OFF} using RF and GB regressors is displayed, respectively, in Figures 8.5 and 8.6. The GB regressor is selected for testing because it outperforms the RF regressor in training.

Figures from Figure 8.7a to c display the V_{TH} test results. Figure 8.7a illustrates how the genuine value of the V_{TH} under test and the anticipated V_{TH} overlap. The prediction error is displayed in Figure 8.7b for each test sample. It can be observed that most samples have an error of zero. As seen in Figure 8.7c, a total of 77.03% of test samples (1073 out of 1393 samples under test) have a prediction error of ±9 µV. Of the 320 test samples that are left, 310 have an inaccuracy of less than ±98 µV. The forecast of SS under test is shown in Figure 8.8a. When SS is between 250 mV/decade and 600 mV/decade in a short-channel domain, there is a small departure from the prediction. As illustrated in Figure 8.8b, a maximum prediction error

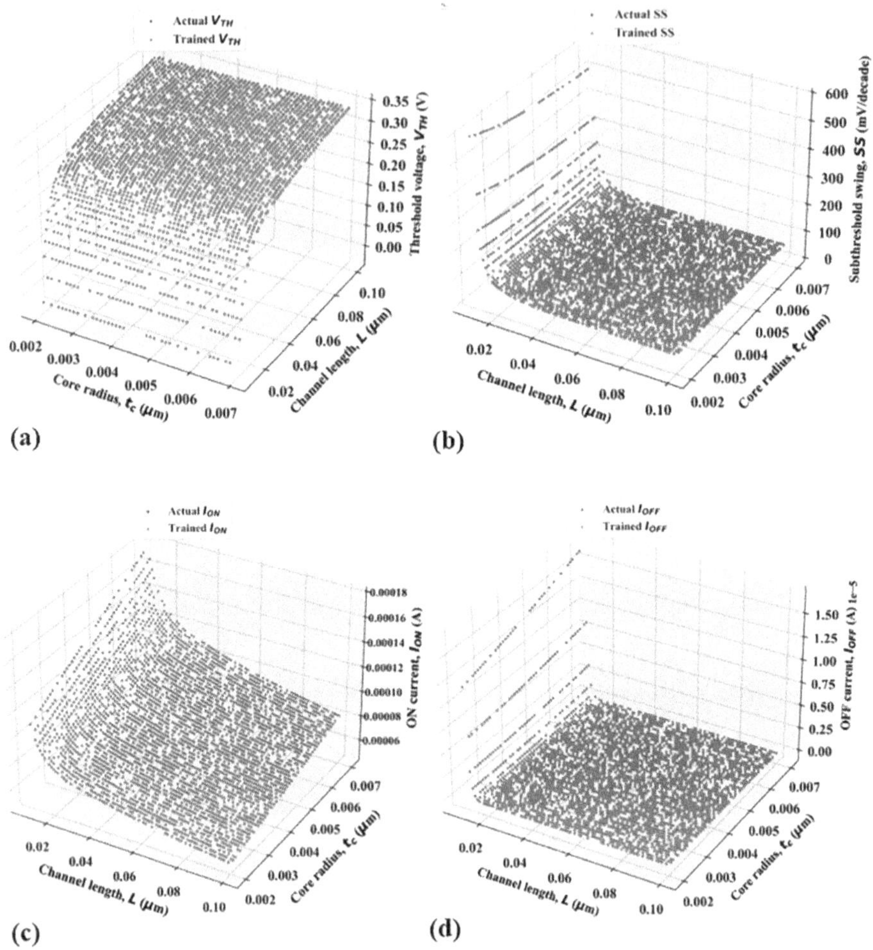

FIGURE 8.6 Training results: (a) V_{TH}, (b) SS, (c) I_{ON}, and (d) I_{OFF} with GB regression.

of ±3 mV/decade to ±4 mV/decade is observed for five test samples in the short-channel range. When compared to the previously described values of SS in the short-channel region, this mistake is essentially insignificant. Figure 8.8c illustrates that 1330 test samples have a prediction error of <0.274 mV/decade. Out of the 63 test samples that were left, 58 samples had a prediction error of less than ±3 mV/decade.

Figure 8.9a shows the prediction of I_{ON} and comparison with test data samples. The profile shows that the I_{ON} is rising with t_c as a result of the SiNTFET's changing effective channel thickness. Figure 8.9b makes clear that most test samples had a prediction error of less than ±1 μA. Figure 8.9c shows that 1055 out of 1393 test samples, or 75.73%, had prediction errors of less than ±0.67 μA. Three hundred and twenty-six test samples from the remaining 338 samples provided an inaccuracy between <±2.6 and >±0.67 μA. As seen in Figure 8.9c, there is only one test sample with an error of magnitude >6 μA. Figure 8.10a shows the prediction and comparison of I_{OFF} with

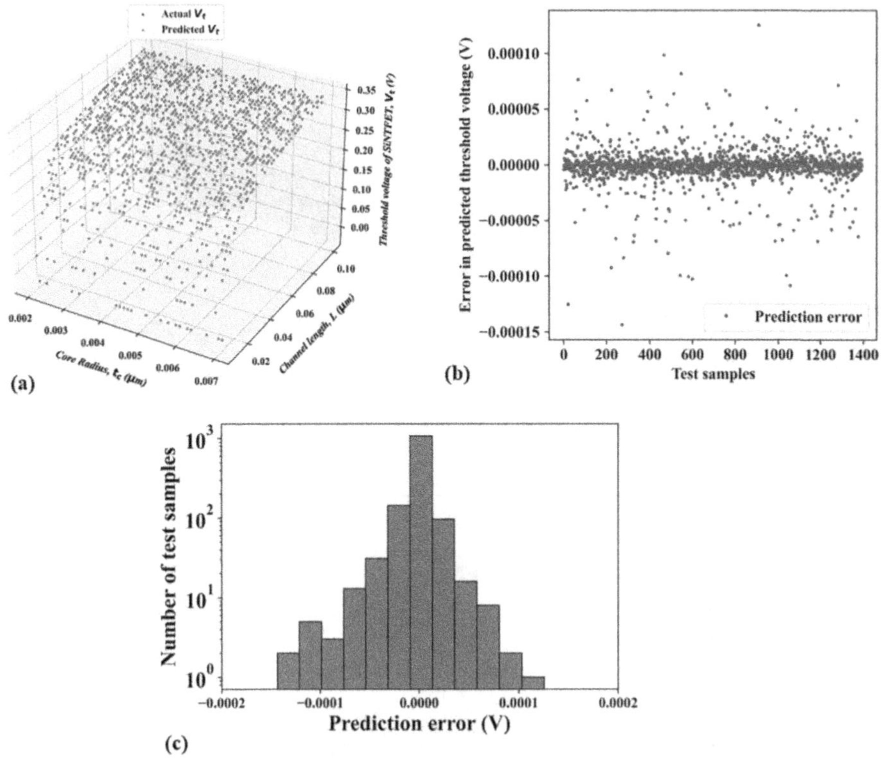

FIGURE 8.7 Testing results for V_{TH}. (a) Prediction and comparison, (b) Error of individual samples, (c) Error distribution.

test data samples. At $L < 20\,\text{nm}$, I_{OFF} has a very strong non-linearity. Figure 8.10b illustrates a fairly strong agreement between the test and projected values of I_{OFF} despite this non-linearity. Thirteen hundred and forty-four test samples (or 97.2%) have a prediction error of less than 5.9 nA. Of the 39 test samples that are still available, 27 of them exhibit an error that falls between $<\pm$ 88 and $>\pm$ 5.9 nA, as shown in Figure 8.10c. The testing performance measures for the GB regressor, MSE and ME, are listed in Table 8.4. In comparison to V_{TH}, I_{ON}, and I_{OFF}, SS provided poorer prediction based on the absolute values of MSE and ME. But as Figure 8.8b illustrates and discusses, the ME provided by SS, i.e., 4.1942, is derived from the short-channel regime and is extremely little in comparison to the order of SS under that regime.

8.3.2 Section 2 Enhanced TCAD-ML Dataset Design

In the conventional ML model, train dataset does not allow the use of input parameters other than those used in the train dataset during the testing of the trained algorithm and the prediction of outputs. The dataset used in the previous section is designed as an array of two input variables L and t_c with three outputs V_{TH}, SS, I_{ON}, and I_{OFF}. The previous dataset does not allow input variables other than L and

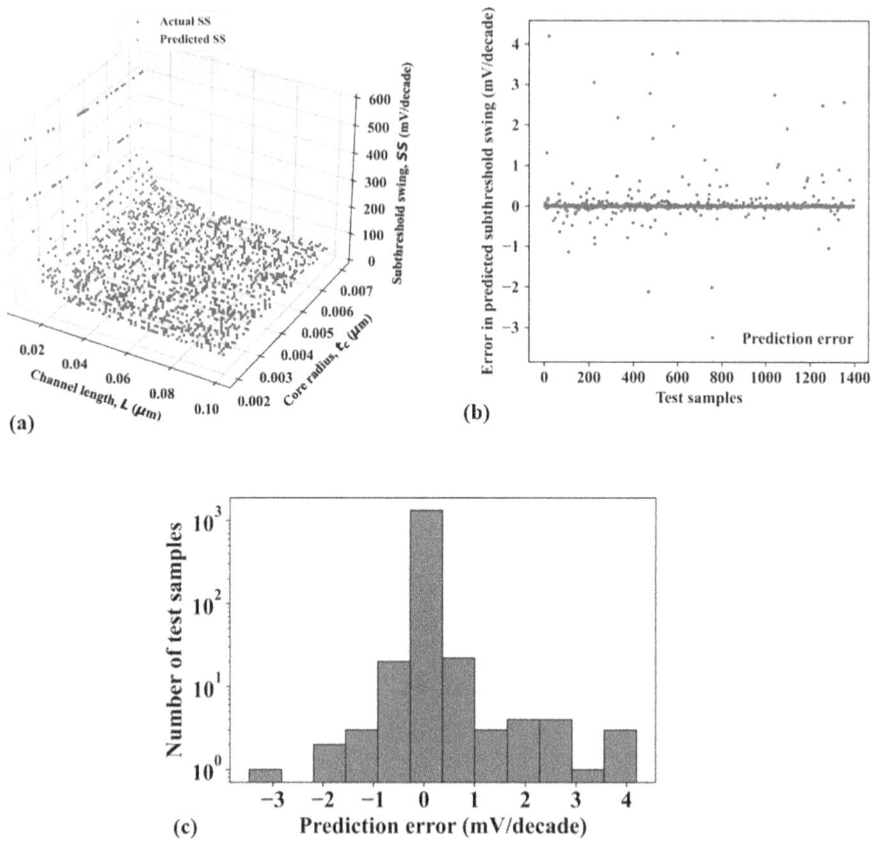

FIGURE 8.8 Testing results for SS. (a) Prediction and comparison, (b) Error of individual samples, (c) Error distribution.

t_c during the algorithm testing for the prediction of output. Besides this, the limits of extrapolation associated with the ML model imply that the prediction of output fails if the given values of the input variables (L and t_c) are outside the range of train dataset. As these algorithms cannot extrapolate (out-of-train range prediction), a large dataset is required for the algorithm training to cover the entire range and the combinations of all the considered inputs. The extraction of such huge data with several input combinations results in a drastic increase in the TCAD simulation time and size of the train dataset. The above-mentioned challenges which are involved in the ML model and dataset design in turn diminish the viability and furtherance of TCAD-aided ML. To enhance the practicality of TCAD-aided ML by eluding the above-mentioned challenges, this section of results proposes an alternate design of the previous dataset for the algorithm training and testing. The range and step size of parameters L and t_c and models used in TCAD simulation remain unchanged.

A new parameter $[(L * t_c)/(L + t_c)] + \log(L)$ is defined from the values of L and t_c in the considered range. The new parameter is chosen to overcome the redundancy

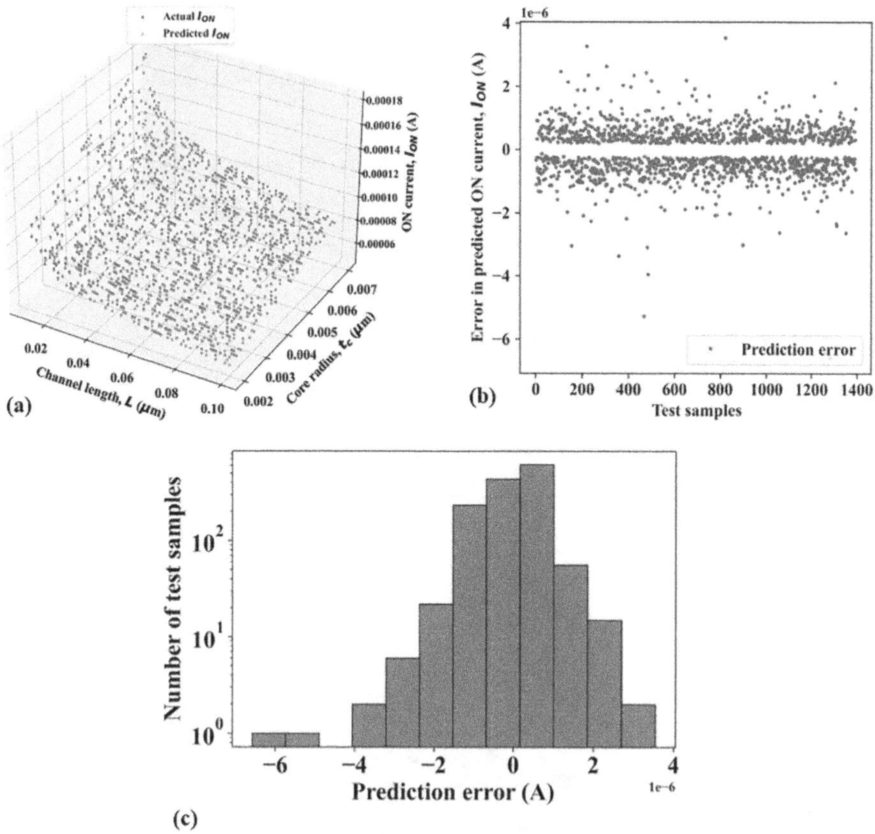

FIGURE 8.9 Testing results for I_{ON}. (a) Prediction and comparison, (b) Error of individual samples, (c) Error distribution.

caused by various combinations of L and t_c values in the given range. Furthermore, $[(L * t_c)/(L+t_c)]+\log(L)$ is multiplied with a factor 10^9 (the resultant parameter is denoted as M) to amplify the difference between its adjacent values which results in better training of the algorithm. This is due to the fact that the values of $[(L * t_c)/(L+t_c)]+\log(L)$ are very small and are in the order of 10^{-9}. The two input parameters, i.e., L and t_c in the previous dataset, are replaced with the newly defined input parameter, i.e., M. The inclusion of M in the proposed new dataset allows the selection of L and t_c values in any range such that M must be within the range of the proposed train dataset. Hence, the boundaries of L and t_c which are used for dataset generation can be surpassed during prediction. This in turn eludes the limit of extrapolation in the ML model. Furthermore, two extracted device characteristics I_{OFF} and I_{ON}/I_{OFF} are also included in the proposed new dataset as input parameters. The inclusion of I_{OFF} and I_{ON}/I_{OFF} as input parameters in the proposed new dataset permits the use of any device parameter (along with L and t_c) as a variable during testing/prediction. This can extend the test range beyond train range and elude the limitation of inputs

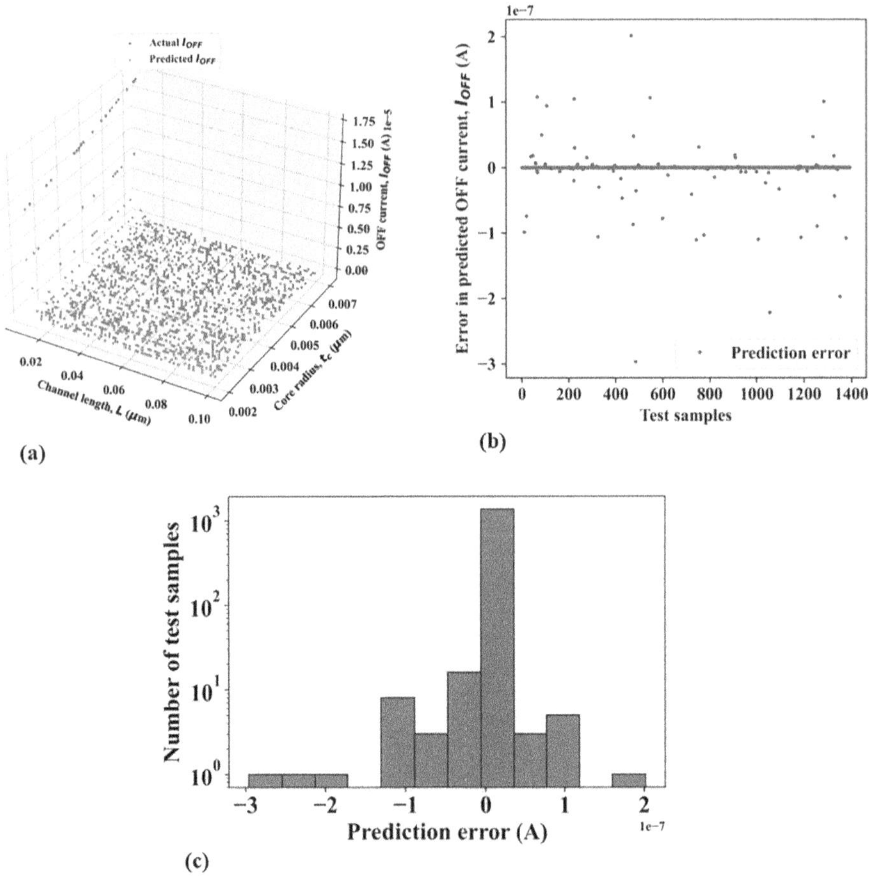

FIGURE 8.10 Testing results for I_{OFF}. (a) Prediction and comparison, (b) Error of individual samples, (c) Error distribution.

TABLE 8.4
Performance Metrics of Testing

GB Regressor	V_{TH} (V)	SS mV/decade	I_{ON} (A)	I_{OFF} (A)
MSE	3.12E-10	0.102	6.39E-13	2.84E-16
ME	0.000144	4.1941	6.57E-06	2.95E-07

associated with conventional train datasets as mentioned. Hence, the proposed new train dataset which is used for the algorithm training has three input variables, i.e., M, I_{OFF}, and I_{ON}/I_{OFF} along with output, i.e., V_{TH}. All the three input parameters of the new train dataset are normalized. The range of M, I_{OFF}, and I_{ON}/I_{OFF} in the train dataset is from 1E+09 to 8542056074, 3.8E-14 to 1.69E-05, and 7.95833 to 754122000, respectively, for the considered range of variable parameters L and t_c.

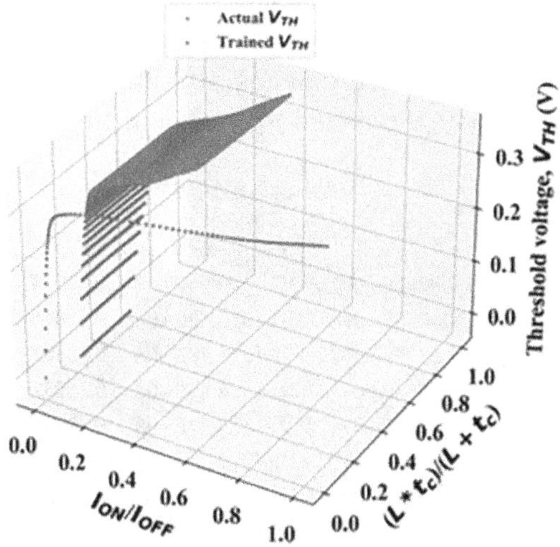

FIGURE 8.11 Training of GB regressor with modified train dataset.

Figure 8.11 depicts the comparison of actual V_{TH} and trained V_{TH} which is presented with only two inputs, i.e., M and I_{ON}/I_{OFF} for the 3D visualization of the trend. The overlapping of both actual and trained outputs in Figure 8.11 is evidence for the successful training of the algorithm. The performance metrics MSE and ME are calculated between actual and trained V_{TH} to estimate the training accuracy of the GB regressor. The obtained MSE and ME under training are 4.18E-15 and 8.65E-7, respectively. As mentioned, the new train dataset permits the use of any of the device parameters along with L and t_c as input variables during testing/prediction. Moreover, the new train dataset also permits the use of L and t_c values out of the train range during testing/prediction. To demonstrate these merits, this study considered channel thickness (t_{si}), gate-oxide thickness (t_{ox}), and channel doping concentration (N_A) as variable device parameters along with L and t_c for the generation of the test dataset which is used for testing/prediction. The considered values of these parameters are shown in Table 8.5 (column 1–5) which are used in TCAD simulations to extract the test dataset as shown in Table 8.5 (column 6–9).

8.4 CONCLUSION

Challenges in TCAD-aided ML are addressed to make it viable. A train dataset design approach is proposed to elude the issues of extrapolation and train data size. The proposed approach is demonstrated with Si NTFET by training a GB regressor algorithm. A set of 4641 data samples are extracted from the TCAD simulations by varying the L and t_c of Si NTFET. The merits of the proposed train dataset are tested with an independent test dataset created by varying the t_{si}, t_{ox}, and N_A along with L and t_c. The training of the regressor algorithm with the proposed train dataset permits the use of any device parameters (including the device parameters which are kept constant for training) as variables for the design of the test dataset. The

TABLE 8.5
Considered Parameters for Generation of Test Dataset (Columns 1–5), Test Dataset (Columns 6–9), Predicted V_{TH} (Column 10), and Prediction Error (Column 11)

t_{si} (nm)	t_{ox} (nm)	N_A (cm^{-3})	L (nm)	t_c (nm)	M	I_{OFF}	I_{ON}/I_{OFF}	V_{TH} (V)	Predicted V_{TH} (V)	Prediction error (V)
16	2	1E15	120	5	6.8792E+09	2.15E-13	2.95E+08	0.322967	0.336457	0.01349
15	3	1E17	110	6	7.7310E+09	1.33E-13	3.52E+08	0.328964	0.336402	0.007438
14	2	5E17	50	7.5	8.2207E+09	4.47E-13	1.78E+08	0.336838	0.303128	−0.03371
12.5	1.5	8E17	45	3	4.4657E+09	1.22E-13	4.95E+08	0.35737	0.336551	−0.02082
9	1.5	5E17	40	8	8.2687E+09	1.99E-13	4.32E+08	0.328012	0.336373	0.008361
9	1.5	5E17	28	3	4.1568E+09	2.35E-12	2.71E+07	0.296329	0.275251	−0.02108
12	1.5	1E15	30	2	3.3521E+09	4.26E-11	1.89E+0^6	0.258672	0.258399	−0.00027
12	1.5	1E17	30	5	5.7628E+09	3.36E-11	3.06E+06	0.268709	0.259011	−0.0097
8	1.5	5E17	14	2	2.8961E+09	2.02E-08	3382.72	0.185928	0.207069	0.021141
8	1.5	1E15	9	2.5	2.9108E+09	6.59E-06	14.67	−0.01319	0.038299	0.051488

proposed model can predict the threshold voltage (V_{TH}) of Si NTFET for the given values of M, I_{OFF}, and I_{ON}/I_{OFF}. This dataset design can be applied to any device structure for a different combination of inputs for the prediction of selected outputs.

REFERENCES

1. S. Moparthi, C. Yadav, G. K. Saramekala and P. K. Tiwari, "Machine learning based device simulation using multi-variable non-linear regression to assess the impact of device parameter variability on threshold voltage of double gate-all-around (DGAA) MOSFET," *2020 IEEE 2nd International Conference on Circuits and Systems (ICCS)*, 2020, pp. 64–67, doi: 10.1109/ICCS51219.2020.9336608.
2. S. Moparthi, P. K. Tiwari and G. K. Saramekala, "Genetic algorithm-based threshold voltage prediction of SOI JLT using multi-variable nonlinear regression," *2021 International Symposium on Devices, Circuits and Systems (ISDCS)*, 2021, pp. 1–4, doi: 10.1109/ISDCS52006.2021.9397911.
3. A. C. Müller and S. Guido. (2016). *Introduction to Machine Learning with Python: A Guide for Data Scientists*. United States, O'Reilly Media.
4. Scikit-learn developers, "Scikit-learn: Machine Learning in Python," 2023. Available: https://scikit-learn.org/stable/

9 Opportunities and Challenges for ML-Based FPGA Backend Flow

Saif Ahmad, Mohd Javed Khan, Indrasen Singh, and Zeenat Fatima

9.1 INTRODUCTION

Field-programmable gate arrays (FPGAs) represent a class of programmable semiconductor devices that offer unparalleled flexibility and customization in digital circuit design. Unlike traditional application-specific integrated circuits (ASICs) with fixed functionality, FPGAs can be configured and reconfigured by users to implement a wide variety of digital circuits and applications. FPGAs are distinguished by their reconfigurable nature. Users can change the functionality of the FPGA by modifying its configuration, allowing for rapid prototyping, iterative development, and adaptation to evolving design requirements.

The core building blocks of FPGAs are configurable logic blocks (CLBs). These blocks consist of lookup tables (LUTs), flip-flops, multiplexers, and other elements that can be programmed to implement different logic functions. FPGAs have a highly flexible interconnect structure that enables the connection of CLBs and other components. Users can define interconnections to establish communication paths between different logic elements, facilitating the creation of complex digital circuits. Input/output (I/O) blocks provide the interface between the FPGA and the external environment. These blocks enable communication with other devices, including sensors, memory, and communication interfaces. FPGAs support both sequential and combinatorial logic designs. Sequential logic involves the use of flip-flops for storing state information, while combinatorial logic consists of logic gates for instantaneous data processing [1–3]. FPGAs excel in parallel processing tasks, enabling the execution of multiple operations simultaneously. This parallelism is particularly beneficial for applications such as digital signal processing (DSP), image processing, and parallel computing.

FPGAs are widely used in DSP applications for tasks like filtering, modulation, and signal analysis. The parallel architecture of FPGAs allows for efficient processing of signal data. FPGAs find application in embedded systems, where their reconfigurability allows for the implementation of various functions such as communication interfaces, control systems, and sensor interfacing. FPGAs are employed in networking equipment, routers, and switches for tasks such as packet processing, protocol handling, and data encryption. Their flexibility enables adaptation to different communication standards.

DOI: 10.1201/9781003483038-9

The parallel processing capabilities of FPGAs make them well-suited for image and video processing applications. FPGAs can be configured to perform tasks like image recognition, video compression, and computer vision. FPGAs play a role in automotive electronics for functions such as engine control, driver-assistance systems, and communication within the vehicle's electronic control units (ECUs). FPGAs are extensively used in prototyping and rapid development scenarios. Designers can quickly implement and test digital circuits without the need for custom ASIC fabrication. FPGAs have limited resources in terms of logic elements, memory, and I/O pins. Designers need to optimize their designs to fit within these constraints. Power consumption can be a concern, especially in battery-powered or energy-efficient applications. Designers should carefully manage power usage through efficient design practices. Designing for FPGAs requires expertise in hardware description languages (HDLs), understanding of FPGA architectures, and considerations for timing, placement, and routing. FPGAs may be more expensive than traditional microcontrollers for certain applications. Cost considerations need to be weighed against the benefits of reconfigurability and parallel processing. FPGAs offer a versatile platform for digital circuit design, allowing for reconfigurability and parallel processing. Their applications span various domains, from signal processing to embedded systems, and they continue to be a crucial component in the landscape of programmable logic devices.

In recent years, the incorporation of machine learning (ML) techniques into various phases of the FPGA design cycle has emerged as a potential strategy for addressing difficulties and generating new opportunities [4]. While ML has shown considerable success in frontend tasks such as synthesis and placement, its application in the backend flow, particularly for FPGA architectures, presents both challenges and exciting prospects. The FPGA backend flow encompasses crucial stages of design implementation, including routing, timing optimization, and bitstream generation. Traditionally, these processes have been rule-based and deterministic, but the increasing complexity of designs and the demand for higher performance prompt the exploration of ML-based methodologies to enhance efficiency and robustness.

9.1.1 Challenges

- ML models introduce a level of non-determinism due to their reliance on training data. Handling the inherent unpredictability in the behavior of ML models during backend flow poses a significant challenge.
- The success of ML models heavily relies on the quality and quantity of training data. Generating representative datasets that capture the diverse characteristics of FPGA designs is a non-trivial task.
- ML models are often considered "black boxes," making it challenging for designers to interpret their decisions. Building trust in ML-based backend flows requires addressing the interpretability of models and ensuring designers can understand and validate the decisions made by the ML algorithms.
- FPGAs typically operate under strict resource constraints. Integrating ML models into the backend flow must balance the benefits of improved performance against the additional resource requirements of running ML algorithms on the FPGA itself.

- FPGA designs are subject to frequent changes during the development cycle. ML models in the backend flow need to adapt seamlessly to design updates without requiring extensive retraining or causing disruptions.

9.1.2 Opportunities

- ML models can optimize critical aspects of the backend flow, such as routing and placement, leading to improved performance and reduced time to market. Accelerating these processes can significantly enhance FPGA design productivity.
- ML algorithms can dynamically manage FPGA resources, adapting to changing workloads and optimizing resource allocation for varying design requirements.
- ML-based backend flows can leverage predictive analytics to anticipate and mitigate potential issues, offering designers insights into optimal design decisions and trade-offs.
- ML models can provide adaptive solutions for routing and timing closure, adjusting strategies based on the specific characteristics and challenges of each design.
- ML can contribute to the creation of energy-efficient FPGA designs by optimizing power consumption during backend processes.
- ML algorithms enable automation in design space exploration, helping designers navigate the vast possibilities efficiently and identify optimal design configurations.

The integration of ML into the FPGA backend flow introduces both challenges and transformative opportunities. Addressing these challenges and leveraging the capabilities of ML models have the potential to revolutionize how FPGAs are designed, optimized, and implemented, paving the way for more efficient and adaptive electronic systems.

9.2 CONVENTIONAL CAD FLOW FOR FPGAS

The conventional CAD flow for FPGAs involves a series of steps to design, implement, and program an FPGA. Here is an overview of the typical CAD flow for FPGAs:

a. **Specification and High-Level Design:**
   a. Define the specifications and requirements of the digital system.
   b. Create a high-level design using HDL such as Verilog or VHSIC (Very High-Speed Integrated Circuit) Hardware Description Language (VHDL).
b. **RTL (Register-Transfer-Level) Design:**
   i. Refine the high-level design into RTL, which represents the digital logic and its behavior.
   ii. Write RTL code in Verilog or VHDL.

c. Simulation:

i. Simulate the RTL code using a simulation tool to verify the functionality and correctness of the design.

ii. Fix any issues found during the simulation.

d. Synthesis:

i. Transform the RTL code into a netlist of logic gates using a synthesis tool.

ii. Map the design to the target FPGA architecture.

e. Technology Mapping:

Perform technology mapping to map the synthesized netlist to the specific FPGA architecture, taking into account the resources and constraints of the target device.

f. Place and Route:

i. Place the logical elements (gates, flip-flops, etc.) onto the physical locations of the FPGA.

ii. Route the interconnections between the placed elements while considering the timing and resource constraints.

g. Static Timing Analysis (STA):

i. Perform static timing analysis to ensure that the design meets the required timing constraints.

ii. Address any timing violations by adjusting the design or constraints.

h. Bitstream Generation:

i. Generate the bitstream file, which is a binary file that contains configuration information for the FPGA.

ii. The bitstream is used to program the FPGA and configure its logic.

i. Verification:

i. Verify the functionality of the design on the FPGA using simulation or hardware verification techniques.

ii. Debug and fix any issues that may arise during this phase.

j. Programming the FPGA:

i. Program the FPGA using the generated bitstream file.

ii. The bitstream configures the FPGA to implement the desired digital logic.

k. Testing and Debugging:

i. Perform extensive testing to ensure that the implemented design meets all specifications.

ii. Debug and fix any issues that may arise during testing.

l. Documentation:

Create documentation that includes design specifications, test plans, and implementation details for future reference and maintenance. This CAD flow provides a systematic approach to design and implement digital circuits on FPGAs, allowing for flexibility in hardware design and rapid prototyping. Keep in mind that specific tools and methodologies may vary based on the FPGA vendor and the complexity of the design.

9.2.1 High-Level Synthesis

High-level synthesis (HLS) is a design process that transforms a high-level programming language, such as C, C++, or SystemC, into an RTL (register-transfer-level) hardware description. HLS allows designers to describe their algorithms and functionalities at a higher level of abstraction, which can improve productivity and reduce design time. Here is an overview of the conventional HLS process:

1. Start with a high-level algorithmic description of the desired functionality in a high-level programming language (e.g., C, C++, and SystemC).
2. Use an HLS tool to compile the high-level code. Popular HLS tools include Vivado HLS, Intel HLS Compiler, and Catapult C. The HLS tool analyzes the high-level code and generates an RTL representation.
3. Optionally, the designer can use HLS directives to guide the synthesis process. The HLS tool performs synthesis to generate an RTL design from the high-level code. During this phase, the tool applies various optimizations to improve performance, reduce area, and meet specified constraints.
4. The HLS tool generates synthesizable RTL code in a HDL (typically Verilog or VHDL) based on the high-level description.
5. Simulate the generated RTL code to verify its functional correctness. This step helps catch any discrepancies between the high-level description and the synthesized hardware.
6. In some cases, designers may perform co-simulation with the software to verify that the hardware and software components work seamlessly together.
7. Use traditional RTL simulation and debugging techniques to verify the RTL code and address any issues that may arise.
8. Iterate between the HLS tool and design refinement to optimize the hardware implementation further. Designers may adjust directives or modify the high-level code to achieve the desired balance between performance, area, and power.
9. Once satisfied with the RTL code, the generated hardware description is used in the conventional FPGA or ASIC design flow. This includes synthesis, place and route, static timing analysis, and bitstream generation for FPGAs or ASIC synthesis and backend flows for ASICs.
10. Test the implemented design on the target hardware, ensuring that it meets performance and functionality requirements.

    The HLS process provides an efficient way to convert algorithmic descriptions into hardware implementations while allowing designers to work at a higher level of abstraction compared to traditional RTL design. The HLS methodology is particularly beneficial for complex algorithms and applications where manual RTL design would be time-consuming and error-prone (Figure 9.1).

9.2.2 Logic Synthesis

Logic synthesis is a crucial step in the design of digital circuits, including ASICs and FPGAs. It involves the transformation of a high-level hardware description (usually

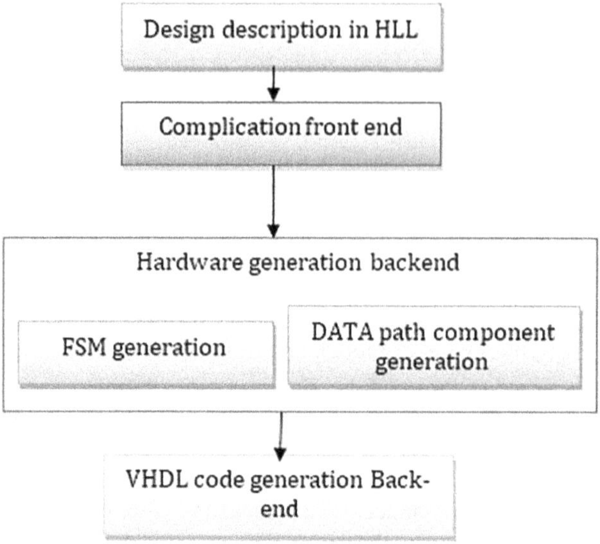

FIGURE 9.1 Traditional HLS process.

in the form of RTL) into a netlist of logical gates that implement the desired functionality. Here is an overview of the logic synthesis process:

- Begin with a high-level hardware description using an HDL such as Verilog or VHDL. This description typically includes the desired behavior of the digital circuit.
- Define design constraints, which may include timing requirements, area constraints, and power constraints. These constraints guide the synthesis tool in optimizing the design.
- Choose an appropriate synthesis tool based on the target technology (ASIC or FPGA) and the design constraints. Common synthesis tools include Synopsys Design Compiler, Cadence Genus, and Xilinx Vivado (for FPGA synthesis).
- The synthesis tool analyzes the RTL code, performs optimizations, and elaborates the design hierarchy. During this stage, the tool may identify and resolve issues such as incomplete or ambiguous descriptions.
- Map the high-level constructs from the RTL code to specific cells in the technology library. These cells represent basic logic gates, flip-flops, and other standard elements available in the target technology.
- The synthesis tool maps the logic functions described in the RTL to the specific logic cells available in the target technology. This process considers the timing, area, and power constraints.
- Apply various optimizations to improve the design in terms of speed, area, and power consumption. Common optimization techniques include logic restructuring, pipelining, and redundancy removal.

- Perform static timing analysis to ensure that the design meets the specified timing constraints. This involves analyzing the critical paths in the circuit and optimizing them to meet the required clock frequency.
- Iterate through the synthesis process, adjusting constraints and design parameters as needed to achieve the desired balance between performance, area, and power.
- Generate a gate-level netlist that represents the synthesized design. This netlist includes information about the logical and physical connections between the various gates and flip-flops.
- Simulate the synthesized netlist to verify that it behaves correctly and meets the design specifications. This may involve functional simulation and timing simulation.
- The synthesized netlist is handed off to backend tools for further processing. For ASICs, this includes place-and-route tools, clock tree synthesis, and physical verification. For FPGAs, the netlist is used in the place-and-route process.

Logic synthesis is a critical step in the overall digital design flow, and the quality of the synthesized netlist significantly influences the performance and efficiency of the final implemented circuit. The synthesis process may vary slightly depending on the target technology and the specific features of the synthesis tool being used.

The mapping and packing procedures are essential steps in preparing the hardware design for the specific target FPGA device. These steps involve mapping the logical design onto the physical resources of the FPGA and packing the design into the available resources efficiently. Below is an overview of the mapping and packing procedures in the FPGA backend flow:

9.2.2.1 Mapping

The logical design, often described in a high-level HDL such as Verilog or VHDL, needs to be mapped onto the physical resources of the target FPGA. This involves associating logical elements, such as gates or flip-flops, with specific physical resources on the FPGA, such as LUTs, multiplexers, and flip-flop cells.

During technology mapping, the high-level logic is mapped to specific elements available in the FPGA architecture. This involves selecting appropriate resources for the implementation of logic functions, taking into account the FPGA's architecture and constraints.

9.2.2.2 Packing

After mapping, the next step is logic packing, where the mapped logic elements are efficiently packed into the available resources of the FPGA. This involves determining how different logical elements will be placed within the CLBs and how the interconnections will be established. Resource allocation involves determining how the various components of the design, such as registers, LUTs, and interconnections, will be allocated within the FPGA. The goal is to maximize resource utilization and minimize resource conflicts. In addition to logic packing, the backend flow also includes routing resource packing. This step involves determining how the physical routing resources, such as programmable interconnects and switch matrices, will be utilized to establish connections between different logic elements.

9.2.2.3 Placement

The physical placement step involves determining the specific locations of logic elements on the FPGA. Placement decisions take into account factors such as signal timing, signal integrity, and the physical proximity of related logic elements. Efficient placement contributes to better overall performance. Global placement involves placing large functional blocks, while local placement focuses on the detailed placement of individual logic elements within these blocks.

9.2.2.4 Optimization

Once the initial placement is done, timing optimization is performed to meet the required performance constraints. This may involve adjusting the placement of critical paths to meet timing requirements. Area optimization aims to minimize the overall area used by the design. Techniques such as logic restructuring, technology mapping, and resource sharing may be applied to achieve better area efficiency.

9.2.2.5 Floor Planning

Floor planning is the process of allocating specific regions of the FPGA for different functional blocks. It involves defining the physical boundaries of regions, considering the placement of I/O pads, clock regions, and other critical components.

9.2.2.6 Iterative Process

Mapping, packing, and placement are often iterative processes. Designers may need to iterate through these steps multiple times, adjusting the mapping and packing strategies to meet performance, area, and timing requirements.

9.2.2.7 Constraint Handling

Constraint Handling: Constraints, such as timing constraints and placement constraints, are applied during the mapping and packing processes. Ensuring that the design adheres to these constraints is crucial for meeting the desired performance goals. The mapping and packing procedures in the FPGA backend flow involve translating the logical design into physical resources, efficiently allocating logic elements, optimizing for timing and area, and iteratively refining the placement for optimal performance. These steps are integral to the successful implementation of FPGA designs.

9.2.2.8 Placement

Simulated annealing is a probabilistic optimization algorithm that is often used for solving combinatorial optimization problems, including the placement of components on a chip. In the context of integrated circuit design, placement refers to determining the physical locations of components (such as gates and flip-flops) on the chip's surface to meet certain design objectives, such as minimizing wirelength or optimizing for timing constraints [5,6].

Here's an overview of the simulated annealing approach for placement:

- Start with an initial random or heuristic-based placement of components on the chip.
- Define an objective function that quantifies the quality of the placement. The objective function could include metrics like wirelength, timing violations, and congestion.
- Set an initial temperature for the simulated annealing algorithm. The temperature parameter controls the likelihood of accepting a worse solution, allowing the algorithm to explore the solution space.
- Define an annealing schedule, which is a scheme for gradually reducing the temperature over iterations. The annealing schedule influences the exploration-exploitation trade-off during the optimization process.
- Generate a neighboring solution by perturbing the current placement. This could involve swapping, moving, or reordering components.
- Compute the objective function value for the new placement.
- Decide whether to accept the new placement based on the acceptance probability, which considers the difference in objective function values and the current temperature.
- If the new placement is accepted, update the current solution. If not, retain the current solution.
- Update the temperature according to the annealing schedule. The temperature is gradually reduced, allowing the algorithm to transition from exploration to exploitation.
- Define stopping criteria based on the number of iterations, a target temperature, or a satisfactory placement quality.
- Return the final placement obtained at the end of the simulated annealing process.

The key idea behind simulated annealing is that the algorithm starts with a high probability of accepting worse solutions (exploration phase) and gradually decreases this probability (exploitation phase) as the temperature decreases. This allows the algorithm to escape local minima and explore the solution space more effectively (Figure 9.2).

9.3 CAD PROCESS FOR FPGAS: AN EMPHASIS ON MACHINE LEARNING

The use of ML techniques in the CAD flow for FPGAs is an evolving area, and researchers are exploring ways to leverage ML to optimize various stages of the FPGA design process. Here's an overview of the CAD flow for FPGAs with an ML perspective:

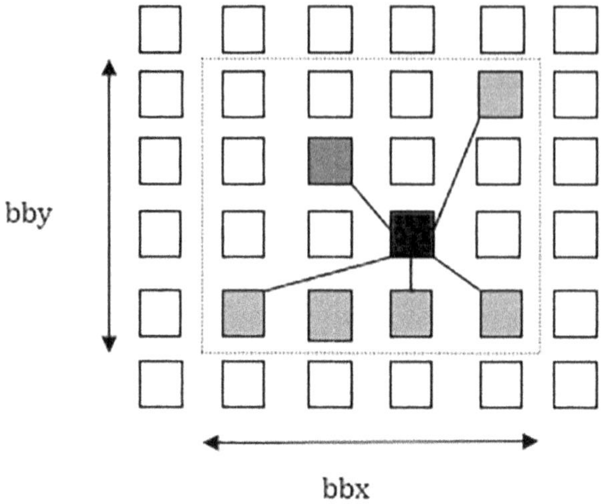

FIGURE 9.2 Placement using the simulated annealing approach.

- Identify specific challenges or optimization problems within the FPGA design flow that can benefit from ML. Examples include optimization of placement and routing, power consumption reduction, and performance improvement.
- Gather data from previous FPGA design projects or simulations. This data may include information about designs, resource utilization, timing, and other relevant metrics. Additionally, data from hardware measurements or simulations can be used.
- Identify relevant features or parameters from the collected data that can be used to describe the characteristics of FPGA designs. This could include features related to circuit structure, timing constraints, and resource utilization.
- For certain tasks, such as optimization or prediction, create labeled datasets where the labels represent the desired outcome or target metric. For example, labels could indicate whether a design meets timing constraints or the level of power consumption.
- Use ML algorithms to train models based on the labeled datasets. Common ML techniques include supervised learning algorithms like decision trees, support vector machines, or neural networks. The model should learn the relationships between the input features and the target metric.
- Deploy the trained models to optimize specific aspects of the FPGA design. For example:
- Predict optimal placement or routing configurations based on historical data.
- Predict the likelihood of a design meeting timing constraints.
- Optimize designs to reduce power consumption based on learned patterns.

- In certain scenarios, reinforcement learning can be used to optimize the FPGA design process. For example, reinforcement learning agents can learn to make decisions during the placement and routing process to maximize performance or minimize power consumption.
- Apply transfer learning techniques to leverage knowledge gained from one FPGA design project to improve the performance of models on new, related projects.
- Validate the performance of the ML models using new designs or datasets not seen during training. Ensure that the models generalize well to different design scenarios.

Establish a feedback loop where the ML models' performance is continuously monitored and refined based on new data and insights. This iterative process can lead to improved models over time.

Each topic focuses on a certain CAD flow stage, such as HLS, logic synthesis, and routing and placement.

9.3.1 HLS through ML

HLS is a process that translates high-level programming languages, such as C or C++, into register-transfer-level hardware descriptions. While conventional HLS relies on algorithms and heuristics to perform this translation, researchers are exploring the application of machine learning (ML) techniques to enhance and optimize the HLS process. Here's an overview of how HLS can be approached using ML [7]:

- Gather a dataset comprising pairs of high-level code and corresponding synthesized RTL descriptions. These pairs represent examples of the input-output relationship that the ML model will aim to learn.
- Identify relevant features or representations from the high-level code that can be used as input to the ML model. This could include chapter syntax trees, control-flow graphs, or other representations that capture the structure and semantics of the code.
- Choose an ML model architecture suitable for the HLS task. Recurrent neural networks (RNNs) [8], long short-term memory networks (LSTMs), graph neural networks (GNNs), or transformer-based models are potential candidates depending on the nature of the input data.
- Train the ML model using the dataset. The model learns to map high-level code features to corresponding RTL descriptions. The loss function can be defined based on the dissimilarity between the synthesized RTL and the ground truth.
- Design an objective function that captures the quality of the synthesized RTL in terms of metrics like resource utilization, power consumption, or performance. The ML model can be trained to optimize this objective function during the synthesis process.

- Apply transfer learning techniques to leverage knowledge gained from one HLS task to improve the performance of the model on new, related tasks. This can help the model generalize well to different application domains.
- Fine-tune the ML model based on specific hardware targets or constraints. This could involve adjusting the model to optimize for FPGA or ASIC architectures, considering factors like clock frequency, area constraints, and power consumption.
- Establish a feedback loop where the ML model's performance is continuously monitored and refined based on new data and insights. Iterative training and fine-tuning can lead to improved models over time.
- Integrate the trained ML model into the HLS toolchain. This involves incorporating the model as a module in the overall HLS flow to assist in the translation of high-level code to RTL.
- Validate the performance of the machine learning-assisted HLS against traditional HLS approaches and benchmarks. Ensure that the generated RTL meets design constraints and exhibits efficient resource utilization (Figure 9.3).

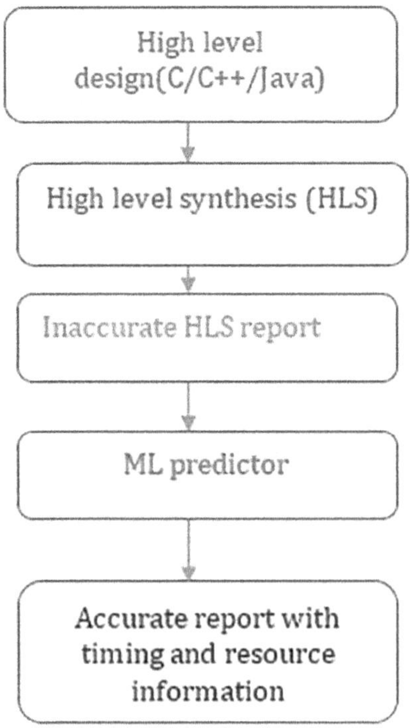

FIGURE 9.3 HLS using ML.

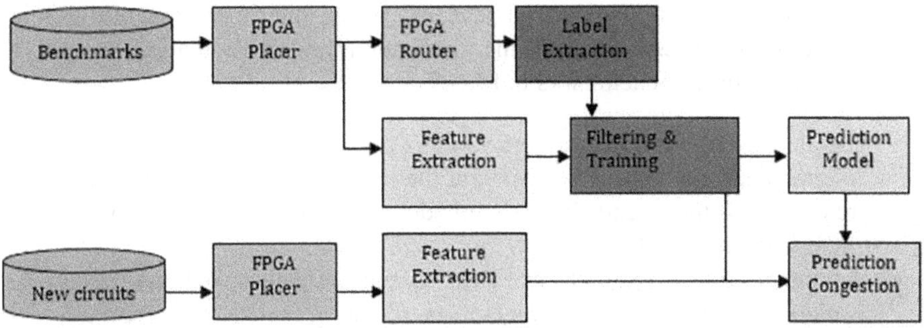

FIGURE 9.4 Placement using machine learning.

9.3.2 Placement

Placement, which aims to assess two objectives, congestion assessment and routability prediction, is one of the CAD flow's most time-consuming operations [9]. Congestion in placement refers to the potential for routing difficulties due to high local demand for routing resources.

- Extract features from the design, such as the density of logic cells in a region, historical congestion data, and proximity to high-demand areas.
- Label data points based on congestion levels (e.g., low, medium, and high).
- Congestion levels and prior placement data can be used to train ML models (such as regression or classification) [10,11].
- Use the trained model to predict congestion levels for new placements, allowing designers to anticipate potential routing challenges.
- Routability prediction assesses how easy it will be to route connections between placed components on the chip.

Extract features related to the spatial arrangement of components, the distance between connected elements, and the available routing resources. Label data points based on routability (e.g., easy, moderate, and difficult). Use the trained model to predict routability for new placements, assisting designers in making informed decisions during the design process (Figure 9.4).

9.4 RESEARCH CHALLENGES AND OPPORTUNITIES

Integrating ML algorithms into CAD for FPGAs is crucial since it reduces time to market by considerably reducing compilation time. There are various gaps in the use of ML techniques across the FPGA CAD backend pipeline [12,13]. The six research possibilities, which we have identified with the assistance of the work detailed in the above section, are mentioned below:

9.4.1 Lack of Open-Source Benchmarks

The lack of open-source benchmarks specific to FPGA design is a notable challenge in the field of hardware development. Open-source benchmarks for FPGAs are essential for evaluating the performance, efficiency, and capabilities of different FPGA designs and tools. Here are some implications and considerations related to the lack of open-source benchmarks in FPGA development [14]:

- The absence of standardized benchmarks makes it challenging for FPGA designers and researchers to compare the performance of different FPGA architectures, tools, and design methodologies. This lack of comparison hinders progress in the field.
- Reproducibility is crucial in FPGA design research. Without openly available benchmarks, it becomes difficult for researchers to reproduce experiments and validate the effectiveness of their FPGA designs against established standards.
- Open-source FPGA benchmarks foster collaboration within the FPGA design community. The lack of such benchmarks may limit the ability of researchers and designers to collaborate, share insights, and collectively address challenges in FPGA design.
- Innovators in FPGA design may face challenges in benchmarking their designs against widely accepted standards. This lack of benchmarking resources can impede the innovation process by making it harder to assess the impact of new FPGA architectures or design techniques.
- Designers working on novel FPGA architectures or design methodologies may find it challenging to benchmark their designs against established norms. Open benchmarks provide a reference point for evaluating the effectiveness of new solutions [15].
- A diverse set of benchmarks is essential to assess FPGA designs under various scenarios and workloads. The absence of diverse and representative benchmarks may result in evaluations that are biased or limited to specific use cases.
- Standardized benchmarks contribute to the establishment of best practices and common evaluation metrics in FPGA design. The lack of open-source benchmarks may lead to a lack of standardization in evaluation practices, hindering the adoption of common metrics [16].
- Open-source benchmarks encourage community participation, allowing FPGA designers to contribute datasets, scenarios, and evaluation metrics. This collaborative effort helps improve the quality and relevance of benchmarks over time.
- FPGA architectures can vary significantly, and benchmarks that account for this heterogeneity are crucial. Open benchmarks can address the challenge of designing evaluations that consider the diverse characteristics of different FPGA architectures.
- Recognizing the importance of open benchmarks in FPGA design, there may be initiatives and organizations working to create and promote standardized benchmarks. These efforts aim to fill the gap by providing accessible and widely adopted benchmarks.

9.4.2 ML-Based Complete Backend Flow

A machine learning-based complete backend flow for FPGA design involves incorporating ML techniques into various stages of the backend design process.

The backend flow typically includes synthesis, place-and-route, and optimization stages. Here's an overview of how ML can be integrated into each stage of the FPGA backend flow [17]:

- **Synthesis:** Use ML models to predict optimal logic synthesis configurations based on input RTL (register-transfer-level) code. This could involve predicting optimal settings for synthesis tools, optimization levels, and other synthesis parameters. Explore the use of ML to automatically generate optimized RTL code. ML models can be trained on datasets of RTL code and corresponding optimized versions, learning patterns that lead to improved synthesis results.
- **Place and Route:** Train ML models to predict optimal placements of logic elements on the FPGA. These models can consider factors such as signal delay, resource utilization, and power consumption. Leverage ML to optimize routing resources by predicting optimal routing paths. This can lead to improved timing, reduced congestion, and overall better performance.
- **Optimization:** Use ML to predict whether a design will meet timing constraints. The model can analyze design characteristics and historical data to provide early insights into potential timing issues.

  Train models to optimize power consumption by making informed decisions on clock gating, power gating, and other power-related optimizations. Employ ML models to predict optimal resource utilization, guiding the selection of design parameters to achieve efficient use of FPGA resources.
- **Security Considerations:** Integrate ML models for anomaly detection during the backend flow. Detect irregularities in the design that may indicate security vulnerabilities or potential attacks. Develop ML models capable of identifying potential hardware Trojans by analyzing patterns in the synthesized and placed-and-routed designs.
- **Customization for Specific Architectures:** Train ML models that are specific to the target FPGA architecture. This customization ensures that the models are well-suited for the unique features and constraints of the hardware.
- **Iterative Learning and Improvement:** Establish a feedback loop where the ML models continuously learn from new designs and adapt to changes in the FPGA architecture or design practices. This iterative learning process helps improve the accuracy and effectiveness of the models over time.
- **User Interaction and Guidance:** Develop machine learning-driven interactive design assistants that provide guidance to designers during the backend flow. These assistants can suggest optimizations, recommend parameter settings, and assist in decision-making.
- **Verification and Validation:** Ensure the reliability of ML models by subjecting them to rigorous verification processes. This involves validating the

models against diverse datasets and testing scenarios. Compare the results and performance of the machine learning-based backend flow with traditional approaches to validate the efficacy and benefits of the ML integration.

9.4.3 Multi-FPGA Prototyping Using Machine Learning

Multi-FPGA prototyping involves the use of multiple FPGAs to create a scalable and efficient platform for prototyping complex digital designs, such as system-on-chip (SoC) designs. Integrating ML techniques into multi-FPGA prototyping can offer several benefits, including optimization of resource utilization, dynamic workload balancing, and improved overall performance. Here are some ways ML can be applied in multi-FPGA prototyping [18].

9.4.3.1 Resource Allocation and Load Balancing

Efficiently distribute the workload across multiple FPGAs while considering resource constraints and avoiding congestion. Train models to predict the resource requirements of different modules based on their characteristics. Develop algorithms that use these predictions to dynamically allocate tasks to FPGAs, ensuring optimal resource utilization and load balancing.

9.4.3.2 Communication Optimization

Minimize communication delays and optimize data transfer between FPGAs to enhance overall system performance. Predict communication patterns based on historical data and application behavior. Develop algorithms to dynamically optimize communication paths, reducing latency and improving overall communication efficiency.

9.4.3.3 Run-Time Adaptation

Adapt to changing workloads and reconfigure the multi-FPGA system dynamically for optimal performance. Train models to recognize patterns in workload changes and system behavior. Implement adaptive control mechanisms that use ML predictions to reconfigure the system on the fly, optimizing for current workloads.

9.4.3.4 Power Consumption Optimization

Minimize power consumption while maintaining performance in the multi-FPGA system. Train models to predict power consumption patterns based on different operating scenarios. Develop algorithms that dynamically adjust FPGA configurations and task allocations to minimize power consumption without sacrificing performance.

9.4.3.5 Fault Detection and Recovery

Detect faults in individual FPGAs and implement recovery strategies to maintain system reliability. Train models to recognize patterns indicative of FPGA faults. Develop algorithms for real-time fault detection and recovery, using ML predictions to initiate corrective actions.

9.4.3.6 Task Scheduling and Prioritization

Schedule tasks efficiently on multiple FPGAs to meet design constraints and deadlines. Train models to predict task execution times and potential bottlenecks. Develop algorithms for intelligent task scheduling and prioritization, taking into account design constraints and deadlines.

9.4.3.7 Design Space Exploration

Explore the vast design space to find optimal configurations for multi-FPGA prototyping. Utilize ML to guide design space exploration by predicting the performance and resource utilization of different configurations. Implement algorithms that automate the search for optimal configurations based on ML insights. It's important to note that the success of ML in multi-FPGA prototyping depends on the availability of representative training data and collaboration between domain experts in FPGA design and ML practitioners. Additionally, the use of ML in this context can be highly application-specific, and the design goals and constraints should guide the selection and development of ML models and algorithms [19].

9.4.4 Domain-Specific Flow

A domain-specific flow (DSF) refers to a specialized design and development process tailored for a specific application domain. In the context of FPGA design, a DSF is crafted to address the unique requirements and characteristics of a particular application area. Here's an overview of the key aspects and considerations in a domain-specific FPGA design flow:

- Understanding the application domain
- Customized HLS
- Application-specific hardware accelerators
- Specialized Intellectual Property (IP) blocks
- Application-specific constraints
- Customized placement and routing
- Custom memory hierarchy
- Domain-specific language extensions
- Performance profiling and optimization
- Verification and validation strategies
- Documentation and knowledge capture
- Iterative refinement

9.4.5 Federated Learning for FPGA

Federated learning is an ML method that allows models to be trained across several dispersed devices or servers while keeping data local. It enables collaborative model training without revealing raw data, making it ideal for privacy-sensitive applications. When considering federated learning for FPGAs, the focus is on leveraging the computational capabilities of FPGAs in a federated learning setting. Here's an overview of how federated learning can be applied to FPGA environments:

A decentralized ML paradigm called federated learning (FL) allows model training across many edge devices or nodes without requiring the exchange of raw data. This approach aligns well with the resource-constrained nature of FPGAs, making it an intriguing framework for collaborative learning on distributed FPGA devices.

FPGAs have emerged as powerful accelerators, providing hardware customization to meet the growing demands of computational workloads. FPGA acceleration involves leveraging the reconfigurable nature of FPGAs to offload and accelerate specific tasks, offering a balance between flexibility and performance [20].

Model parallelism is a crucial strategy in the field of artificial intelligence (AI) that involves distributing the computational load of a neural network across multiple processing units. On FPGA, model parallelism can be strategically employed to tailor hardware configurations, maximizing the efficiency of AI model execution.

Privacy preservation is a paramount concern in the era of data-driven technologies. FPGAs, known for their reconfigurable nature and parallel processing capabilities, present unique challenges and opportunities in the context of preserving privacy while enabling accelerated computations. Communication efficiency is a critical aspect in the design and deployment of systems utilizing FPGAs. FPGAs, known for their parallel processing capabilities, require optimized communication strategies to ensure efficient data exchange between components, both on-chip and across distributed systems. Customized FPGA designs involve the creation of hardware configurations that are specifically tailored to meet the unique requirements of a particular application or use case. FPGAs offer a versatile platform for designing custom hardware solutions, allowing for optimization, acceleration, and adaptability in a wide range of applications. Distributed learning across FPGAs involves the parallelized training or inference of ML models across multiple FPGA devices. This collaborative approach enables scalable and efficient processing, making it well-suited for applications demanding large-scale data analysis and complex model training. Resource optimization for FPGAs involves the systematic management and efficient utilization of the available hardware resources. By carefully designing and configuring the FPGA logic, memory, and routing resources, developers can achieve optimal performance for a given application while minimizing resource consumption. FPGAs present unique security challenges due to their reconfigurable nature and widespread use in various applications. Ensuring the security of FPGA-based systems involves addressing vulnerabilities related to configuration integrity, unauthorized access, and potential attacks on the hardware.

9.4.6 Lack of Explainability in ML-Based Flow

The difficulty in understanding and comprehending a machine learning model's decision-making process is referred to as a lack of explainability in ML processes. This lack of transparency may be a major issue, especially in essential applications, where knowing why a given decision was taken is critical. Here are some important aspects to consider about the lack of explainability in ML-based flows [21–25]:

Many advanced ML models, particularly deep neural networks, are complicated and include several parameters. Understanding how these models make particular decisions might be difficult because of their non-linear and complex architecture.

Some ML models are considered "black boxes" because the internal workings are not easily interpretable. The relationship between input features and model predictions may not be transparent, making it difficult to explain why a particular decision was made.

In some cases, the ML model may perform feature transformations or learn chapter representations that are not directly interpretable by humans. This can further obscure the understanding of how input features contribute to the output. ML models trained on high-dimensional data, such as images or text, may learn complex patterns that are not readily understandable by humans. Interpreting decisions based on such learned features becomes challenging. Ensemble models, which combine multiple individual models for improved performance, can add another layer of complexity. Understanding the contribution of each sub-model to the final decision becomes intricate. Model performance and explainability are frequently weighed against one another. Simpler models may be easier to understand, but they may not capture complicated patterns as well as more advanced models. In regulated domains or applications where ethical considerations are paramount (such as healthcare or finance), there is a growing need for transparency and explainability to ensure accountability and compliance with regulations.

Explainable AI (XAI) is a subfield of ML focused on developing models and techniques that inherently provide explanations for their decisions. XAI aims to strike a balance between model performance and transparency. The development of user-friendly tools that visualize and explain the decision-making process of ML models is an area of ongoing research. Such tools can help users, including non-experts, understand model predictions.

9.5 CONCLUSIONS AND FUTURE WORK

The backend flow of FPGAs comprises lengthy and complicated phases, and the final design's performance is heavily dependent on the tools' quality and the designer's knowledge and proficiency with them. ML algorithms have recently been integrated into almost every layer of the FPGA backend. The major purpose of these algorithms is to minimize implementation time while retaining the final design quality and reducing dependency on designers. This chapter gives a thorough description of different ML methods and their applications across the FPGA backend flow. According to a review of recent research, the use of ML algorithms such as random forest, SVM, CNN, and Reinforcement Learning (RL) has greatly decreased implementation time while preserving design quality.

However, the use of ML algorithms in the FPGA backend flow is still in its early phases, and further study is required. For example, the bulk of the studies analyzed in this chapter are based on a small set of criteria that are not available as an open-source resource for academic research. To accelerate research in this area, it is important to construct a set of open-source and complicated benchmarks akin to the typical FPGA CAD pipeline. Furthermore, many studies focus on a single phase of the backend flow, with no end-to-end flow that employs ML techniques from HLS to routing. Furthermore, the complexity of ML algorithms might lead to inexplicable results. This restricts their use to a certain domain, and extensive effort is required to make them generally applicable in Electronic Design Automation (EDA) in general and FPGA CAD flow in particular. Cloud-based solutions, such as CAOS, show promise and require further exploration in the context of the issues raised in this chapter. To close the gaps in the FPGA backend flow, the research community must prioritize the aforementioned concerns and opportunities.

The exploration of an ML-based FPGA backend flow represents a significant stride toward revolutionizing the traditional methodologies in electronic system development. Through the challenges and opportunities unveiled in this endeavor, several key conclusions emerge. The integration of ML introduces a non-deterministic element into the FPGA backend flow, challenging the conventional deterministic design processes. Addressing this challenge requires a nuanced understanding of how to effectively manage and harness the benefits of unpredictability.

The success of ML models crucially hinges on the quality and diversity of training data. Efforts to improve data quality and ensure diversity become imperative. Additionally, enhancing the interpretability of ML models is vital for establishing trust among designers, enabling them to comprehend and validate decision-making processes. Balancing the potential benefits of ML-based optimizations with the resource constraints inherent in FPGA designs poses a significant struggle. Future work should delve into refining algorithms and strategies to optimize resource utilization without compromising efficiency. The dynamic nature of FPGA designs necessitates ML models that can seamlessly adapt to changes without extensive retraining. The ability to dynamically adjust to evolving design requirements is critical for the continued success of ML-based FPGA backend flows.

Future work should focus on the generation of high-quality, diverse datasets that capture the intricacies of FPGA designs. This involves exploring techniques for data augmentation, synthesis, and curation to ensure comprehensive coverage. Addressing the interpretability challenge is a key aspect of future work. Developing techniques to enhance the interpretability and explainability of ML models will be crucial for fostering trust among designers and stakeholders. Future efforts should be directed toward the development of resource-aware ML algorithms that optimize FPGA backend processes while adhering to strict resource constraints. This includes exploring techniques for lightweight models and efficient computations. Research should delve into the creation of ML models that exhibit adaptability to dynamic changes in FPGA designs. This involves investigating techniques for incremental learning and transfer learning to facilitate continuous adjustment without extensive retraining. The seamless integration of ML into existing design automation tools is an area for future development. Enabling ML models to work collaboratively with conventional tools will streamline the FPGA design process. Future work should include comprehensive validation and benchmarking of ML-based FPGA backend flows across a diverse set of designs. This involves developing standardized metrics and methodologies to evaluate the effectiveness and efficiency of these approaches.

The journey toward a machine learning-based FPGA backend flow has laid the groundwork for transformative changes. Addressing the identified challenges and pursuing the outlined future work will propel the field toward a new era of adaptive, efficient, and intelligent FPGA design methodologies.

REFERENCES

1. Shah, D.; Hung, E.; Wolf, C.; Bazanski, S.; Gisselquist, D.; Milanovic, M. Yosys+ nextpnr: an open source framework from verilog to bitstream for commercial fpgas. In *Proceedings of the 2019 IEEE 27th Annual International Symposium on Field-Programmable Custom Computing Machines (FCCM)*, San Diego, CA, USA, 28 April–1 May **2019**; pp. 1–4.
2. Fiduccia, C.M.; Mattheyses, R.M. A linear-time heuristic for improving network partitions. In *Papers on Twenty-Five Years of Electronic Design Automation*, ACM, Las Vegas, NV, USA, **1988**; pp. 241–247.
3. Bui, T.N.; Chaudhuri, S.; Leighton, F.T.; Sipser, M. Graph bisection algorithms with good average case behavior. *Combinatorica* **1987**, 7, 171–191.
4. Taj, I.; Zaman, N. Towards industrial revolution 5.0 and explainable artificial intelligence: challenges and opportunities. *Int. J. Comput. Digit. Syst.* **2022**, 12, 295–320.
5. Chhabria, V.A.; Kahng, A.B.; Kim, M.; Mallappa, U.; Sapatnekar, S.S.; Xu, B. Template-based PDN synthesis in floorplan and placement using classifier and CNN techniques. In *Proceedings of the 2020 25th Asia and South Pacific Design Automation Conference (ASP-DAC)*, Beijing, China, 13–16 January **2020**; pp. 44–49.
6. Al-Hyari, A.; Areibi, S. Design space exploration of convolutional neural networks based on evolutionary algorithms. *J. Comput. Vis. Imaging Syst.* **2017**, 3, https://doi.org/10.15353/vsnl.v3i1.162.
7. Wang, Z.; Schafer, B.C. Machine learning to set meta-heuristic specific parameters for high-level synthesis design space exploration. In *Proceedings of the 2020 57th ACM/IEEE Design Automation Conference (DAC)*, San Francisco, CA, USA, 20–24 July **2020**; pp. 1–6.
8. Chen, T.; Guestrin, C. Xgboost: a scalable tree boosting system. In *Proceedings of the 22nd ACM Sigkdd International Conference on Knowledge Discovery and Data Mining*, San Francisco, CA, USA, 13–17 August **2016**; pp. 785–794.
9. Chen, S.C.; Chang, Y.W. FPGA placement and routing. In *Proceedings of the 2017 IEEE/ACM International Conference on Computer-Aided Design (ICCAD)*, Irvine, CA, USA, 13–16 November **2017**; pp. 914–921.
10. Liaw, A.; Wiener, M. Classification and regression by randomForest. *R News* **2002**, 2, 18–22.
11. Fix, E.; Hodges, J.L. Discriminatory analysis. Nonparametric discrimination: consistency properties. *Int. Stat. Rev./Rev. Int. Stat.* **1989**, 57, 238–247.
12. Al-Hyari, A.; Abuowaimer, Z.; Martin, T.; Gréwal, G.; Areibi, S.; Vannelli, A. Novel congestion-estimation and routability- prediction methods based on machine learning for modern FPGAs. *ACM Trans. Reconfigurable Technol. Syst. (TRETS)* **2019**, 12, 1–25.
13. Zhou, Q.; Wang, X.; Qi, Z.; Chen, Z.; Zhou, Q.; Cai, Y. An accurate detailed routing routability prediction model in placement. In *Proceedings of the 2015 6th Asia Symposium on Quality Electronic Design (ASQED)*, Kula Lumpur, Malaysia, 4–5 August **2015**; pp. 119–122.
14. Roorda, E.; Rasoulinezhad, S.; Leong, P.H.; Wilton, S.J. FPGA architecture exploration for DNN acceleration. *ACM Trans. Reconfigurable Technol. Syst. (TRETS)* **2022**, 15, 1–37.
15. Ward, S.; Ding, D.; Pan, D.Z. PADE: A high-performance placer with automatic datapath extraction and evaluation through high-dimensional data learning. In *Proceedings of the DAC Design Automation Conference 2012*, San Francisco, CA, USA, 3–7 June **2012**; pp. 756–761.

16. Wang, Z.; Schafer, B.C. Learning from the past: efficient high-level synthesis design space exploration for FPGAs. *ACM Trans. Des. Autom. Electron. Syst. (TODAES)* **2022**, 27, 1–23.
17. Tang, Y.; Zhang, L.; Bao, G.; Ren, F.; Pedrycz, W. Symmetric implicational algorithm derived from intuitionistic fuzzy entropy. *Iran. J. Fuzzy Syst.* **2022**, 19, 27–44.
18. Schafer, B.C.; Wang, Z. High-level synthesis design space exploration: past, present, and future. *IEEE Trans. Comput.-Aided Des. Integr. Circuits Syst.* **2019**, 39, 2628–2639.
19. Brayton, R.K. The decomposition and factorization of Boolean expressions. In *Proceedings of the International Symposium on Circuits and Systems (ISCAS 82)*, Rome, Italy, 10–12 May **1982**.
20. Zhuo, C.; Yu, B.; Gao, D. Accelerating chip design with machine learning: from pre-silicon to post-silicon. In *Proceedings of the 2017 30th IEEE International System-on-Chip Conference (SOCC)*, Munich, Germany, 5–8 September **2017**; pp. 227–232.
21. Reyes Fernandez de Bulnes, D.; Maldonado, Y.; Trujillo, L. Development of multiobjective high-level synthesis for FPGAs. *Sci. Program.* **2020**, 8, 1–25.
22. Cong, J.; Ding, Y. FlowMap: an optimal technology mapping algorithm for delay optimization in lookup-table based FPGA designs. *IEEE Trans. Comput.-Aided Des. Integr. Circuits Syst.* **1994**, 13, 1–12.
23. Cong, J.; Ding, Y. On area/depth trade-off in LUT-based FPGA technology mapping. In *Proceedings of the 30th International Design Automation Conference*, Dallas, TX, USA, 14–18 June **1993**; pp. 213–218.
24. Cong, J.; Hwang, Y.Y. Simultaneous depth and area minimization in LUT-based FPGA mapping. In *Proceedings of the Third International ACM Symposium on Field-Programmable Gate Arrays*, Monterey, CA, USA, 12–14 February **1995**; pp. 68–74.
25. Cong, J.; Hwang, Y.Y. Structural gate decomposition for depth-optimal technology mapping in LUT-based FPGA designs. *ACM Trans. Des. Autom. Electron. Syst. (TODAES)* **2000**, 5, 193–225.

10 Role of Machine Learning Applications in VLSI Design

Mohd Javed Khan, Pooja Gaur, Indrasen Singh, and Saif Ahmad

10.1 INTRODUCTION

10.1.1 Overview of VLSI Design Challenges

The technique of integrating millions or billions of transistors on a single chip is known as "very-large-scale integration (VLSI)." Microprocessors, memory chips, communication systems, and other advanced electronic devices can be developed by employing VLSI design. However, the growing complexity, power consumption, timing constraints, manufacturing problems, and design verification hurdles also pose significant challenges to VLSI design. Some of the major challenges associated with VLSI design are as follows [1]:

1. **Complexity**: With millions of components and connections, VLSI designs have grown remarkably complex. In order to ensure that the design satisfies power, area, and reliability restrictions, designers must handle the complexity of architectures, data pathways, memory hierarchies, as well as performance specifications. Designers employ a variety of strategies, including abstraction, hierarchy, modularity, and reuse, to handle the complexity. To help them with the design process, they also use design automation technologies and processes. But these methods and instruments also bring with them additional challenges like complicated design validation, verification, and debugging [2].
2. **Power Management:** The power consumption rises with the number of transistors. The requirement to strike a balance between energy efficiency and performance, as well as lower cooling and heat dissipation expenses, presents power management issues. Power gating, voltage scaling, and clock gating are just a few of the strategies that designers must use to maximize power efficiency without compromising reliability or performance. They must also take into account how power management strategies affect other design elements including area, time, noise, and variability [3].
3. **Timing Closure**: This technique guarantees that all signals arrive at their designated locations within predetermined time constraints. Meeting timing

DOI: 10.1201/9781003483038-10

requirements is significantly hampered by the declining clock cycles and rising operating frequencies. To achieve timing closure, designers must carefully plan and optimize the design, use time-driven approaches, and make use of sophisticated place-and-route and advanced synthesis tools. They must additionally take into consideration how temperature, aging, and process variables affect the design's timing behavior [4].

4. **Design for Manufacturability (DFM):** Design for manufacturability (DFM) is essential in order to guarantee that the design can be produced in large quantities with a satisfactory yield and quality. Additional challenges including yield optimization, lithography constraints, and process variability appear when technology nodes get smaller. To overcome DFM obstacles and guarantee good manufacturing yield, designers must optimize layouts, carry out statistical analysis, and use strategies like design-technology co-optimization (DTCO). To make sure the design is feasible and compatible, they must also work with the foundry and the fabrication process [5].
5. **Physical Design Challenges**: Millions of components must be placed and routed on a chip while adhering to timing, power, and area restrictions during the physical design stage. To create a physically robust design, designers must maximize floor layout, make use of sophisticated algorithms for placement and routing, and deal with problems like signal integrity and congestion. Additionally, they must handle the difficulties brought on by interconnect scaling, which include increasing resistance, capacitance, and crosstalk and which have an impact on the design's power, performance, and reliability [6].
6. **Verification Complexity**: Verification makes sure the design fulfills performance and functional requirements. The size and complexity of the design cause the verification complexity to increase exponentially. To test the design at multiple levels of abstraction and ensure its accuracy and quality, designers must use a variety of verification approaches, including simulation, formal methods, emulation, and prototyping. They must also overcome the difficulties involved in validating intricate architectures, including those with many cores, heterogeneous systems, and reconfigurable components [7].

10.1.2 Introduction to the Role of Machine Learning in Addressing These Challenges

A specialized branch of artificial intelligence called machine learning (ML) gives computers the ability to learn from data and become more efficient without the need for explicit programming. By offering automated and data-driven solutions, ML can be extremely helpful in tackling the problems associated with VLSI design. The following are some instances of how ML can support various facets of VLSI design [8]:

1. **Complexity:** By automating the creation and optimization of architectures, data pathways, and memory hierarchies, ML can assist in lowering the complexity of VLSI design [9]. ML algorithms have the ability to learn from current designs and produce new ones that satisfy power and performance

constraints. By employing strategies like formal methods, emulation, and prototyping, ML may also help with the verification, validation, and debugging of intricate designs.

2. **Power Management:** By anticipating and optimizing the power consumption of various components and modules, ML can assist in raising the power efficiency of VLSI circuits. With the help of previous data, ML algorithms are able to precisely calculate a circuit's power dissipation and leakage under various operating scenarios. ML can assist with low-power circuit design by employing strategies including voltage scaling, clock gating, and power gating [10].
3. **Timing Closure:** Through the analysis and inference of past data, ML has the potential to assist in timing closure by precisely predicting a circuit's timing behavior under various operating scenarios. ML algorithms have the ability to learn from past timing reports and optimize a circuit's timing margins and limitations. By utilizing methods like timing-driven approaches and sophisticated algorithms, ML can also assist in the synthesis and place-and-route of VLSI circuits [11].
4. **Design for Manufacturability (DFM):** Through the analysis and learning of process variations, lithography constraints, and yield optimization data, ML can contribute to the improvement of VLSI circuits' manufacturability and yield. In order to reduce the impact of variations and faults, ML algorithms can optimize a circuit's layouts and masks by learning from prior fabrication data. ML can assist in VLSI circuit layout optimization and mask synthesis through the use of methods like statistical analysis and DTCO [12].
5. **Physical Design challenges:** Through data analysis and learning from interconnect scaling, congestion, and signal integrity, ML can assist in overcoming the physical design issues associated with VLSI circuits. The placement and routing of a circuit can be optimized by ML algorithms to reduce interconnect delay and noise by learning from prior physical design data. By applying strategies like sophisticated algorithms and optimization approaches, ML can also assist in the placement and routing of VLSI circuits [13].
6. **Verification Complexity:** By evaluating and learning from enormous amounts of circuit data as well as functional and performance specifications, ML can assist in addressing the verification complexity of VLSI circuits. More effectively and precisely, ML algorithms can test a circuit's performance and functionality by learning from past verification data. Pattern recognition and anomaly detection are two methods that ML can use to aid in the diagnostic and fault detection of VLSI circuits [14].

10.1.3 Objectives

A subfield of artificial intelligence called ML allows computers to learn without explicit programming by using data and experience. The process of assembling millions of transistors and other components on a single chip to create integrated circuits (ICs) is known as VLSI. The art and science of creating and refining ICs for a range of uses, including sensing, computation, and communication, is known as VLSI design [15]. Applying ML to VLSI design aims to increase the efficiency, quality, and reliability of IC design and manufacture by automating data-intensive and complicated processes like:

1. **Device Modeling**: ML can assist in developing concise and precise models, based on simulations or measurements, of the physical behavior and functionality of transistors and other devices [13,16]. For example, ML can be used to discover the nonlinear relationship between the electrical characteristics, including capacitance, voltage, and current, and the device parameters [13]. Through the use of methods like dimensionality reduction, sparsification, and compression, ML) can also assist in lowering the computational cost and memory requirements of device simulations [16].
2. **Layout Verification:** ML can assist in confirming that an IC's layout is robust and correct by identifying and fixing mistakes such as lithography defects, design rule violations, and parasitic effects [16,17]. In order to apply various verification techniques appropriately, ML can be used, for instance, to categorize layout patterns into groups like crucial, non-critical, and ambiguous [17]. By employing strategies including feature extraction, anomaly detection, and optimization, ML can also aid in enhancing the precision and effectiveness of layout verification [16].
3. **Yield Prediction:** By examining how process variations and defects affect circuit performance, ML can assist in estimating the probability that an IC will meet requirements and operate as intended [13,16]. For example, ML can be used to determine the statistical distribution of circuit characteristics like power, delay, and noise, and then compute the yield while taking into account the specified constraints [13]. Additionally, by utilizing methods like regression, classification, and clustering, ML can improve yield prediction [16].
4. **Post-Silicon Validation**: By locating and analyzing errors and optimizing the testing and debugging procedures, ML can assist in verifying the quality and operation of an IC after it has been fabricated. To create the best test responses and sequences, for instance, ML can be used to develop test patterns and fault models. By applying methods like fault injection, fault localization, and fault rectification, ML can also assist in lowering the duration and expense of post-silicon validation [16].
5. **Reliability Analysis:** Through modeling and mitigating the impacts of aging, stress, and environmental conditions, ML can assist in evaluating and enhancing an IC's lifetime and reliability. For example, ML can be used to understand the failure modes and degradation mechanisms, and then forecast the IC's reliability and remaining useful life. By utilizing methods like reliability modeling, reliability monitoring, and reliability management, ML can also improve the reliability analysis [16].
6. **Analog and Digital Synthesis:** Through design space exploration, the generation of potential solutions, and an assessment of the trade-offs between

performance, power, area, and cost, ML can assist in the synthesis and optimization of analog and digital circuit design. For instance, ML can be used to understand the design constraints and objectives, after which the synthesis problem can be formulated as an optimization problem. By utilizing methods like reinforcement learning (RL), evolutionary algorithms, and neural networks, ML can also aid in enhancing the effectiveness and quality of analog and digital synthesis [16].

10.2 FUNDAMENTALS OF VLSI DESIGN

10.2.1 Brief Explanation of VLSI Design Process

The technique of assembling several thousand or millions of transistors and other electronic components on a single chip to create ICs is known as VLSI [18]. VLSI design is an iterative process that includes various steps, including:

1. **Specification and Architecture**: The key components of the IC or system-on-a-chip (SoC), such as processors, memory, peripherals, and interfaces, are defined in this step by translating the system-level specifications into a high-level architecture [19]. The data flow, control signals, and communication protocols between the components are all specified in the architecture. Typically, block diagrams, data flow graphs, or state machines are used to depict the architecture. The architecture is assessed using a range of analysis and estimating methods for cost, power, performance, and practicality [20].
2. **Design Entry**: Using a graphical tool like a schematic editor or a hardware description language (HDL) like Verilog or VHSIC (Very High-Speed Integrated Circuit) Hardware Description Language (VHDL), the architecture is implemented in this step [21]. At various abstraction levels, such as the behavioral, structural, or register-transfer level (RTL), the HDL or schematic represents the structure and functioning of the IC or SoC [22]. In order to verify the design's functionality, testbenches and stimuli must also be created for the design entry. An integrated development environment (IDE) with editing, debugging, and simulation functions is typically used for the design entry [23].
3. **Logic Design:** A netlist, or representation of the IC or SoC in terms of logic gates and connections, is created from the HDL or schematic in this step [24]. Then, using logic synthesis tools, the netlist is optimized for power, area, and performance [25]. To enhance the quality of the netlist, the logic synthesis tools carry out a number of modifications, including retiming, technology mapping, logic replication, and logic minimization [26]. Logic simulation tools are also used to verify the timing and functionality of the netlist [27]. The outputs of the netlist are compared to the anticipated results from the testbenches and stimuli using the logic simulation tools. Additionally, reports on the netlist's coverage, power, and delay are produced by the logic simulation tools.

4. **Circuit Design:** The netlist is mapped to a particular technology library in this step, which includes models and properties of the transistors and other devices that are accessible during the fabrication process [28]. In order to meet the performance, area, and power constraints, the devices must also be sized, biased, and placed as part of the circuit design [29]. Circuit simulation tools, like Simulation Program with Integrated Circuit Emphasis (SPICE), are used to verify the circuit design by analyzing the electrical behavior of the circuit. Reports on the circuit's voltage, current, power, and noise are also produced by the circuit simulation tools.
5. **Physical Design:** The circuit design is transformed into a layout in this step, which is a representation of the SoC or IC in terms of layers and geometric shapes. In addition to placing and routing the wires and devices, the layout includes DFM elements like metal density, antenna diodes, and dummy fill [30]. Physical verification technologies including parasitic extraction (PEX), layout versus schematic (LVS), and design rule check (DRC) are used to verify the layout [31]. The DRC ensures that the layout complies with the fabrication specifications, including minimal overlap, width, and spacing. The LVS ensures the consistency of the layout with the netlist, which includes aspects such as orientation, polarity, and connectivity. The PEX removes the layout's parasitic capacitance, resistance, and inductance, which have an impact on the circuit's power, noise, and performance [32].
6. **Fabrication and Testing:** The layout is transmitted to a foundry during this phase, where a series of photolithography, etching, deposition, and doping processes are used to manufacture the IC or SoC [33]. Using a variety of test equipment and techniques, including automatic test pattern generation (ATPG), scan chain, built-in self-test (BIST), and fault diagnosis, the fabricated IC or SoC is subsequently evaluated for quality, functionality, and performance [34]. Circuit faults, including stuck-at, bridging, and delay faults, can be identified by the test patterns generated by the ATPG. A tester can access the test points inserted into the circuit via the scan chain in order to apply test patterns and observe the outputs. In order to enable self-testing and self-diagnosis, the BIST embedded test logic inside the circuit. The circuit faults are identified by the fault diagnosis in terms of their location and nature [35].
7. **Verification and Validation:** This crucial stage guarantees the correctness and quality of the design throughout each phase of the VLSI design flow. Validation is the process of determining whether the design satisfies the demands and expectations of the end-users, whereas verification examines whether the design conforms to the specifications and requirements [36]. Formal methods, simulation, emulation, prototyping, and testing are all viable approaches for conducting verification and validation (Figure 10.1).

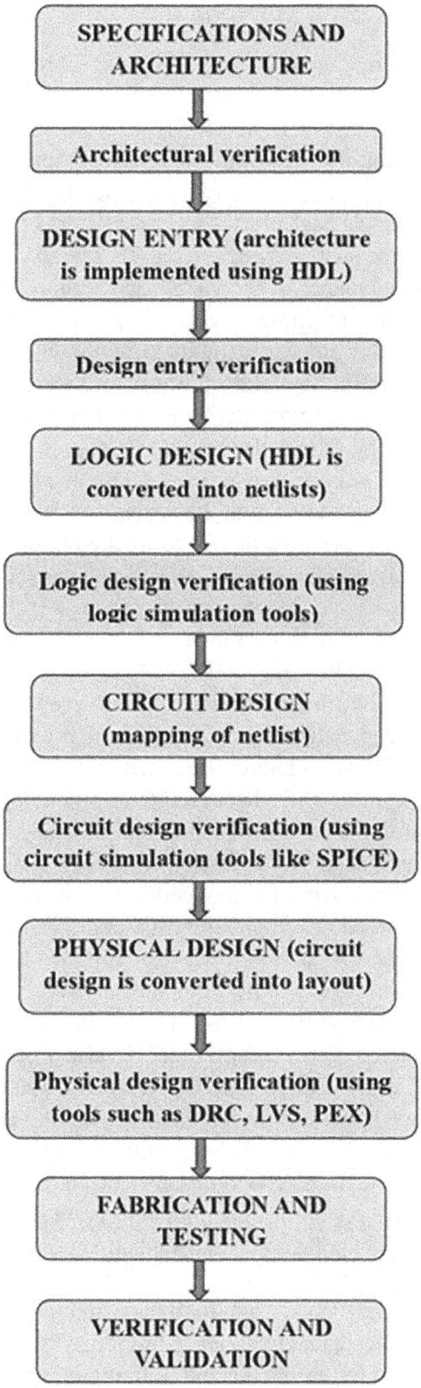

FIGURE 10.1 VLSI design process.

10.2.2 Key Parameters of VLSI Design

The key parameters of fundamentals of VLSI design are the characteristics that impact IC performance, cost, and reliability. A few of these parameters include:

1. **Power Consumption:** The result of multiplying supply voltage by current is power consumption. The current relies on the circuit's switching activity and capacitance. During each clock cycle, transistors change their states (on and off), resulting in switching activity [37]. The total amount of charge held in the circuit elements is known as capacitance. VLSI designers can minimize capacitance, minimize switching activity, lower the supply voltage, and use low-power process methods to cut down on power usage [38].
2. **Area:** The product of the size of each transistor and the number of transistors yields the area. The size of each transistor relies on the design rules and the process technology. The collection of techniques and materials needed to fabricate the chip is known as process technology. The constraints and instructions for placing and routing the transistors are referred to as design rules. VLSI designers can employ methods including scaling, layout optimization, logic minimization, and high-density process technologies to minimize area [39].
3. **Performance:** Performance is defined as the inverse of the circuit's execution time or delay. The delay relies on the interconnect delay and the propagation delay. The time it takes a signal to propagate through a logic gate is known as propagation delay [40]. The time required for a signal to travel via a wire is known as interconnect delay. VLSI designers can employ strategies like pipelining the stages, reducing the delay, parallelizing the computation, raising the clock frequency, and utilizing high-speed process technologies for enhanced performance [41].
4. **Design Time:** The amount of time needed to create, check, and test the semiconductor, reduced development costs, and a quicker time to market are the results of a shorter design time. The chip's design time relies on the functionality and complexity of the chip, the design methodology and tools, and the testing and debugging process. The quality and performance of the chip are also impacted by the design time. A variety of methods, including modular design, hierarchical design, design automation, and design for testability, are employed by VLSI designers to shorten the design time [42].
5. **Testability:** Testability is the chip's capacity to detect and correct errors with ease. Greater yield and reliability are correlated with increased testability [43]. A chip's testability is contingent upon its fault coverage, test pattern creation, test application, and fault diagnosis. The chip's performance and cost are also impacted by its testability. A variety of strategies, including scan design, BIST, design for debug, and fault-tolerant design, are employed by VLSI designers to enhance testability [44].
6. **Cost:** The entire cost for producing and utilizing the chip, profit and affordability both increase with reduced costs. The chip's expense relies on the fabrication cost, the packaging cost, the testing cost, and the operating cost [45].

The market and consumer demand for the chip are also impacted by cost. Yield improvement, design optimization, process improvement, and cost-benefit analysis are just a few of the methods VLSI designers employ to cut costs [46].

10.3 MACHINE LEARNING BASICS AND TECHNIQUES IN VLSI DESIGN

10.3.1 Overview of Machine Learning Concepts Relevant to VLSI Design

Complex and difficult processes like device modeling, layout verification, yield prediction, post-silicon validation, and reliability analysis are all part of VLSI design. ML can assist VLSI designers in automating these processes, enhancing their efficiency and precision, and managing the growing intricacy and unpredictability of contemporary IC technology [47].

The following are some ML ideas that are pertinent to VLSI design:

1. **Supervised Learning:** Supervised learning is a sort of ML in which the computer learns from labeled data in the form of input–output pairs. The objective is to identify a function, like a regression or classification function, that maps the input to the output [48]. For instance, you can use supervised learning to identify a function that estimates a house's price based on its features if you have a data collection of house prices and features, such as size, location, number of rooms, etc. The output of this problem is continuous, making it a regression problem. On the other hand, supervised learning can be used to identify a function that categorizes an image according to its label if you have a data collection of photos and labels, such as cat, dog, bird, etc. The output of this task is discrete, making it a classification problem.

   Supervised learning can be applied to VLSI design tasks like yield prediction, where the input is the design parameters and the output is the probability of passing manufacturing tests, or device modeling, where the input is the device geometry and the output is the electrical behavior [49].
2. **Unsupervised Learning:** Unsupervised learning is a sort of ML in which the computer learns from unlabeled data that simply contains inputs. Using techniques like dimensionality reduction or clustering functions, the objective is to identify structures, patterns, or clusters in the data [50]. For instance, unsupervised learning can be used to identify a function that lowers the dimensionality of an image data collection by compressing the images or removing the most significant features. Since the output is a lower-dimensional representation of the input, this is an issue involving dimensionality reduction. On the other hand, unsupervised learning can be employed to create a function that organizes the images into similar clusters, such as by identifying common qualities or themes among the images. This is a clustering problem, as the output consists of a set of labels indicating the input's cluster membership.

For VLSI design tasks like layout verification—where the layout geometry is the input and the output is the identification of possible errors or violations—or post-silicon validation—where the chip measurements are the input and the detection of faults or anomalies are the output—unsupervised learning can be applied [51].

3. **Semi-Supervised Learning:** This is a technique that integrates unlabeled and labeled data. It employs a reduced quantity of labeled data (in which the output labels are specified) and an increased quantity of unlabeled data (in which the output labels are unknown) [52]. Analogous to supervised learning, the objective is to develop a model that provides precise predictions of the output variable using the input features. However, the model is trained on a dataset that includes both labeled and unlabeled data. However, the model is trained on a dataset that includes both labeled and unlabeled data., for instance, anomaly detection, performance prediction, image classification, and text classification [53]. This method has numerous applications, including the classification of web content, speech analysis, and more. This method offers numerous advantages, including enhanced model generalization in comparison to solely supervised approaches and decreased annotation expenses.
4. **Reinforcement Learning:** This form of ML involves the computer acquiring knowledge through its own actions and feedback, which is comprised of rewards, states, and actions. By utilizing a control or optimization function, the objective is to identify a policy that optimizes the expected cumulative reward. Logic synthesis, in which the input is a behavioral description and the output is the optimal gate-level implementation, and analog design, in which the input consists of design specifications and the output is the optimal circuit configuration, are two VLSI design tasks that can benefit from RL [54,55] (Figure 10.2).

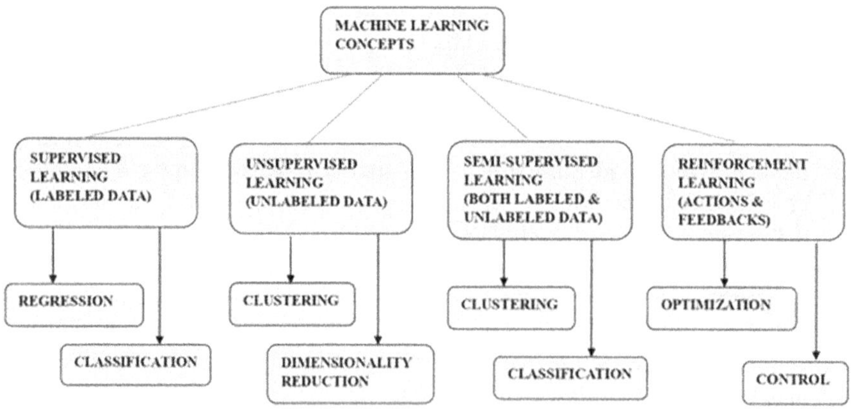

FIGURE 10.2 Machine learning concepts in VLSI design.

10.3.2 Introduction to Feature Selection and Dimensionality Reduction

Two methods, feature selection and dimensionality reduction, are intended to minimize the quantity of input variables in an ML model, particularly in the case of high-dimensional data. For VLSI design, high-dimensional data can present difficulties by driving up the cost of computing and data storage, affecting the speed and accuracy of problem-solving, and contributing to the "curse of dimensionality" [56].

1. **Feature Selection:** The process of choosing a subset of pertinent features from the original data and eliminating the redundant or irrelevant ones is known as feature selection [57]. In addition to lowering computational complexity and the risk of overfitting, feature selection can enhance the performance, interpretability, and generalization of ML models. Filter methods, wrapper methods, and embedding methods are the three categories into which feature selection can be classified [57]. The features are selected using filter methods according to certain statistical criteria like the chi-square test, information gain, or correlation, and the top-ranked features are selected [58]. Using an ML model as a black box, wrapper techniques assess the features according to the model's performance metrics, such as accuracy, precision, or recall, and then choose the optimal subset of features [57]. Embedded approaches, such as regularization, decision trees (DTs), or neural networks, incorporate the feature selection process into the training of an ML model and select the features with the greatest impact on the model's output.
2. **Dimensionality Reduction:** The process of converting original high-dimensional data into a lower-dimensional representation while maintaining the data's structure or key information is known as dimensionality reduction [59]. In addition to lowering noise and redundancy in the data, dimensionality reduction can help with classification, clustering, and visualization. There are two types of dimensionality reduction techniques: linear techniques and nonlinear techniques [60]. For reducing dimensionality, linear techniques such as principal component analysis (PCA), linear discriminant analysis (LDA), or singular value decomposition (SVD) employ linear projections or transformations on the assumption that the data is on or near a linear subspace. In nonlinear techniques, the data is assumed to be on or near a nonlinear manifold; to reduce the dimensionality, nonlinear mappings or embeddings are applied, such as multidimensional scaling (MDS), kernel PCA, or t-distributed stochastic neighbor embedding (t-SNE) [61].

10.3.3 Machine Learning Techniques

With the use of ML approaches, computers may learn from data and carry out operations that are challenging or impossible for them to accomplish using traditional algorithms. ML approaches can be used for fault tolerance, optimization, verification, testing, diagnostics, and other elements of VLSI design (Figure 10.3).

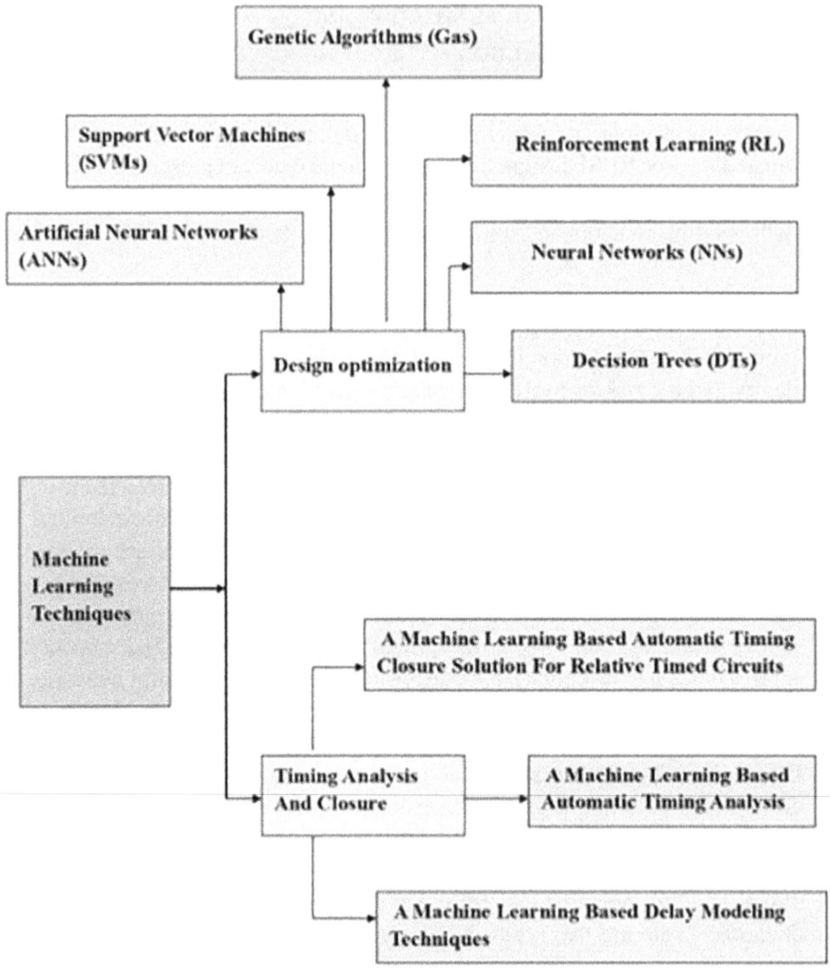

FIGURE 10.3 Machine learning techniques in VLSI design.

10.3.3.1 Design Optimization

The process of using ML techniques to enhance the area, power, and performance of VLSI circuits is known as design optimization of ML techniques in VLSI design [62]. ML approaches have the potential to automate and accelerate a number of processes in the VLSI design flow, including yield prediction, reliability analysis, logic synthesis, placement, routing, timing analysis, power estimation, and defect detection and fault diagnostics. In order to determine the optimum trade-offs between competing goals, ML techniques can also be used to explore and optimize the design space of VLSI circuits. VLSI circuit design optimization commonly uses a number of ML approaches, including:

1. **Artificial Neural Networks (ANNs):** ANNs are computer models that can learn intricate and nonlinear functions from data. They are useful for a number of optimization tasks, including timing analysis, logic synthesis, placement, routing, and power estimation for VLSI circuits. ANNs can also be implemented in hardware by means of VLSI circuits that mimic the functionality and behavior of biological neurons and synapses, such as memristors, spintronics, and neuromorphic chips [63].
    a. Artificial neurons are arranged in layers and connected by weights to form ANNs. To generate an output, each neuron calculates the weighted sum of its inputs and applies a nonlinear activation function. Throughout the learning process, the weights are modified in accordance with the error between the desired and actual outputs.
    b. With enough information and hidden neurons, ANNs can learn to approximate any function. The loss function, which estimates the discrepancy between the outputs and the targets, can be minimized using a variety of algorithms, including stochastic gradient descent, gradient descent, and backpropagation, to train ANNs [64].
    c. Depending on whether or not the connections are directed, ANNs can be categorized as feedforward or recurrent networks. Feedforward networks can be utilized for static input–output mapping because they don't have any cycles in their connections. Recurrent networks can be utilized for sequential and dynamic data processing because they contain cycles in their connections.
    d. ANNs can be applied to a range of VLSI design tasks, including:
        a. **Logic Synthesis:** A Boolean function can be mapped to a network of logic gates using ANNs, which minimizes the circuit's area, delay, and power consumption [65].
        b. **Placement:** ANNs can be employed to determine the best locations for the components on the chip, reducing wire length and congestion.
        c. **Routing:** By preventing overlaps and crossovers, ANNs can be utilized to determine the best paths for the interconnections between the components.
        d. **Timing Analysis:** By considering the impacts of temperature, voltage, and process variations, ANNs can be utilized to predict the circuit delay [66].
        e. **Power Estimation:** ANNs can be utilized to estimate the power consumptions in circuits by taking into account the wires' resistance, capacitance, and switching activity.
2. **Support Vector Machines:** Support vector machines (SVMs) are supervised learning models that are capable of performing regression and binary or multiclass classification. SVMs are capable of handling high-dimensional and nonlinear data and can determine the best hyperplane to divide the data into distinct classes. SVMs can be applied to a variety of optimization problems, including yield prediction, fault diagnosis, defect detection, and VLSI

circuit reliability analysis [67]. SVMs are additionally suitable to optimize building energy, as they are able to predict a building's energy demand and consumption depending on a range of parameters, including occupancy, weather, and mode of operation.

a. SVMs employ a kernel function to map the data to a higher-dimensional space, from which a linear hyperplane is found to divide the classes. A subset of the data points that are on the margin of the classes—referred to as support vectors—define the hyperplane [68].
b. Depending on the nature of the issue and the data, SVMs can employ a variety of kernel functions, including sigmoid, polynomial, linear, and radial basis functions. In the feature space, the kernel function calculates the distance or similarity between the data points.
c. The trade-off between the model's accuracy and complexity can be managed by SVMs using a variety of regularization parameters, such as gamma or C. The width of the margin and the penalty for the misclassified points are established by the regularization parameters.
d. SVMs can be applied to a range of VLSI design tasks, including:
    i. **Defect Detection:** SVMs can reduce false positives and false negatives by classifying the chips as defective or non-defective based on the test results.
    ii. **Fault Diagnosis:** SVMs can be employed for fault detection in order to pinpoint the kind and position of circuit faults by analyzing the malfunctioning outputs, which lowers the expense and time required for diagnosis.
    iii. **Yield Prediction:** Using process parameters, design attributes, and test outcomes, SVMs can be employed to predict the chip yield, which is the percentage of functional chips within a batch.
    iv. **Reliability Analysis:** SVMs can be employed to analyze a chip's reliability, which is the probability that it will continue to function correctly over time, taking into account stress factors, aging effects, and failure mechanisms [69].

3. **Genetic Algorithms (GAs)**: Genetic algorithms (GAs) are methods of optimization based on the principles of natural process of evolution. GAs are able to adapt to changing environments and discover nearly optimal solutions to complicated and multimodal issues. Several optimization activities in VLSI circuits, including test pattern generation, circuit design, layout optimization, and fault tolerance VLSI circuit, can be addressed with GAs [70]. Through the identification of optimum parameters for population size, crossover rate, and mutation rate, GAs can also be utilized to maximize the energy usage and performance of parallel ML algorithms.
    a. GAs employ a population consisting of individuals who, when compared to a fitness function that evaluates their quality, represent potential solutions to the problem. To create new generations of individuals who are more fit than the previous ones, GAs employ operators like selection, crossover, and mutation.

b. GAs can select individuals for reproduction using a variety of techniques, including tournaments, roulette wheels, and rank-based systems [71]. The population's quality and diversity are intended to be maintained by the selection procedures.
c. The way the individuals are merged to produce offspring is determined by the crossover techniques that GAs employ, which include one-point, two-point, and uniform [72]. The crossover techniques seek to enlarge the search space and produce novel solutions.
d. Different mutation techniques, like bit-flip, swap, or invert, can be used by GAs to determine how the individuals are changed to add variation. The goal of the mutation techniques is to avoid local optima and stop premature convergence.
e. GAs can be applied to a range of VLSI design tasks, including:
    i. **Circuit Design:** GAs can be applied to optimize the performance, area, and power of analog and digital circuits by designing their topology, size, and biasing.
    ii. **Layout Optimization:** By reducing the area, wire length, and parasitic of the circuit, GAs can be utilized to optimize the chip's layout [70].
    iii. **Test Pattern Generation:** Using GAs, test patterns can be created that identify the maximum number of faults present in the circuit, cutting down on test duration and expense.
    iv. **Fault Tolerance:** Using GAs, circuits can be designed to be fault-tolerant, meaning they can either tolerate or retrieve from faults, hence enhancing the circuit's availability and reliability [70].

4. **Reinforcement Learning (RL):** Reinforcement learning (RL) is an ML technique that allows a system to learn from its own actions and rewards. The design space of VLSI circuits can be explored and optimized using RL to determine the optimum trade-offs between area, power, and performance. Adaptive control techniques for VLSI systems, including error correction, thermal management, and voltage and frequency scaling, can also be learned by RL [73].
    a. In RL, an agent interacts with the environment and gains knowledge from the feedback it gets. The value function calculates the expected return, the reward function estimates the results, and the policy dictates the agent's course of action.
    b. There are two types of RL methods: model-based and model-free, based on whether the agent has access to an environment model or not. Model-free techniques depend on trial-and-error learning, but model-based techniques can employ simulation and planning to enhance the policy [74].
    c. RL can be divided into value-based and policy-based approaches, contingent on whether the agent directly learns the policy or the value function. In policy-based approaches, actions are sampled using the policy, but in value-based methods, the optimal action is chosen using the value function.
    d. RL can be applied to a range of VLSI design tasks, including:

i. **Design Space Exploration:** Resistant logic can be utilized to determine the optimum possible configuration of circuit parameters, like transistor sizes, wire widths, and supply voltages, in order to maximize power, area, and performance objectives.
ii. **Adaptive Control:** RL can be utilized to train dynamic control policies for VLSI systems, which include controlling the thermal dissipation, modifying the voltage and frequency levels, and correcting errors based on temperature, workload, and reliability conditions [75].

5. **Neural Networks (NNs):** Neural networks (NNs) are a unique class of ANNs that draw inspiration from the functions and structures of biological brains. NNs are useful for a variety of VLSI applications, including computer vision, speech recognition, image processing, and natural language processing. They are also capable of learning complex and nonlinear functions from data. Neuromorphic devices, spintronics, memristors, and other VLSI circuits can be employed to implement NNs in hardware [76].
   a. Neurons in NNs are arranged in layers and joined by synapses. Every neuron generates an output by applying a nonlinear activation function to the weighted sum of its inputs. Based on the error between the desired and actual outputs, the synapses retain the weights that are learned during the training phase.
   b. Depending on whether or not the connections are directed, NNs can be categorized into feedforward and recurrent networks. Feedforward networks can be utilized for static input–output mapping because they don't have any cycles in their connections. Recurrent networks can be utilized for sequential and dynamic data processing because they contain cycles in their connections.
   c. Based on the number of hidden layers, NNs can also be categorized into shallow and deep networks. Shallow networks are useful for simple, linear issues and can contain one or more hidden layers. Complex and nonlinear issues can be addressed by deep networks, which have numerous hidden layers [77].
   d. NNs can be applied to a range of VLSI design tasks, including:
      i. **Image Processing:** NNs can be applied to a range of image processing applications, including restoration, enhancement, segmentation, compression, and recognition [76].
      ii. **Speech Recognition:** NNs have the ability to identify speech signals and translate them into commands or text.
      iii. **Natural Language Processing:** NNs are capable of understanding and generating texts in natural language for tasks including sentiment analysis, question answering, translation, and summarization3.
      iv. **Computer Vision:** NNs are capable of handling a wide range of computer vision applications, including action recognition, object recognition, scene understanding, and face detection [76].
      v. **Hardware Implementation:** Implementing NNs in hardware can be implemented through the application of VLSI circuits that mimic the functionality and behavior of biological neurons and synapses, like memristors, spintronics, and neuromorphic chips.

6. **Decision Trees (DTs):** Regression and classification can be carried out by DTs, which are supervised learning models. Based on specific criteria, DTs can divide the data into smaller subsets, creating a structure similar to a tree [78]. DTs can be applied to VLSI activities such as test pattern generation, fault diagnosis, reliability analysis, and yield prediction [79].
    a. DTs employ a recursive partitioning algorithm, which divides the data into two or more child nodes according to a feature and a threshold, beginning at the root node. The process of splitting keeps going until a predetermined threshold—such as a minimum quantity of samples, a maximum depth, or a purity level—is reached.
    b. Based on the nature of the problem and the data, DTs can employ a variety of splitting criteria, including information gain, the Gini index, or mean squared error. The objective of the splitting criteria is to reduce the impurity or increase the homogeneity of the child nodes.
    c. DTs can enhance the generalization performance and prevent overfitting by employing various pruning strategies, such as minimum error pruning, cost complexity pruning, or reduced error pruning [80]. The goal of pruning procedures is to eliminate branches or nodes from the tree that are hazardous or superfluous.
    d. DTs can be used for VLSI jobs like:
        i. **Test Pattern Generation:** By employing fault coverage, fault detection probability, or fault activation probability as the splitting criterion, DTs can be utilized to generate test patterns that are capable of detecting circuit faults.
        ii. **Fault Diagnosis:** By classifying fault types or locations as classes and employing faulty outputs, fault models, or fault signatures as features, DTs can be used to diagnose circuit faults.
        iii. **Yield Prediction:** Utilizing process parameters, design features, or test outcomes as features and the yield rate or yield loss as the target, DTs can be employed in predicting the chip's yield.
        iv. **Reliability Analysis:** By utilizing stress factors, aging effects, or failure mechanisms as features and the failure rate or failure mode as the target, DTs can be utilized to analyze the chip's reliability [79].

10.3.3.2 Timing Analysis and Closure

Artificial intelligence and data-driven techniques are used in ML models for timing analysis and closure in VLSI design to maximize the reliability and performance of ICs. Automating the processes of determining and enhancing circuit delays, meeting timing constraints, and cutting down on design complexity and turnaround time are all possible with these techniques.

The following are some instances of ML models used in VLSI design for time analysis and closure:

1. **A Machine Learning-based Automatic Timing Closure Solution for Relative Timed Circuits** [81]:
    a. One type of asynchronous circuit that employs relative timing constraints rather than absolute timing limitations is known as a relative

timed circuit. This indicates that rather than the precise time values, the behavior of the circuit depends on the relative order of events.

b. The machine learning-based approach minimizes the total run-time to timing close a design by using a RL algorithm to learn from the features throughout each iteration. The circuit topology, timing constraints, slack values, gate sizes, and buffer locations are some of the features.
c. To determine the optimal action, such as resizing, buffering, or rewiring, for each gate or net, the RL method employs a policy network. A reward function that determines improvements in timing closure as well as decreases in area and power consumption is used to train the policy network.
d. A value network is also used by the RL algorithm to evaluate the expected future reward for each state and direct the trade-off between exploration and exploitation [82]. Using temporal difference learning, the value network is trained by updating the value function according to the variation between the predicted and actual rewards.

2. **A Machine Learning-based Static Timing Analysis Framework:**
   a. A technique for validating a circuit's timing performance is static timing analysis, which involves computing the worst-case delays of all feasible paths [83]. The paths with the biggest delays or the smallest slacks are known as the critical paths, and they can be identified through static timing analysis.
   b. Based on parameters including gate types, fan-in and fan-out, wire lengths, and process variations, the machine learning-based framework employs a deep neural network for predicting the slack violations and path delays of a circuit.
   c. A traditional static timing analysis tool provides the target values for the supervised learning approach that is employed to train the deep neural network. In comparison to the traditional tool, the deep neural network is capable of achieving excellent accuracy and efficiency, particularly for large-scale circuits with millions of paths.
   d. In addition, the machine learning-based framework optimizes gate sizing and buffer insertion using a GA. These two methods allow circuit timing performance to be enhanced by varying drive strengths and lowering interconnection delays [84]. The deep neural network is used by the GA as a fitness function for evaluating each solution's quality, and crossover and mutation operators are applied to generate novel solutions.
3. **A Machine Learning-based Delay Modeling Technique:**
   a. A method for estimating a circuit's interconnect delays—that is, the delays brought on by the cables connecting the gates—is called delay modeling [85]. Delays at interconnects rely on the coupling capacitance, loop resistance, transportation topology, and wire geometry.
   b. Utilizing a support vector regression model, the ML approach estimates a circuit's interconnect delays by analyzing several characteristics such as wire diameter, wire length, wire spacing, number of bends, and slew rate [86].

c. The training process for the support vector regression model involves supervised learning. The target values utilized in this approach are determined through a parasitic extraction tool, which extracts the resistance and capacitance values of the wires. In contrast to the parasitic extraction tool, the support vector regression model can provide superior accuracy and efficiency, particularly when applied to intricate routing topologies comprising numerous wires and vias.
d. The technique that is based on ML additionally optimizes the routing topology, which is the configuration of the wires connecting the gates, using a gradient boosting algorithm [87]. The gradient boosting algorithm measures the interconnect delays using the support vector regression model as a loss function. It then employs a greedy search technique to identify the optimal routing topology that reduces the delays.

The vast quantity of data produced throughout the VLSI design process can be utilized by these ML models, which can offer quick and precise timing analysis and closure solutions. They are also capable of enhancing the yield and quality of the VLSI circuits by adjusting to variations and uncertainties in the fabrication process.

10.4 CHALLENGES AND CONSIDERATIONS

10.4.1 Data Quality and Quantity

Numerous processes related to VLSI design and technology, including layout synthesis, placement and routing, timing closure, DFM, verification, and testing, can be automated and optimized through the use of ML. Nevertheless, the implementation of ML in the design and technology of VLSI presents numerous challenges, most notably concerning the availability and quality of training data.

The data utilized to train an ML model to carry out a particular task is known as training data. The quantity and quality of training data have a direct impact on the ML model's accuracy and performance [88]. Several challenges pertain to the availability and quality of training data in ML models for VLSI design and technology. These include:

1. **Data Scarcity:** This refers to the insufficient amount of data that is available to train ML models for VLSI design and technology. This may occur as a result of the complexity and specialization of VLSI design and technology, which call for a high level of domain knowledge and expertise [89]. As a result, it is challenging and expensive to find and collect relevant data. Furthermore, access to and exchange of data between various parties may be restricted by the possibility that certain data is proprietary or confidential. The accuracy and performance of the ML model might be impacted by a scarcity of data since the model may not be able to adequately represent the complexity and variability of the data and make good generalizations to new situations [90].
2. **Data Quality:** This refers to the degree of noise, incompleteness, unpredictability, or imbalance in the data that is available for training ML models

for VLSI design and technology. This may occur due to the fact that VLSI design and technology deal with a number of different sources of variation and uncertainty, including environmental factors, device aging, and process variations. These may introduce the data with errors and noise. Furthermore, faulty measurement or transmission could result in some data being lost or corrupted. Furthermore, an unequal distribution of samples or labels may lead to bias or skewing of some data. Since the ML model may be overly sensitive to noise and outliers in the data and perform poorly on unseen data, data quality might have an impact on the model's robustness and reliability [91].

3. **Data Labeling:** This means that precise and consistent labels are needed for supervised learning in order to use the data that is available for training ML models for VLSI design and technology. This can occur because labeling data—such as analog signals, circuit defects, or design errors—requires a great deal of human participation and judgment in VLSI design and technology. Nonetheless, data labeling can be labor-intensive, time-consuming, and error-prone, particularly when dealing with big and complicated datasets [92]. Furthermore, labeling certain data may be challenging or ambiguous, which introduces subjectivity and human bias into the data. Data labeling may not accurately reflect the true nature and properties of the data, which might have an impact on the validity and generalizability of the ML model.
4. **Data Diversity:** This indicates that the data used to train ML models for VLSI design and technology is not diverse and representative of the various scenarios and conditions the model may come across. This is possible because there are extensive number of tasks, objectives, and constraints associated with VLSI design and technology, which vary between design stages, abstraction levels, and technology nodes. As a result, it might be difficult to find and collect diverse data because it may necessitate numerous resources, experiments, and simulations. Diversity of data may prevent the ML model from being able to handle the various requirements and scenarios that it may encounter, which could have an impact on its scalability and adaptability [93].

These are a few of the major challenges with training data quality and availability in ML models for VLSI design and technology. Here are a few potential solutions to address these challenges:

1. **Data Augmentation:** This method creates new data by transforming existing data using transformations like rotation, scaling, cropping, flipping, or noise addition. Reducing overfitting and enhancing the quantity and diversity of data are two benefits of data augmentation. For instance, in the field of VLSI design and technology, artificial data can be created by data augmentation and generative adversarial networks (GANs) for predicting the performance of circuit. One kind of neural network called GAN is able to learn to create realistic data from a provided distribution, like circuit data. ML models that were trained on a small training dataset may become more accurate with the use of GANs for data augmentation [94].

2. **Data Cleaning:** This is a process that uses techniques like filtering, interpolation, normalization, or resampling to eliminate or correct noisy, incomplete, inconsistent, or imbalanced data [95]. Reducing underfitting or overfitting and enhancing data quality and reliability are two benefits of data cleaning. For instance, data cleaning can be used in VLSI design and technology to address a variety of causes of variation and uncertainty, including process variations, device aging, and environmental factors that may introduce errors and noise into the data. Reducing the influence of these issues on the design and testing of VLSI circuits can be achieved by data cleaning [13].
3. **Data Annotation:** This is a methodology that uses automatic, semi-automatic, or manual annotation to apply labels to data for supervised learning. Data annotation may minimize human bias and error while assisting to give consistent and accurate labeling for data. For instance, data annotation can be employed in VLSI design and technology to label data for various design tasks, including physical design verification, layout generation, and optimization. ML models trained on labeled data may benefit from data annotation in terms of validity and generalizability [96]
4. **Data Synthesis**: This method uses techniques like generative models, simulation, and emulation to generate novel data through scratch. In order to alleviate data scarcity and creating sufficient and relevant data for training ML models, data synthesis can be beneficial [97]. Data synthesis, for instance, can be employed in VLSI design and technology to generate data for various design scenarios and conditions, such as various design stages, abstraction levels, and technology nodes. The scalability and adaptability of ML models trained on synthetic data can be enhanced through data synthesis [13].

10.4.2 Interpretability and Explainability

Due to the fact that ML is being implemented more and more in VLSI design for tasks like layout optimization, fault detection, and performance prediction, this is a highly relevant and intricate topic. Nevertheless, the perception of ML models as "black boxes" frequently makes it challenging to comprehend how they decide what to do and what factors influence their results [98]. This presents a number of challenges for VLSI designers, including:

1. **Reliability and Robustness:** In the design of VLSI, ML models must be robust and reliable. This entails the ability of the models to consistently generate accurate results amid a variety of scenarios and conditions [99]. As an illustration, a circuit performance prediction model based on ML should possess the capability to deal with fluctuations in the input data, the fabrication process, and environmental factors. Nevertheless, the reliability and robustness of ML models may be compromised by factors including bias, noise, overfitting, and underfitting. Consequently, it is imperative that VLSI designers comprehend the origins of these complications and devise

methods to alleviate them through the implementation of regularization, cross-validation, and ensemble methods [100].

2. **Validation and Verification:** In the design of VLSI, ML models must be validated and verified, which entails testing and inspecting them to ensure their accuracy and correctness. When maximizing a circuit layout, for instance, an ML model should be able to adhere to design constraints and specifications including area, power, and time. However, errors and uncertainties in ML models may exist, thereby impacting validation and verification processes. Consequently, VLSI designers must comprehend how to utilize metrics including accuracy, precision, recall, and F1-score to determine and evaluate the quality and performance of ML models [101].
3. **Communication and Justification:** In the design of VLSI, ML models must be justifiable and communicable, which means they must be capable of providing explanations and insights regarding the reasoning behind their recommendations and decisions. An illustration of this can be seen in the capability of an ML model to identify faults in a circuit, as well as to pinpoint their locations and root causes, as well as to propose potential solutions. In contrast, the communication and justification of ML models may be compromised due to their complexity and opaqueness [102]. Hence, it is imperative for VLSI designers to possess the knowledge and skills necessary to derive and display pertinent data and evidence from ML models by employing methodologies like feature importance, DTs, and saliency maps.
4. **Scalability and Efficiency:** Efficiency and scalability are critical considerations in the design of VLSIs; ML models must possess the capability to process large and intricate datasets and problems [103]. An instance of this would be an ML model tasked with generating optimal solutions within a reasonable timeframe and capable of handling millions of transistors and cables when synthesizing the layout of a chip. However, the high memory and computational costs associated with ML models may impact their scalability and efficiency [104]. Hence, it is imperative for VLSI designers to comprehend the methods by which ML models' resource and time demands can be diminished and optimized through the implementation of approximation, parallelization, and compression.
5. **Privacy and Security:** In order to safeguard the sensitive and confidential information and data associated with the design process, ML models utilized in VLSI design must be both private and secure. As an illustration, an ML model that acquires knowledge from a corporation's design data ought to refrain from disclosing or otherwise exposing the company's trade secrets or intellectual property to unauthorized individuals. Nevertheless, ML models could potentially be susceptible to a range of threats that compromise their privacy and security, including adversarial examples, data contamination, and model theft. Consequently, VLSI designers must comprehend how to employ techniques such as adversarial training, homomorphic encryption, differential privacy, and homomorphic encryption to protect and encrypt the data and information utilized by ML models [98].

10.4.3 Hardware Constraints

ML models frequently necessitate substantial quantities of memory and power to operate, resulting in high computational demands. Certain applications, such as embedded systems, Internet of Things devices, or printed circuits, may necessitate the deployment of these models on hardware platforms with constrained resources. In regard to size, weight, power, and performance, these platforms are severely constrained (SWAP) [105]. Consequently, the integration of ML models onto such platforms presents considerable challenges and necessitates extremely careful hardware–software co-design and optimization.

The implementation of ML models on resource-constrained platforms is impacted by the following hardware constraints:

1. **Memory Capacity and Bandwidth:** The amount of data that may be stored and moved between a hardware platform's memory and its processing units depends on two factors: memory size and bandwidth. The term "memory size" describes the total amount of memory—either on-chip or off-chip—that is accessible on the platform. In general, on-chip memory (e.g., SRAM or eDRAM) exhibits enhanced speed and energy efficiency; however, it entails a higher cost and a restricted capacity. Off-chip memory, including flash and DRAM, is typically more affordable and has a greater capacity, but it operates slowly and is more energy-hungry. The rate of data transfer between the memory and the processing units is referred to as the memory bandwidth. This rate is contingent upon the bus width, clock frequency, and memory access pattern. The implementation of ML models on resource-constrained platforms may be hindered by memory capacity and bandwidth, which can impact the model's size and complexity, the inference's speed and energy efficiency, as well as the results' accuracy and robustness [106].
2. **Computing Resources and Throughput:** These parameters dictate the amount of computation that a hardware platform's processing units are capable of performing [107]. The quantity and kind of processing units—which might be either general purpose or specialized—that are available on the platform are referred to as computing resources. Central Processing Units (CPUs) and Graphics Processing Units (GPUs) are some examples of general-purpose processing units that are typically more programmable and adaptable, but they are also more intricate and power-hungry. Field-programmable gate arrays (FPGAs) and application-specific integrated circuits (ASICs) are some examples of specialized processing units that are typically more expensive and rigid, but also more efficient and customized. Throughput, which is dependent on clock frequency, latency, and pipelining of the processing units, is the amount of computing that the processing units can accomplish per unit time [108]. The implementation of ML models on resource-constrained platforms may be restricted by computing resources and throughput, which can impact the models' complexity and parallelism, the inference process' speed and energy efficiency, and the accuracy and performance of the results.

3. **Power Consumption and Heat Dissipation:** These two parameters dictate the amount of energy that the hardware platform uses and wastes throughout the inference. The term "power consumption" describes the amount of electrical energy needed for the operation of the hardware platform, which is contingent upon component voltage, current, and switching activity. There are two types of power consumption: static power refers to the power consumed by components when they are inactive and dynamic power concerns the power consumed when the components are in operation (active). The amount of thermal energy that the hardware platform generates and releases throughout the inference is referred to as heat dissipation, and it is dependent upon the temperature, cooling mechanisms, and power consumption of the individual components. Since heat dissipation can lead to component failure, degradation, or thermal stress, it can have an impact on the hardware platform's reliability and lifetime [109]. The implementation of ML models on resource-constrained platforms may be limited by power consumption and heat dissipation, as these factors may impact the platform's sustainability and battery life, as well as the precision and dependability of the outcomes.

Some potential solutions to get around these hardware constraints are as follows:

1. **Model Compression and Quantization:** By eliminating redundant or irrelevant layers, parameters, or operations, or by employing lower-precision arithmetic, like binary or ternary, these strategies seek to minimize the size and complexity of ML models [110]. These methods can keep the models' accuracy and performance within reasonable bounds while lowering their memory footprint, computational complexity, and power consumption.
   a. Here are a few instances of model compression methods:
      i. **Pruning:** Pruning is a method of eliminating weights or neurons from a neural network that is superfluous or irrelevant, based on certain standards like magnitude, gradient, or relevance [111]. Pruning can enhance the neural network's inference speed and energy efficiency by reducing the number of parameters and processes.
      ii. **Distillation:** Through reducing the disparity among their outputs or intermediary representations, distillation is a strategy that moves knowledge from a large and complicated teacher model to a simplified and smaller student model. Distillation can maintain a neural network's accuracy and performance while reducing its size and intricacy.
   b. A few instances of quantization methods are as follows:
      i. **Fixed-Point Quantization**: One technique that is commonly used for representing the weights and activations of a neural network is called fixed-point quantization. It involves taking floating-point numbers and converting them to fixed-point numbers, which have a defined number of bits for the integer and fractional components [112]. In addition to facilitating the use of faster and simplified

arithmetic units, such as multipliers and adders, fixed-point quantization can minimize the bandwidth and memory size.

ii. **Binary or Ternary Quantization:** One way to further minimize the number of bits needed to represent the weights and activations of a neural network is to utilize binary or ternary quantization, which can reduce the number of bits to either one bit (binary) or two bits (ternary) [112]. In addition to the use of bitwise operations, such as Exclusive OR (XOR) and Exclusive NOR (also known as equivalence) (XNOR), which are considerably quicker and more energy-efficient than arithmetic operations, binary or ternary quantization can substantially reduce memory size and bandwidth.

2. **Neural Architecture Search and Hardware-Aware Optimization**: One approach that seeks to identify the optimum neural network architecture and implementation for a specific hardware platform and task is through the utilization of hardware-aware optimization and neural architecture search [113]. In order to explore the design space of neural architecture, hardware implementation, and quantization, as well as to maximize the trade-off among accuracy, latency, and energy efficiency, these methods may employ learning-based approaches like RL or evolutionary algorithms. The following are instances of techniques involving hardware-aware optimization and neural architecture search:

   i. **ProxylessNAS**: ProxylessNAS is a methodology that employs no surrogate or proxy models, including accuracy or latency predictors, in its search for the optimum neural network architecture on the target hardware platform [114]. ProxylessNAS uses gradient-based techniques that simultaneously maximize accuracy and latency goals in order to determine the optimal architecture for each given hardware platform, including smartphones, GPUs, and FPGAs.

   ii. **HAQ:** Hardware-Aware Quantization (HAQ) is a methodology that autonomously ascertains the optimum bit-width for each layer of a neural network, taking into consideration the hardware feedback and the accuracy constraint [115]. HAQ has the capability to discover a quantization policy that reduces the neural network's latency or energy consumption on the target hardware platform, including CPUs, GPUs, or FPGAs, through the utilization of RL.

3. **Non-Conventional Computing Approaches**: The objective of non-conventional computing approaches is to perform ML processes more effectively and scalably by taking benefit of the physical characteristics of innovative materials or devices, including memristors, spintronics, or quantum dots [116]. In order to construct ML models with reduced power consumption, higher density, and enhanced speed, these methods can take advantage of the inherent parallelism, stochasticity, or approximation of these materials or devices. A few instances of non-conventional computing methods are as follows:

   i. **Memristive Computing:** This method employs memristors—resistive devices that can store and alter their resistance state in response to applied voltage or current—to implement neural network functions like activation functions, matrix multiplication, and weight storage. Memristive computing employs crossbar arrays of memristors to accomplish parallel and analog computations, allowing for high-density and low-power neural network implementation [117].

ii. **Spintronic Computing:** Spintronic computing is a neural network technology that leverages spintronics—devices that control the spin of electrons instead of their charge—to perform neural network operations like logic gates, oscillators, and weight storage. By utilizing spin torque nano-oscillators, domain wall magnets, or magnetic tunnel junctions, spintronic computing can implement NNs with reduced power and high speed while carrying out non-volatile and stochastic computations [118].

10.5 FUTURE TRENDS

Emerging technologies and trends in this field include the following:

1. **Machine Learning for the Automation of VLSI Design:** The layout optimization of VLSI circuits, for instance, can be automated, utilizing ML. The process of arranging the circuit components and interconnections on a chip so as to reduce its area, power consumption, and delay is known as layout optimization. This process has traditionally been performed manually or with heuristic algorithms, both of which are time-consuming and prone to error. In contrast, the layout optimization can be accomplished autonomously and effectively through the use of ML techniques, including RL, GAs, and NNs. In order to generate optimized layouts that adhere to the design specifications and constraints, ML algorithms have the capability to acquire knowledge from pre-existing layouts and data [119].
2. **Machine Learning for VLSI Circuit Performance Prediction and Optimization**: ML, for instance, can be utilized for predicting and optimizing VLSI circuit performance in a variety of operating conditions [120]. A number of factors, including temperature, voltage, and process variations, can have an impact on the performance of VLSI circuits. These factors may lead to deviations in the circuits' expected behavior and performance, which may result in failure or degradation. In addition to optimizing circuit parameters like power consumption, speed, and reliability, ML algorithms are also capable of analyzing and modeling the behavior and performance of VLSI circuits under various operating conditions. ML algorithms have the ability to acquire knowledge from historical data and make precise predictions about how a circuit will perform under various operating conditions [121]. This makes it possible for designers to efficiently attain the targeted performance goals and optimize the circuit parameters. These VLSI circuits and systems include, for instance:

i. **Field-Programmable Gate Arrays (FPGAs)** and **Application-Specific Integrated Circuits (ASICs)** are reconfigurable and customizable VLSI circuits and systems that possess the ability to implement particular ML algorithms and architectures with maximum flexibility and efficiency. In addition to supporting dynamic reconfiguration, pipelining, and parallelism, ASICs and FPGAs can offer specific hardware resources, such as multipliers, adders, memory, and logic, for ML computations [122].

3. **Machine Learning for VLSI Circuit Innovation and Exploration**: Through its ability to explore unconventional design spaces and produce creative solutions, ML, for example, can be utilized to generate unique and innovative VLSI architectures and circuits. Neural network accelerators, quantum computing circuits, and nanoelectronic devices are just a few examples of the unique and innovative VLSI architectures and circuits that can be generated by ML algorithms. Additionally, ML techniques have the potential to bridge the gap between VLSI design and forthcoming technologies like nanoelectronics, quantum computing, and neuromorphic computing [123]. In order to provide innovative and optimum solutions that go beyond conventional design paradigms, ML algorithms may acquire knowledge from pre-existing data and designs.
4. **VLSI for ML Inference and Training:** The two primary stages of ML applications are ML inference and training, and VLSI is capable of providing the hardware support for both [124]. The procedure of implementing a trained ML model to novel data is referred to as ML inference, whereas the procedure of learning an ML model from data is known as ML training. VLSI is capable of implementing training and inference engines that can accurately and efficiently carry out ML algorithms. Moreover, VLSI is capable of supporting a variety of ML models, including supervised, unsupervised, and RL.
    a. For instance, inference engines, which are hardware devices that are capable of performing ML inference tasks like speech recognition, image classification, and natural language processing, can be implemented using VLSI [125]. The high-speed, low-power, parallel computing capabilities needed for ML inference can be achieved with VLSI. NNs, convolutional networks, and recurrent networks are just a few examples of the specialized architectures and circuits that VLSI can implement for certain ML models. The accuracy, latency, and energy consumption of the inference engines can all be optimized with VLSI.
    b. One other instance is the application of VLSI for training engines, which are hardware devices that are capable of carrying out machine learning training tasks, like figuring out an ML model's parameters from an extensive dataset. The scalable, high-memory, high-performance computing capabilities needed for ML training can be achieved through VLSI. Stochastic gradient descent, backpropagation, gradient descent, and other ML models and algorithms can be facilitated by a variety of general-purpose architectures and circuits that can be implemented using VLSI. The training engines' stability, convergence, and speed can all be optimized with VLSI [126].

ML has a very wide and promising future in the integration of VLSI design because it can offer a variety of advantages and answers to the issues and challenges that the VLSI industry faces. The following are a few potential future applications of ML in VLSI design:

1. **Machine Learning for Advanced Lithography and Nanofabrication**: The lithography and fabrication processes grow more intricate and challenging as VLSI technology scales descend to the nanometer regime, necessitating a high degree of precision and accuracy. The lithography and fabrication processes, including mask design, optical proximity correction, etching, and deposition, can be modeled and optimized with the assistance of ML [127]. Defects and variations in nanoscale devices can also be detected and corrected with the guidance of ML.
2. **Machine Learning for Hardware Security and Reliability:** These are essential features of VLSI design, particularly for applications including aerospace, biometrics, and cryptography that deal with sensitive and crucial data and operations. ML has the potential to improve the security and reliability of VLSI circuits through the detection and prevention of hardware attacks, including counterfeiting, hardware Trojans, and side-channel attacks, as well as by enhancing the circuits' fault tolerance and error correction capabilities [128].

10.6 CONCLUSION

We have introduced the fundamental concepts and principles of VLSI and ML, as well as their interrelation, in this chapter. The application of ML techniques to diverse facets of VLSI design and manufacturing, including layout optimization, fault detection and diagnosis, performance prediction and optimization, and exploration of novel design possibilities, has been the subject of our discourse. Recent research projects and applications that illustrate the advantages and difficulties of integrating ML into VLSI have been examined. For instance, we have seen how ML can automate the time-consuming and error-prone process of layout optimization, which is traditionally performed manually, by generating optimized layouts that reduce power consumption and enhance signal integrity through learning from existing layouts. Furthermore, we have observed how ML can improve the reliability and robustness of VLSI circuits through the identification of anomalous patterns in vast amounts of circuit data, diagnosis of the faults, and recommendation of potential solutions. Furthermore, we have seen how ML can enhance the power efficiency and performance of VLSI circuits through the optimization of design parameters, including voltage, frequency, and temperature, and by analyzing and learning from historical data to accurately predict the performance of a circuit under various operating conditions. In addition, we have observed how ML can enable the investigation of novel design possibilities by generating inventive circuit architectures and exploring unconventional design spaces, including quantum computing and neuromorphic computing, using ML algorithms.

REFERENCES

1. Zamiri Azar, K., Mardani Kamali, H., Farahmandi, F., & Tehranipoor, M. (2023). Basics of VLSI Design. In Kimia Zamiri Azar, Hadi Mardani Kamali, Farimah Farahmandi, & Mark Tehranipoor (Eds.), *Understanding Logic Locking* (pp. 1–23). Springer International Publishing.
2. Thakur, G., Jain, S., & Sohal, H. (2022). Current issues and emerging techniques for VLSI testing: A review. *Measurement: Sensors, 24*, 100497.
3. Harlow, J. E. (2003, February). Toward design technology in 2020: Trends, issues, and challenges [VLSI design]. In *Proceedings of IEEE Computer Society Annual Symposium on VLSI, 2003.* Taipei, Taiwan. (pp. 3–4). IEEE.
4. Simpson, P. A., & Simpson, P. A. (2015). Timing Closure. In Philip Andrew Simpson (Ed.), *FPGA Design: Best Practices for Team-Based Reuse* (pp. 191–226). Springer.
5. Chauhan, P., & Gupta, S. (2023). Challenges and future perspectives of low-power VLSI circuits: A study. In *Modern Electronics Devices and Communication Systems: Select Proceedings of MEDCOM 2021* (pp. 561–569). Springer Nature Singapore.
6. Li, L., Cai, Y., & Zhou, Q. (2022). A survey on machine learning-based routing for VLSI physical design. *Integration, 86*, 51–56.
7. Ganai, M. K., & Gupta, A. (2007). Design verification challenges. In Malay Ganai, & Aarti Gupta (Eds.), *SAT-Based Scalable Formal Verification Solutions* (pp. 1–16). Springer.
8. Tripathi, A., Singh, A. K., Singh, K. K., Choudhary, P., & Vashist, P. C. (2021). Machine learning architecture and framework. In Krishna Kant Singh, Mohamed Elhoseny, Akansha Singh, & Ahmed A. Elngar (Eds.), *Machine Learning and the Internet of Medical Things in Healthcare* (pp. 1–22). Academic Press.
9. Shafique, M., Hafiz, R., Javed, M. U., Abbas, S., Sekanina, L., Vasicek, Z., & Mrazek, V. (2017, July). Adaptive and energy-efficient architectures for machine learning: Challenges, opportunities, and research roadmap. In *2017 IEEE Computer society annual symposium on VLSI (ISVLSI).* Lahore, Pakistan (pp. 627–632). IEEE.
10. Shreyanth, S., Harshitha, D. S., & Niveditha, S. (2023). Implementation of machine learning in VLSI integrated circuit design. *SN Computer Science, 4*(2), 137.
11. Bai, L., & Chen, L. (2018, October). Machine-learning-based early-stage timing prediction in SoC physical design. In *2018 14th IEEE International Conference on Solid-State and Integrated Circuit Technology (ICSICT).* Beijing, China (pp. 1–3). IEEE.
12. De Chen, Y., Huang, J., Zhao, D., Yim, D., & Ervin, J. (2019, March). Advanced 3D design technology co-optimization for manufacturability. In *2019 China Semiconductor Technology International Conference (CSTIC).* Shanghai, China (pp. 1–3). IEEE.
13. Amuru, D., Zahra, A., Vudumula, H. V., Cherupally, P. K., Gurram, S. R., Ahmad, A., & Abbas, Z. (2023). AI/ML algorithms and applications in VLSI design and technology. *Integration, 93*, 102048.
14. Varambally, B. S., & Sehgal, N. (2020). Optimising design verification using machine learning: An open source solution. *arXiv preprint arXiv:2012.02453.* https://doi.org/10.48550/arXiv.2012.02453.
15. Kirschbaum, M., & Plos, T. (2014). Hardware and VLSI Designs. In Konstantinos Markantonakis, & Keith Mayes (Eds.), *Secure Smart Embedded Devices, Platforms and Applications* (pp. 95–116). Springer.
16. Elfadel, I. M., Boning, D. S., & Li, X. (Eds.) (2019). *Machine Learning in VLSI Computer-Aided Design* (pp. 0278–0070). Springer.
17. Sarker, I. H. (2021). Machine learning: Algorithms, real-world applications and research directions. *SN Computer Science, 2*(3), 160.
18. Iniewski, K. (2008). *VLSI Circuits for Biomedical Applications.* Artech House.

19. Taraate, V. (2022). System and architecture design. In Vaibbhav Taraate (Ed.), *Digital Design from the VLSI Perspective: Concepts for VLSI Beginners* (pp. 27–40). Springer Nature Singapore.
20. Jerraya, A. A., Romdhani, M., Valderrama, C. A., Le Marrec, P., Hessel, F., Marchioro, G. F., & Daveau, J. M. (1997). Languages for system-level specification and design. In Jørgen Staunstrup, & Wayne Wolf (Eds.), *Hardware/Software Co-Design: Principles and Practice* (pp. 235–262). Springer US.
21. Chu, P. P. (2006). *RTL Hardware Design Using VHDL: Coding for Efficiency, Portability, and Scalability.* John Wiley & Sons.
22. Sozzo, E. D., Conficconi, D., Zeni, A., Salaris, M., Sciuto, D., & Santambrogio, M. D. (2022). Pushing the level of abstraction of digital system design: A survey on how to program fpgas. *ACM Computing Surveys*, *55*(5), 1–48.
23. Rodriguez-Andina, J. J., Moure, M. J., & Valdes, M. D. (2007). Features, design tools, and application domains of FPGAs. *IEEE Transactions on Industrial Electronics*, *54*(4), 1810–1823.
24. Chakravarthi, V. S. (2020). *A Practical Approach to VLSI System on Chip (SoC) Design.* Springer International Publishing.
25. Araújo, R. A. (2018). *Study and Implementation of Optimized Solutions for Re-Configurable Logic over ASIC Design Flow* (Doctoral dissertation, Universidade do Porto (Portugal)).
27. Gopakumaran, B. (1997). Gate-array IC design tools. *Computer, 30*(01), 117–122.
28. Lienig, J., Scheible, J., Lienig, J., & Scheible, J. (2020). Steps in physical design: From netlist generation to layout post processing. In Jens Lienig, & Juergen Scheible (Eds.), *Fundamentals of Layout Design for Electronic Circuits* (pp. 165–211). Springer.
29. Gielen, G. G., & Rutenbar, R. A. (2000). Computer-aided design of analog and mixed-signal integrated circuits. *Proceedings of the IEEE*, *88*(12), 1825–1854.
30. Cho, M., Mitra, J., & Pan, D. Z. (2007). *Manufacturability Aware Routing.* ECE Department, UT Austin.
31. Chakraverty, M., Prabhu, K. A., & Harisankar, P. S. (2017, May). Process design kits for RF analog & mixed signal design methodologies enabling silicon success. In *2017 International Conference on Trends in Electronics and Informatics (ICEI).* Tirunelveli, Tamil Nadu, India (pp. 669–676). IEEE.
32. Sen, P., Woods, W. H., Sarkar, S., Pratap, R. J., Dufrene, B. M., Mukhopadhyay, R., … & Laskar, J. (2006). Neural-network-based parasitic modeling and extraction verification for RF/millimeter-wave integrated circuit design. *IEEE Transactions on Microwave theory and Techniques*, *54*(6), 2604–2614.
33. Yang, S. M., Lee, T., & Cong, M. (2010). Design and verification of a thermoelectric energy harvester with stacked polysilicon thermocouples by CMOS process. *Sensors and Actuators A: Physical*, *157*(2), 258–266.
34. Coelho, J. R. D. S. (2013). *Aging monitoring methodology for built-In self-test applications* (Doctoral dissertation). Brazil: Universidade Estadual de Campinas (UNICAMP).
35. Kochte, M. A., & Wunderlich, H. J. (2017). Self-test and diagnosis for self-aware systems. *IEEE Design & Test*, *35*(5), 7–18.
36. Opiyo, E. Z., Horváth, I., & Vergeest, J. S. (2002). Quality assurance of design support software: Review and analysis of the state of the art. *Computers in Industry*, *49*(2), 195–215.
37. Darwish, T., & Bayoumi, M. (2005). Trends in low-power VLSI design. In Wai-Kai Chen (Ed.), *The Electrical Engineering Handbook* (pp. 263–280). Academic Press.
38. Sharma, A., & Sohal, H. (2017). Considerations for ultra-low-power VLSI design: A survey. *Journal of Nanoelectronics and Optoelectronics*, *12*(1), 1–21.
39. Pedram, M., Bhat, N., & Kuh, E. S. (1997). Combining technology mapping with layout. *VLSI Design*, *5*(2), 111–124.

40. Vaidya, S., & Dandekar, D. (2010). Delay-power performance comparison of multipliers in VLSI circuit design. *International Journal of Computer Networks & Communications (IJCNC)*, *2*(4), 47–56.
41. Darwish, T., & Bayoumi, M. (2005). Trends in low-power VLSI design. In Wai-Kai Chen (Ed.), *The Electrical Engineering Handbook* (pp. 263–280). Academic Press, Copyright © 2005 Elsevier.
42. Sait, S. M., & Youssef, H. (1999). *VLSI Physical Design Automation: Theory and Practice* (Vol. 6). World Scientific.
43. Wang, L. T., Stroud, C. E., & Touba, N. A. (2010). *System-on-Chip Test Architectures: Nanometer Design for Testability*. Morgan Kaufmann.
44. Wang, L. T., Chang, Y. W., & Cheng, K. T. T. (Eds.). (2009). *Electronic Design Automation: Synthesis, Verification, and Test*. Morgan Kaufmann.
45. Dong, X., Zhao, J., & Xie, Y. (2010). Fabrication cost analysis and cost-aware design space exploration for 3-D ICs. *IEEE Transactions on Computer-Aided Design of Integrated Circuits and Systems*, *29*(12), 1959–1972.
46. Samson, J. R. (1993, January). Optimizing real-time fault tolerance design in WSI. In *1993 Proceedings Fifth Annual IEEE International Conference on Wafer Scale Integration*. San Diego, California, USA (pp. 147–162). IEEE.
47. Mohammad, A., Das, R., Islam, M. A., & Mahjabeen, F. (2023). AI in VLSI design advances and challenges: Living in the complex nature of integrated devices. *Asian Journal of Mechatronics and Electrical Engineering*, *2*(2), 121–132.
48. Mahesh, B. (2020). Machine learning algorithms: A review. *International Journal of Science and Research* (*IJSR*), *9*(1), 381–386.
49. Rapp, M., Amrouch, H., Lin, Y., Yu, B., Pan, D. Z., Wolf, M., & Henkel, J. (2021). MLCAD: A survey of research in machine learning for CAD keynote paper. *IEEE Transactions on Computer-Aided Design of Integrated Circuits and Systems*, *41*(10), 3162–3181.
50. Usama, M., Qadir, J., Raza, A., Arif, H., Yau, K. L. A., Elkhatib, Y., … & Al-Fuqaha, A. (2019). Unsupervised machine learning for networking: Techniques, applications and research challenges. *IEEE Access*, *7*, 65579–65615.
51. Rapp, M., Amrouch, H., Lin, Y., Yu, B., Pan, D. Z., Wolf, M., & Henkel, J. (2021). MLCAD: A survey of research in machine learning for CAD keynote paper. *IEEE Transactions on Computer-Aided Design of Integrated Circuits and Systems*, *41*(10), 3162–3181.
52. Amini, M. R., & Gallinari, P. (2002, August). The use of unlabeled data to improve supervised learning for text summarization. In *Proceedings of the 25th Annual International ACM SIGIR Conference on Research and Development in Information Retrieval*. Tampere, Finland (pp. 105–112).
53. Duarte, J. M., & Berton, L. (2023). A review of semi-supervised learning for text classification. *Artificial Intelligence Review*, *56*, 9401–9469. https://doi.org/10.1007/s10462-023-10393-8.
54. Settaluri, K., Haj-Ali, A., Huang, Q., Hakhamaneshi, K., & Nikolic, B. (2020, March). Autockt: Deep reinforcement learning of analog circuit designs. In *2020 Design, Automation & Test in Europe Conference & Exhibition (DATE)*. Grenoble, France (pp. 490–495). IEEE.
55. Qian, Y., Zhou, X., Zhou, H., & Wang, L. (2024). An efficient reinforcement learning based framework for exploring logic synthesis. *ACM Transactions on Design Automation of Electronic Systems*, *29*(2), 1–33.
56. Yin, S., Dai, G., & Xing, W. W. (2023, January). High-dimensional yield estimation using shrinkage deep features and maximization of integral entropy reduction. In *Proceedings of the 28th Asia and South Pacific Design Automation Conference*. Singapore (pp. 283–289).

57. Jović, A., Brkić, K., & Bogunović, N. (2015, May). A review of feature selection methods with applications. In *2015 38th International Convention on Information and Communication Technology, Electronics and Microelectronics (MIPRO)*. Opatija, Croatia (pp. 1200–1205). IEEE.
58. Basri, S., Imam, A. A., Alazzawi, A. K., Adeyemo, V. E., & Kumar, G. (2023). Empirical analysis of threshold values for rank-based filter feature selection methods in software defect prediction. *Journal of Engineering Science and Technology, 18*(1), 187–209.
59. Engel, D., Hüttenberger, L., & Hamann, B. (2012). A survey of dimension reduction methods for high-dimensional data analysis and visualization. In *Visualization of Large and Unstructured Data Sets: Applications in Geospatial Planning, Modeling and Engineering-Proceedings of IRTG 1131 Workshop 2011*. Schloss Dagstuhl-Leibniz-Zentrum fuer Informatik.
60. Xu, X., Liang, T., Zhu, J., Zheng, D., & Sun, T. (2019). Review of classical dimensionality reduction and sample selection methods for large-scale data processing. *Neurocomputing, 328*, 5–15.
61. Sumithra, V., & Surendran, S. (2015). A review of various linear and non linear dimensionality reduction techniques. *International Journal of Computer Science and Information Technologies, 6*(3), 2354–2360.
62. Park, S. J., Bae, B., Kim, J., & Swaminathan, M. (2017). Application of machine learning for optimization of 3-D integrated circuits and systems. *IEEE Transactions on Very Large Scale Integration (VLSI) Systems, 25*(6), 1856–1865.
63. Sun, S., Cao, Z., Zhu, H., & Zhao, J. (2019). A survey of optimization methods from a machine learning perspective. *IEEE Transactions on Cybernetics, 50*(8), 3668–3681.
64. Rosa, J. P., Guerra, D. J., Horta, N. C., Martins, R. M., Lourenço, N. C., Rosa, J. P., … & Lourenço, N. C. (2020). Overview of artificial neural networks. In João P. S. Rosa, Daniel J. D. Guerra, Nuno C. G. Horta, Ricardo M. F. Martins, & Nuno C. C. Lourenço (Eds.), *Using Artificial Neural Networks for Analog Integrated Circuit Design Automation* (pp. 21–44). Springer.
66. Ram, O. S., & Saurabh, S. (2020). Modeling multiple-input switching in timing analysis using machine learning. *IEEE Transactions on Computer-Aided Design of Integrated Circuits and Systems, 40*(4), 723–734.
67. Ye, F., Zhang, Z., Chakrabarty, K., & Gu, X. (2013). Board-level functional fault diagnosis using artificial neural networks, support-vector machines, and weighted-majority voting. *IEEE Transactions on Computer-Aided Design of Integrated Circuits and Systems, 32*(5), 723–736.
68. Kecman, V. (2005). Support vector machines: A introduction. In *Support Vector Machines: Theory and Applications* (pp. 1–47). Springer.
69. Vijayan, A., Chakrabarty, K., & Tahoori, M. B. (2019). Machine learning-based aging analysis. In Ibrahim (Abe) M. Elfadel, Duane S. Boning, & Xin Li (Eds.), *Machine Learning in VLSI Computer-Aided Design* (pp. 265–289). Springer.
70. Saini, S., Lata, K., & Sinha, G. R. (Eds.). (2021). *VLSI and Hardware Implementations Using Modern Machine Learning Methods*. CRC Press.
71. Yadav, S. L., & Sohal, A. (2017). Comparative study of different selection techniques in genetic algorithm. *International Journal of Engineering, Science and Mathematics, 6*(3), 174–180.
72. Bao, Z., & Watanabe, T. (2010). Circuit design optimization using genetic algorithm with parameterized uniform crossover. *IEICE Transactions on Fundamentals of Electronics, Communications and Computer Sciences, 93*(1), 281–290.
73. Pagani, S., Manoj, P. S., Jantsch, A., & Henkel, J. (2018). Machine learning for power, energy, and thermal management on multicore processors: A survey. *IEEE Transactions on Computer-Aided Design of Integrated Circuits and Systems, 39*(1), 101–116.

74. Ding, Z., Huang, Y., Yuan, H., & Dong, H. (2020). Introduction to reinforcement learning. In Hao Dong, Zihan Ding, & Shanghang Zhang (Eds.), *Deep Reinforcement Learning: Fundamentals, Research and Applications* (pp. 47–123). Springer.
75. Kim, R. G., Choi, W., Chen, Z., Doppa, J. R., Pande, P. P., Marculescu, D., & Marculescu, R. (2017). Imitation learning for dynamic VFI control in large-scale manycore systems. *IEEE Transactions on Very Large Scale Integration (VLSI) Systems, 25*(9), 2458–2471.
76. Li, Y. (2017). Deep reinforcement learning: An overview. *arXiv preprint arXiv:1701.07274.*
77. Ardakani, A., Leduc-Primeau, F., Onizawa, N., Hanyu, T., & Gross, W. J. (2017). VLSI implementation of deep neural network using integral stochastic computing. *IEEE Transactions on Very Large Scale Integration (VLSI) Systems, 25*(10), 2688–2699.
78. Priyanka, & Kumar, D. (2020). Decision tree classifier: A detailed survey. *International Journal of Information and Decision Sciences, 12*(3), 246–269.
79. Blanton, R. D., Fang, C., Huang, Q., & Mittal, S. (2022). Machine learning in logic circuit diagnosis. In Patrick Girard, Shawn Blanton, & Li-C. Wang (Eds.), *Machine Learning Support for Fault Diagnosis of System-on-Chip* (pp. 135–171). Springer International Publishing.
80. Soto, V., García-Moratilla, S., Martínez-Muñoz, G., Hernández-Lobato, D., & Suárez, A. (2014). A double pruning scheme for boosting ensembles. *IEEE Transactions on Cybernetics, 44*(12), 2682–2695.
81. Sharma, T., Kolluru, S., & Stevens, K. S. (2021). Learning based timing closure on relative timed design. In *VLSI-SoC: Design Trends: 28th IFIP WG 10.5/IEEE International Conference on Very Large Scale Integration, VLSI-SoC 2020*, Salt Lake City, UT, USA, October 6–9, 2020, Revised and Extended Selected Papers 28 (pp. 133–148). Springer International Publishing.
82. Murray, K. E., & Betz, V. (2019, September). Adaptive FPGA placement optimization via reinforcement learning. In *2019 ACM/IEEE 1st Workshop on Machine Learning for CAD (MLCAD).* San Francisco, California, USA (pp. 1–6). IEEE.
83. Bhasker, J., & Chadha, R. (2009). *Static Timing Analysis for Nanometer Designs: A Practical Approach.* Springer Science & Business Media.
84. Blaauw, D., Chopra, K., Srivastava, A., & Scheffer, L. (2008). Statistical timing analysis: From basic principles to state of the art. *IEEE Transactions on Computer-Aided Design of Integrated Circuits and Systems, 27*(4), 589–607.
85. Dartu, F., Menezes, N., Qian, J., & Pillage, L. T. (1994, June). A gate-delay model for high-speed CMOS circuits. In *Proceedings of the 31st annual Design Automation Conference.* San Diego, California, USA (pp. 576–580).
86. Nath, S. (2016). *New Applications of Learning-Based Modeling in Nanoscale Integrated-Circuit Design.* University of California.
87. Huang, G., Hu, J., He, Y., Liu, J., Ma, M., Shen, Z., … & Wang, Y. (2021). Machine learning for electronic design automation: A survey. *ACM Transactions on Design Automation of Electronic Systems (TODAES), 26*(5), 1–46.
88. Zhou, L., Pan, S., Wang, J., & Vasilakos, A. V. (2017). Machine learning on big data: Opportunities and challenges. *Neurocomputing, 237*, 350–361.
89. Wu, N., & Xie, Y. (2022). A survey of machine learning for computer architecture and systems. *ACM Computing Surveys (CSUR), 55*(3), 1–39.
90. Bansal, M. A., Sharma, D. R., & Kathuria, D. M. (2022). A systematic review on data scarcity problem in deep learning: Solution and applications. *ACM Computing Surveys (CSUR), 54*(10s), 1–29.
91. Gudivada, V., Apon, A., & Ding, J. (2017). Data quality considerations for big data and machine learning: Going beyond data cleaning and transformations. *International Journal on Advances in Software, 10*(1), 1–20.
92. Barboza, E. M. C. (2021). *Machine Learning Techniques for Performance Prediction and Diagnosis of VLSI Designs* (Doctoral dissertation, Texas A&M University).

93. Lwakatare, L. E., Raj, A., Crnkovic, I., Bosch, J., & Olsson, H. H. (2020). Large-scale machine learning systems in real-world industrial settings: A review of challenges and solutions. *Information and Software Technology, 127,* 106368.
94. Srivastava, P., Kumar, P., & Abbas, Z. (2023). Qualitative data augmentation for performance prediction in VLSI circuits. *arXiv preprint arXiv:2302.07566.*
95. Dhawas, P., Dhore, A., Bhagat, D., Pawar, R. D., Kukade, A., & Kalbande, K. (2024). Big data preprocessing, techniques, integration, transformation, normalisation, cleaning, discretization, and binning. In Dina Darwish (Ed.), *Big Data Analytics Techniques for Market Intelligence* (pp. 159–182). IGI Global.
96. Chiueh, T. C., & Katz, R. (1991, January). An annotation system for VLSI design. In *1991 Tenth Annual International Phoenix Conference on Computers and Communications,* Phoenix, Arizona, USA (pp. 755–756). IEEE Computer Society.
97. Lu, Y., Shen, M., Wang, H., Wang, X., van Rechem, C., & Wei, W. (2023). Machine learning for synthetic data generation: A review. *arXiv preprint arXiv:2302.04062.*
98. Tsang, W. K., & Benoit, D. F. (2022). Interpretability and explainability in machine learning. In Yukio Ohsawa (Ed.), *Living Beyond Data: Toward Sustainable Value Creation* (pp. 89–100). Springer International Publishing.
99. Amrouch, H., Chowdhury, A. B., Jin, W., Karri, R., Khorrami, F., Krishnamurthy, P., … & Tan, S. X. D. (2021, April). Special session: Machine learning for semiconductor test and reliability. In *2021 IEEE 39th VLSI Test Symposium (VTS).* San Diego, CA, USA (pp. 1–11). IEEE.
100. He, H., & Garcia, E. A. (2009). Learning from imbalanced data. *IEEE Transactions on Knowledge and Data Engineering, 21*(9), 1263–1284.
101. Chowdhury, A. B., Tan, B., Garg, S., & Karri, R. (2021). Robust deep learning for IC test problems. *IEEE Transactions on Computer-Aided Design of Integrated Circuits and Systems, 41*(1), 183–195.
102. Burrell, J. (2016). How the machine 'thinks': Understanding opacity in machine learning algorithms. *Big Data & Society, 3*(1), 2053951715622512.
103. Sze, V., Chen, Y. H., Emer, J., Suleiman, A., & Zhang, Z. (2017, April). Hardware for machine learning: Challenges and opportunities. In *2017 IEEE Custom Integrated Circuits Conference (CICC),* Austin, TX, USA (pp. 1–8). IEEE.
104. Al-Jarrah, O. Y., Yoo, P. D., Muhaidat, S., Karagiannidis, G. K., & Taha, K. (2015). Efficient machine learning for big data: A review. *Big Data Research, 2*(3), 87–93.
105. Lu, Q., Jiang, W., Xu, X., Shi, Y., & Hu, J. (2019). On neural architecture search for resource-constrained hardware platforms. *arXiv preprint arXiv:1911.00105.*
106. Shuvo, M. M. H., Islam, S. K., Cheng, J., & Morshed, B. I. (2022). Efficient acceleration of deep learning inference on resource-constrained edge devices: A review. *Proceedings of the IEEE, 111*(1), 42–91.
107. Che, S., Li, J., Sheaffer, J. W., Skadron, K., & Lach, J. (2008, June). Accelerating compute-intensive applications with GPUs and FPGAs. In *2008 Symposium on Application Specific Processors*, San Jose, CA, USA (pp. 101–107). IEEE.
108. Neuman, S. M., Plancher, B., Duisterhof, B. P., Krishnan, S., Banbury, C., Mazumder, M., … & Reddi, V. J. (2022, June). Tiny robot learning: Challenges and directions for machine learning in resource-constrained robots. In *2022 IEEE 4th International Conference on Artificial Intelligence Circuits and Systems (AICAS).* New York, NY, USA (pp. 296–299). IEEE.
109. Zervakis, G., Tahoori, M. B., & Henkel, J. (2023). Hardware–software co-design for ultra-resource-constrained embedded machine learning inference: A printed electronics use case. In *Embedded Machine Learning for Cyber-Physical, IoT, and Edge Computing: Software Optimizations and Hardware/Software Codesign* (pp. 201–224). Springer Nature Switzerland.
110. Polino, A., Pascanu, R., & Alistarh, D. (2018). Model compression via distillation and quantization. *arXiv preprint arXiv:1802.05668.*

111. Vadera, S., & Ameen, S. (2022). Methods for pruning deep neural networks. *IEEE Access, 10*, 63280–63300.
112. Guo, Y. (2018). A survey on methods and theories of quantized neural networks. *arXiv preprint arXiv:1808.04752*.
113. Sekanina, L. (2021). Neural architecture search and hardware accelerator co-search: A survey. *IEEE Access, 9*, 151337–151362.
114. Liu, S., Zhang, H., & Jin, Y. (2022). A survey on computationally efficient neural architecture search. *Journal of Automation and Intelligence, 1*(1), 100002.
115. Rezk, N. M., Nordström, T., Stathis, D., Ul-Abdin, Z., Aksoy, E. E., & Hemani, A. (2022). MOHAQ: Multi-objective hardware-aware quantization of recurrent neural networks. *Journal of Systems Architecture, 133*, 102778.
116. Li, Y., & Ang, K. W. (2021). Hardware implementation of neuromorphic computing using large-scale memristor crossbar arrays. *Advanced Intelligent Systems, 3*(1), 2000137.
117. Grollier, J., Querlioz, D., Camsari, K. Y., Everschor-Sitte, K., Fukami, S., & Stiles, M. D. (2020). Neuromorphic spintronics. *Nature Electronics, 3*(7), 360–370.
118. Ielmini, D., & Wong, H. S. P. (2018). In-memory computing with resistive switching devices. *Nature Electronics, 1*(6), 333–343.
119. Kahng, A. B. (2018, March). Machine learning applications in physical design: Recent results and directions. In *Proceedings of the 2018 International Symposium on Physical Design*. San Francisco, CA, USA (pp. 68–73).
120. Devi, S., Tilwankar, G., & Zele, R. (2021, September). Automated design of analog circuits using machine learning techniques. In *2021 25th International Symposium on VLSI Design and Test (VDAT)*. Vijayawada, India (pp. 1–6). IEEE.
121. Zheng, X., Wang, H., Jiang, K., & He, B. (2019). A novel machine learning-based short-circuit current prediction method for active distribution networks. *Energies, 12*(19), 3793.
122. Molanes, R. F., Amarasinghe, K., Rodriguez-Andina, J., & Manic, M. (2018). Deep learning and reconfigurable platforms in the internet of things: Challenges and opportunities in algorithms and hardware. *IEEE Industrial Electronics Magazine, 12*(2), 36–49.
123. Mohammad, A., Das, R., Islam, M. A., & Mahjabeen, F. (2023). AI in VLSI design advances and challenges: Living in the complex nature of integrated devices. *Asian Journal of Mechatronics and Electrical Engineering*, 2(2), 121–132.
124. Abreu, B. A., Grellert, M., & Bampi, S. (2020, October). VLSI design of tree-based inference for low-power learning applications. In *2020 IEEE International Symposium on Circuits and Systems (ISCAS)*. Seoul, South Korea (pp. 1–5). IEEE.
125. Bavikadi, S., Sutradhar, P. R., Khasawneh, K. N., Ganguly, A., & Pudukotai Dinakarrao, S. M. (2020, September). A review of in-memory computing architectures for machine learning applications. In *Proceedings of the 2020 on Great Lakes Symposium on VLSI*. Chicago, Illinois, USA (pp. 89–94).
126. Lu, J., Ni, C., & Wang, Z. (2022). ETA: An efficient training accelerator for DNNs based on hardware-algorithm co-optimization. *IEEE Transactions on Neural Networks and Learning Systems, 33*(5), 2158–2169.
127. Shiely, J. P. (2019). Machine learning for compact lithographic process models. In Ibrahim (Abe) M. Elfadel, Duane S. Boning, & Xin Li (Eds.), *Machine Learning in VLSI Computer-Aided Design* (pp. 19–68). Springer International Publishing.
128. Pan, Z., & Mishra, P. (2022). A survey on hardware vulnerability analysis using machine learning. *IEEE Access, 10*, 49508–49527.

11 Application of Artificial Intelligence/Machine Learning in VLSI Design

Abhay Pratap Singh, Vimal Kumar Mishra, and Shamim Akhter

11.1 INTRODUCTION

Historically, computers were operated by large vacuum tubes. At the time, they had the title of being the most rapid, but the advent of semiconductor transistors in 1948 by Bell Laboratories ushered in a new era for the entire computing system. This allows the devices to be smaller, more affordable, and more efficient. In the 1960s, we employed a method called small-scale integration (SSI) that utilized approximately ten transistors, using a low-density integration approach. During the late 1960s, we achieved the integration of approximately 100 transistors on a single chip, a technological advancement referred to as medium-scale integration (MSI). In the early 1970s, this integration increased with the help of transistor-transistor logic (TTL), which enables space for higher density integration. In the late 1980s, the integration of transistors crossed the mark of 1000 transistors on a single chip and was called VLSI. Due to power consumption and the need for a higher number of gates on a single chip, Metal-Oxide-Semiconductor (MOS) gradually replaced TTL. These integration evaluations follow Moore's law, which predicts that the number of transistors on a single chip will increase exponentially and double in nearly two years. For the last five decades, this industry has followed suit [1,2]. Reducing the transistor size achieves this phenomenon. As the transistor size decreases, the transistor density on the chip increases and the transistor's performance improves [3]. This process will help the semiconductor industry grow as it faces the need for portable devices. Scaling VLSI circuits can achieve the small size and efficient power consumption necessary for portability. This continuous scaling growth is bringing our semiconductor industries to a new high. Devices are currently reducing in size to sub-3 nm gate size regimes and even smaller. Currently, Complementary Metal-Oxide-Semiconductor (CMOS) technology is the dominant and revolutionary technology in the field of microelectronics. As CMOS technology downscaling increases, it enables short-channel effects (SCEs), which include drain-induced barrier lowering (DIBL), leakage, a decrease in gain, channel length modulation, and threshold voltage roll-off [4,5]. As a result, as miniaturization increases, the dependence of device characteristics on various parameters increases, and the sensitivity of process variation in manufacturing also increases, ultimately influencing chip yield [6].

 DOI: 10.1201/9781003483038-11

Therefore, we must use advanced and affordable design techniques in our VLSI design processes that are sustainable for future growth in the semiconductor industry. The VLSI Computer-Aided Design (CAD) tool offers an essential role in multiple stages of the chip design process. Modern chip design and performance evaluation are dependent on the capabilities of these CAD tools. The semiconductor chips' turnout time heavily varies on the effectiveness of the Electronic Design Automation (EDA) tools to overcome the aforementioned limitations. The conventional EDA works on rule-driven methodologies, and when the complexity of the IC increases, it results in process variation and environmental variables. The result is a time delay [7]. As the count of transistors on a chip increases, advancements in CAD tools are required. Many studies show that we can use an artificial network-based solution to our EDA process to achieve quick convergence, which provides lesser time consumption and high efficiency [8]. To provide a solution to these constraints, many researchers are looking toward artificial intelligence (AI). AI mirrors human intelligence and can perform complex tasks. AI includes machine learning (ML), which can readily identify trends and patterns in massive datasets and forecast the output of the desired input. This field has witnessed significant expansion in recent times, mostly due to its resilience and extensive range of applications. Recently, ML has found extensive applications in various disciplines such as mathematics, physics, chemistry, computer vision, medicine, and image processing [9–12]. These ML methods operate at high computational rates and are capable of handling heterogeneous and multivariant data. Because of these advantages, researchers are incorporating AI and ML into various design processes within the VLSI design flow. This chapter summarizes the various research studies conducted in VLSI design at various levels.

11.2 VLSI DESIGN FLOW

The typical integrated circuit (IC) design process consists of multiple hierarchical layers, as shown in Figure 11.1. This encompasses the back- and front-end stages of IC design. The design flow delineates the functions, interface, and overarching structure of the IC being examined. Design specifications provide a detailed description of a digital circuit's functionality, interface, and architecture. This includes information like block diagrams, propagation lags, packaging type, and design limitations. Design specifications also serve as a formal contract between the customer and the manufacturer. The architectural planning layer encompasses the fundamental structure of a system, including the functionality descriptions of its subsystem elements. The behavioral design level, also known as the functional design level, is typically described using Verilog HDL/VHDL. The functionality is described in a general way, while the specific details of how it is implemented are hidden. A high-level synthesis (HLS) tool can convert system specifications written in C or C++ into high-level hardware description language (HDL) or create a gate-level interpretation based on the high-level behavioral interpretation. Test bench simulation is used to verify the logic, while design for testability (DFT) is used for formal verification and scan insertion. These methods are employed to investigate the Register Transfer Level (RTL) mapping [13]. The process of system partitioning is used to break down large and complex systems into smaller modules. This is followed by floor layout, placement,

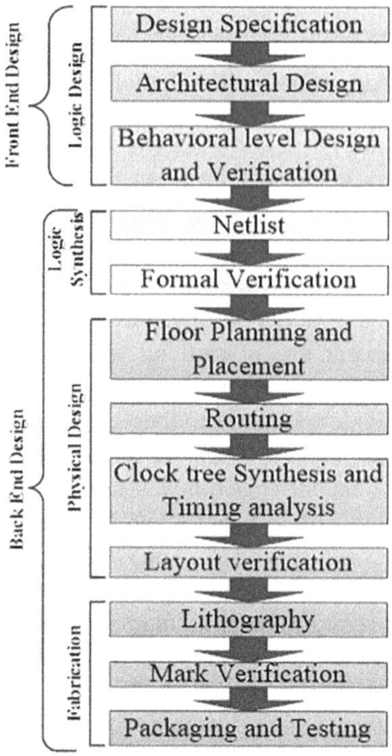

FIGURE 11.1 Modern chip design flow.

and routing. The floor planner is responsible for estimating the necessary IC space for implementing standard cell/module design and optimizing design performance. The place-and-route tools are accountable for the positioning of sub-modules, gates, and flip-flops. Verifying the layout is essential to ensure that the design aligns with the source schematic and adheres to the electrical and physical design principles.

Once the post-layout simulation is finished, the chip undergoes the sign-off stage, where parasitic resistance and capacitance are extracted and verified [14]. The fabrication of ICs includes a multitude of intricate chemical and physical processes, including lithography, which encompasses the creation, verification, and delineation of masks and materials in specific locations. Repeating this step multiple times is crucial due to alterations in the process and the downsizing of technology nodes. Once the chips have been manufactured, we proceed to divide the wafers into smaller sections, separate the individual chips, package each chip, and conduct thorough testing to ensure that they function properly and meet the specified design criteria [15,16].

11.3 ARTIFICIAL INTELLIGENCE/MACHINE LEARNING

Statistical learning is an important tool in many scientific and technological disciplines because it enables the study of huge volumes of data to make future predictions. AI, a technology that simulates human behavior, has a subset of ML and

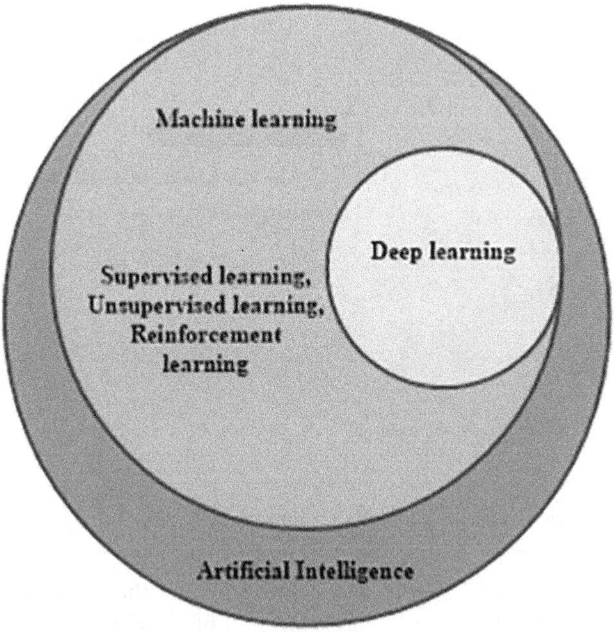

FIGURE 11.2 Overview of artificial intelligence techniques.

deep learning, as shown in Figure 11.2. We may categorize ML under three types: supervised learning, unsupervised learning, and reinforcement learning. Supervised learning assigns output labels to all elements in the given data, while unsupervised learning simply uses input variables, and semi-supervised learning uses data in which some are labeled examples and the remaining are unlabeled [17].

11.3.1 Supervised Learning

Supervised learning is an ML technique that involves algorithm learning patterns using labeled information. It entails training a model on a set of data in which every input is linked to a corresponding output. Throughout the training process, the model optimizes its parameters to reduce the discrepancy between its predictions and the actual labels. This allows the model to generalize well and make precise predictions on fresh, unseen data. Supervised learning can be categorized into two primary forms: classification and regression. Classification is a supervised learning task where algorithms assign input data to predefined categories or classes. Models learn from labeled examples to provide predictions on previously unobserved data. These models, known as classifiers, provide predictions about discrete categorical class labels. Regression is a supervised learning task that predicts continuous numerical values by analyzing the relationship between a dependent variable and one or more independent variables. We use this fundamental statistical technique to forecast and understand data trends [18,19].

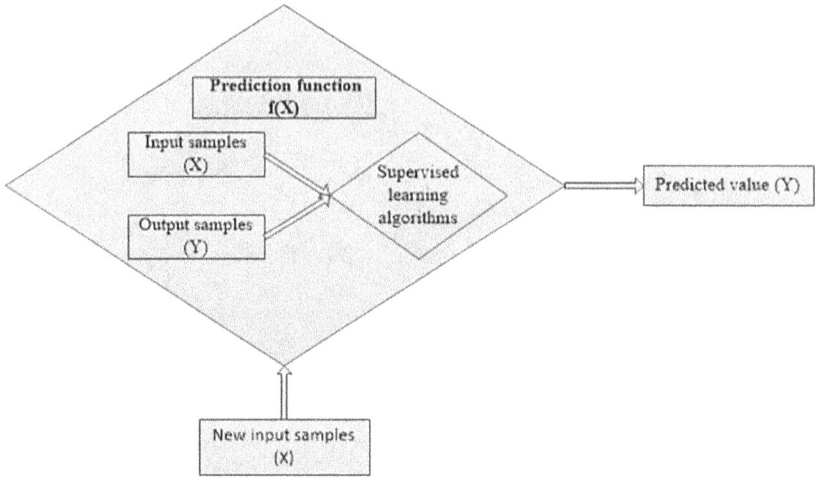

FIGURE 11.3 Learning function of classification/regression algorithms.

These processes are obtainable as a learning function that aims to anticipate a cartography of $Y = f(X)$, where Y represents a set of variables that are outputs for a given set of input variables X. We calculate the cartography function to forecast the corresponding class label Y for a given set of new input variables X, as shown in Figure 11.3.

Some of the most common methods for regression and classification are decision trees (DTs), gradient boosting machines (GBMs), random forests (RFs), support vector machines (SVMs), and ensemble learning [20].

11.3.2 Unsupervised Learning

Unsupervised learning is a type of ML where the model is trained using input data that does not have labeled responses. The algorithm autonomously analyzes the data to reveal patterns or structures without any explicit instructions on what to search for. Instead, its objective is to identify intrinsic links or groupings within the data, typically using methods such as clustering, dimensionality reduction, or density estimation. Unsupervised learning is especially beneficial for problems with limited or nonexistent labeled data, as it enables the discovery of the underlying structure of the dataset [21]. Some common methods used in unsupervised learning are clustering, association rule learning, dimensionality reduction, and anomaly detection. Clustering is the process of grouping together data points that are alike, while dimensionality reduction techniques reduce feature complexity. Association rule learning identifies relationships between variables, while anomaly detection flags outliers. Density estimation estimates the distribution of data, and generative modeling learns to generate new samples. Common algorithms include K-nearest neighbors, Apriori algorithm, Isolation Forest, kernel density estimation, K-means clustering, and principal component analysis [22].

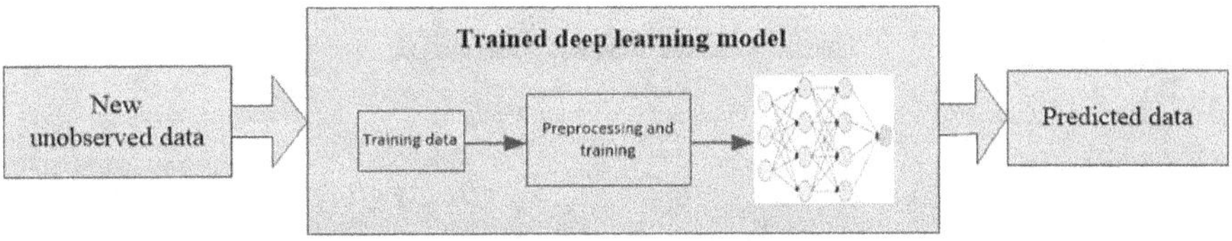

FIGURE 11.4 Overview of deep learning model flow.

11.3.3 Reinforcement Learning

Reinforcement learning is a subfield of ML that focuses on finding the best actions to take in different scenarios to achieve the maximum numerical reward. Its primary emphasis is on learning that is oriented toward achieving certain goals through interactions. Reinforcement learning is the method of acquiring knowledge via experience and identifying the most optimal approach that optimizes a reward signal [23].

11.3.4 Deep Learning

Deep learning is a kind of ML that employs artificial neural networks having numerous layers to uncover intricate patterns across huge datasets. These neural networks, which are designed based on the anatomy and functioning of the brain of a person, have the capacity to automatically learn representations of data [24]. Multilayer perceptrons (MLPs) are very important parts of systems for deep learning. There are many levels of nodes (neurons) in an MLP. There is an input layer, one or multiple hidden layers, and an output layer. As shown in Figure 11.4, we refer to it as a densely connected network because each neuron in one layer connects to every neuron in the layer below it. Various deep learning models are shallow neural networks, deep reinforcement learning, generative adversarial networks, recurrent neural networks, and deep neural networks.

11.4 APPLICATION OF AI/ML AT CIRCUIT SIMULATION LEVEL

Simulation is critical in IC device modeling because performance evaluation is difficult in the nano-scale zone due to process and environmental variables [25,26]. To get the appropriate characteristics, the device needs to be carefully examined. Early detection of functional and electrical performance changes can increase IC yield. AI and ML algorithms can assist, which improves turnaround time and performance, resulting in less design effort. Many researchers presented substitute methods for defining device characterization parameters [27]. Researchers proposed numerous learning algorithms that characterize devices at numerous layers of abstraction, including device and gate levels. Many researchers worked on different stages of simulation, including device, gate, circuit, RTL, and post-layout simulation levels. Many researchers are developing ML/AI models to predict the performance of devices using small sets of simulation results. The ML/AI algorithms help to analyze

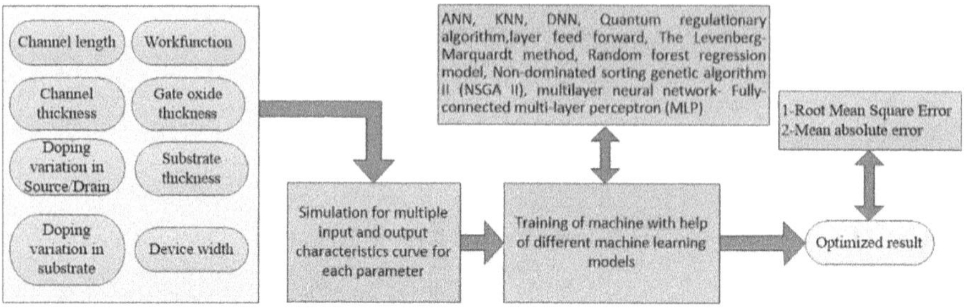

FIGURE 11.5 ML/AI-based device simulation flow process.

and optimize the characteristics of the device, computing the parametric yield with extremely high speed with an accuracy greater than 98%. Figure 11.5 shows the basic building blocks of the application of ML and AI at the device-level simulation process.

In 2019, Carrillo et al. developed an NN model for TCAD device modeling, utilizing 1380 training data points from junction-less nanowire transistors. The model demonstrated that ML techniques could reduce a simulation worktime of 5520 to 2 minutes [28]. In 2021, Yoon and team used TCAD and ML to maximize the efficiency and effectiveness of vertical nanowire field-effect transistors. The completely linked multilayer perceptron neural network accurately predicted input parameters, minimizing RC delay [29]. In 2022, Akbar and team used an ML model to find an efficient solution for sophisticated device simulations of line TFETs. The model had an R^2-score greater than 99% and below 1% of error rate, reducing computing expenses by nearly 99% [30]. Stillmaker et al. [31] offered a dataset and a method for modeling the performance of transistors at technology nodes spanning from 180 nanometers to 7 nanometers. In 2019, Amuru and colleagues [32] presented a regression model for estimating static leakage power and total power dissipation in a sub-nanodevice. They suggested multivariate linear regression (MLR) and multivariate polynomial regression (MPR), and their models had less than 1% inaccuracy. Khang et al. (2015) demonstrated an ML-based methodology that can predict incremental delay induced by SI and path delay, and this methodology got the worst-case error for these two as 5.2 and 8.2 ps each [33]. The paper [34] proposed a precise power estimation method for CMOS VLSI circuits using random forest, which outperforms neural networks. The results demonstrated a strong correlation with the ISCAS'89 benchmark circuits. The paper [35] introduced a ResNet-based digital circuit optimization framework that is both quick and efficient. The framework focuses on minimizing leakage and delay. The findings from the study of 22 nano-meter Metal Gate High-K digital cells demonstrated a significant decrease of 36.7% in delay and 18.8% in leakage when employing a genetic algorithm.

11.5 APPLICATION OF AI/ML AT ARCHITECTURAL LEVEL

The modern era inspires most architectural improvements with the need for high processing speed, a dependable system, and reduced development costs and time. AI and ML significantly enhance architectural level VLSI design by enabling

HLS and swift exploration of architectural options. AI models help designers quickly assess a wide range of design options by predicting parameters related to performance, power usage, and area. ML techniques enhance the quality of outputs regarding performance, area, and power efficiency by optimizing the translation of high-level code (such as C/C++) into HDLs. Architects may create more effective and efficient VLSI designs by using AI and ML to make better-informed judgments, cut down on design iterations, and speed up the entire process. The inference phase of neural networks was sped up by Neural Network Processing-In-Memory (NNPIM), which was suggested by Gupta and his team [36] in 2019. In order to make the inference phase of neural networks work more quickly, they used crossbar memory-based architecture and weight sharing to lower the amount of computation that neural networks need to do. They achieved an energy efficiency that is 131.5 times higher and a speedup that is 48.2 times faster than conventional neural networks. Yang et al. [37] examined the energy usage of data transfer in the von Neumann architecture using a computing-in-memory architecture based on Static Random-Access Memory (SRAM). The architecture is transformed from being focused on computing to being focused on data, resulting in an energy-efficient design that fully eliminates the computing bottleneck of the von Neumann architecture. They addressed the issue of read/write stability in SRAM. The study demonstrated that a computing-in-memory architecture consisting of 96000 SRAM arrays successfully resolved the problem of read/write stability in 6T SRAM.

In the VLSI architecture design, systolic arrays use a grid of processing elements (PEs) to handle data in parallel and in a pipeline, which speeds up computations as much as possible. This design makes tasks like matrix multiplication and neural network operations faster. It also lowers the amount of extra work that needs to be done when sending data, and it makes semiconductor devices more scalable and energy-efficient. In 2019, Asgari and team [38] demonstrated that systolic array solutions can accelerate DNN inference and are extremely valuable for dense linear algebra. However, implementing sparsification mechanisms during or after training makes the issue sparse. To circumvent this issue, they proposed an Eridanus method for structured pruning that generates DNNs that operate with a constant flow of data from memory to systolic arrays. In 2020, Parmar et al. [39] introduced a hardware-effective and non-multiplier method for convolutions in CNNs. Furthermore, in comparison with their time structures, they suggested a novel dataflow model featuring a reconfigurable systolic ring architecture that increases resource efficiency by 55% and decreases power consumption by the same amount. In 2024, Inayat et al. [40] presented two novel systolic arrays that use factoring and radix-8 multipliers. The first model reduced the delay and area by increasing the number of registers, but this was balanced by the lower requirement of registers in the pipeline of radix-8 offsets. This model gets better results by 15%, 13%, and 23%, respectively, for delay, area, and power. The second one can compress interconnects with two steps of hard-multiple addition; this achieves maximum reductions in area of 11.7% and maximum reduction in power of 16.7%.

11.6 APPLICATION OF AI/ML AT PHYSICAL DESIGN LEVEL

AI and ML greatly improve the physical design stage of VLSI by simplifying critical operations, including placement, routing, and layout verification. ML algorithms have the capability to forecast and enhance placement tactics, resulting in the reduction of wire length and improvement of timing. AI tools in routing aid in the effective exploration of route paths to manage congestion and minimize delays. AI-powered design rule checking (DRC) utilizes AI to anticipate and address possible rule infractions, hence enhancing the feasibility of manufacturing processes. In addition, ML models optimize layouts to improve power distribution and thermal management, hence boosting the overall performance of the design. These technological breakthroughs decrease the amount of time required for designing, increase the productivity of VLSI circuits, and improve their dependability.

In 2021, A. Mirhoseini et al. demonstrated a deep reinforcement learning model for chip floor layout and developed an edge-based graph convolutional neural network architecture capable of learning rich and transferable chip representations. This architecture can save hundreds of hours on modern accelerator design. Google used this approach to create their processors [41]. In 2021, Xie et al. [42] established a deep RL solution for an ML model by utilizing a constraint solver. This methodology improves random search and simulated annealing by 6.11% and 5.85%, respectively, while also cutting search time from 3 hours to 9 minutes. Tsung-Wei Huang [43] focused on the optimization of GPU acceleration in electronic design automation through placement and timing analysis. In order to accomplish this, he utilized ML to implement a novel heterogeneous parallel EDA solution. He was able to reach speeds that are 500 times faster for GPU acceleration path-based timing analysis. In 2021, Huang et al. [44] addressed the solution to global routing congestion in physical design with the help of an ML model. This model generates predictions of routing violations and provides them back to the system, which creates new placements with improvements.

In 2022, Yeh et al. [45] employed supervised ML techniques to tackle the challenges associated with substrate signal routing in high-density packages. The researchers introduced an innovative rerouting framework that can anticipate inefficient routing and yield improvements of 36.5% and 1.6% for wire band count and wire length, respectively.

11.7 APPLICATION OF AI/ML AT MANUFACTURING AND RELIABILITY LEVEL

AI and ML are transforming VLSI production by improving process control, defect prediction, and reliability analysis. AI improves process parameters in real time, increasing yield and eliminating faults in semiconductor manufacturing. ML methods forecast potential faults based on historical data, allowing for proactive corrective actions and reducing production losses. Furthermore, AI-driven reliability analysis uses stress test data to estimate long-term device performance, resulting in robust and reliable devices. These applications streamline manufacturing, cut costs, and improve overall VLSI device quality and yield, thereby promoting semiconductor industry efficiency and innovation. Mansi Phute and her team [46] emphasized the significance of lithography in VLSI fabrication. In addition, they examined the utilization of different ML models, including artificial neural networks (ANNs), convolutional neural networks (CNNs), and SVMs, for the purpose of identifying hotspots. The authors illustrated the potential for false-negative outcomes resulting from imbalanced datasets and proposed the use of Gaussian process assurance in order to address this problem. In 2021, Meng and team [47] suggested innovative ML models for etch bias applications and a unique data structure for these models. Additionally, they presented a new theory based on the edge placement error. Ying Chen and his colleagues [48] addressed the problem of acquiring a large dataset for ML applications in hotspot detection. The authors provided a method for detecting hotspots using a self-paced, multi-task learning approach in a semi-supervised manner. They achieved a higher accuracy of 4.6%–6.5% compared to conventional ML methods, while maintaining the same false alarm levels. This improvement is shown even when using only 10%–50% of the training data.

AI and ML improve the testing process of VLSI design by optimizing the development of test patterns, diagnosing faults, and predicting yield. ML algorithms produce optimized test patterns that enhance fault coverage and minimize test duration, hence guaranteeing thorough testing with little resource utilization. AI-powered fault diagnosis rapidly detects and pinpoints flaws, hence minimizing the duration required for troubleshooting and fixing. In addition, ML algorithms make predictions about manufacturing yield and detect possible problem sites by analyzing test data and historical patterns. These developments raise the precision of tests, decrease the expenses associated with testing, and improve the overall dependability and excellence of VLSI devices. Soham Roy and his colleagues [49] demonstrated a system for generating test patterns automatically using ML. They combined an unsupervised ML model with a decision-making system that is focused on paths. This approach minimized the need for backtracking and significantly decreased the amount of Central Processing Unit (CPU) time required. In 2021, Liu and Ou [50] developed a method using ML to increase the efficiency of system-level testing, reduce testing duration, and manage the number of defective components per million. The pseudo-probability of a chip's failure in the system-level test procedure was estimated using SVM and auto-encode. They proved that their approach has the capacity to decrease the expense of system-level testing by 40%.

11.8 CONCLUSION

This chapter demonstrated the utilization of AI and ML in multiple areas of VLSI logical and physical planning, which include circuit simulation, architecture, physical design, testing, and production. These examples illustrate the broad range of uses of AI and ML in the initial phases of VLSI design. AI/ML at the system level assists in architecture discovery and HLS. These technologies assist in optimizing code and detecting bugs for RTL design, thereby improving code quality and reducing the need for verification. AI/ML enhances the test generation, formal verification, and simulation processes during verification, ensuring comprehensive and efficient validation. AI in logic synthesis optimizes synthesis parameters and assists in obtaining timing closure, hence improving design quality. AI/ML techniques enhance the placement and routing of components, conduct design rule verification, and optimize the layout during the physical design phase to ensure efficient manufacturing and high yield. DFT utilizes AI to produce optimized test patterns, identify flaws, and forecast manufacturing yield, thereby guaranteeing reliable testing and achieving high production yields. AI/ML plays a crucial role in manufacturing by improving process control, forecasting faults, and evaluating reliability. This ultimately leads to enhanced production quality and an increased gadget lifespan. Integrating AI and ML into the VLSI design pipeline has substantial prospects for optimizing efficiency, decreasing time to market, and improving the overall quality of semiconductor designs. We anticipate that the ongoing progress of AI and ML technologies will broaden their utilization in VLSI design, thereby bringing about a further revolution in the sector. To achieve the desired implementation, it is crucial to possess interdisciplinary competence, which involves integrating an understanding of VLSI design principles with modern AI and ML methodologies. Furthermore, a persistent partnership between academics and industry would stimulate advancements and tackle the changing obstacles in VLSI design.

REFERENCES

1. J.-A. Carballo, W.-T. J. Chan, P. A. Gargini, A. B. Kahng, and S. Nath, "ITRS 2.0: toward a re-framing of the semiconductor technology roadmap," In *2014 IEEE 32nd International Conference on Computer Design (ICCD)*, Oct. 2014, doi: 10.1109/iccd.2014.6974673.
2. G. E. Moore, "Cramming more components onto integrated circuits," *Proceedings of the IEEE*, vol. 86, no. 1, pp. 82–85, Jan. 1998, doi: 10.1109/jproc.1998.658762.
3. H.-S. P. Wong, D. J. Frank, P. M. Solomon, C. H. J. Wann, and J. J. Welser, "Nanoscale CMOS," *Proceedings of the IEEE*, vol. 87, no. 4, pp. 537–570, Apr. 1999, doi: 10.1109/5.752515.
4. A. P. Singh, V. K. Mishra, and S. Akhter, "A perspective view of silicon based classical to non-classical MOS transistors and their extension in machine learning," *Silicon*, vol. 15, no. 16, pp. 6763–6784, Jun. 2023, doi: 10.1007/s12633-023-02541-0.
5. H. Iwai, "Logic LSI technology roadmap for 22 nm and beyond," In *2009 16th IEEE International Symposium on the Physical and Failure Analysis of Integrated Circuits*, Jul. 2009, doi: 10.1109/ipfa.2009.5232710.
6. M. H. Abu-Rahma and M. Anis, "Variability in VLSI circuits: sources and design considerations," In *2007 IEEE International Symposium on Circuits and Systems (ISCAS)*, May 2007, doi: 10.1109/iscas.2007.378156.
7. M. Belleville, O. Thomas, A. Valentian, and F. Clermidy, "Designing digital circuits with nano-scale devices: challenges and opportunities," *Solid-State Electronics*, vol. 84, pp. 38–45, Jun. 2013, doi: 10.1016/j.sse.2013.02.030.
8. A. P. Singh, V. K. Mishra, and S. Akhter, "Investigating machine learning applications for FDSOI MOS-based computer-aided design," In *2023 9th International Conference on Signal Processing and Communication (ICSC)*, Dec. 2023, doi: 10.1109/icsc60394.2023.10441540.
9. L. Wang and M. Luo, "Machine learning applications and opportunities in IC design flow," In *2019 International Symposium on VLSI Design, Automation and Test (VLSI-DAT)*, Apr. 2019, doi: 10.1109/vlsi-dat.2019.8742073.

10. C. C. John, V. Ponnusamy, S. Krishnan Chandrasekaran, and N. R, "A survey on mathematical, machine learning and deep learning models for COVID-19 transmission and diagnosis," *IEEE Reviews in Biomedical Engineering*, vol. 15, pp. 325–340, 2022, doi: 10.1109/rbme.2021.3069213.
11. R. Rai and C. K. Sahu, "Driven by data or derived through physics? A review of hybrid physics guided machine learning techniques with cyber-physical system (CPS) focus," *IEEE Access*, vol. 8, pp. 71050–71073, 2020, doi: 10.1109/access.2020.2987324.
12. A. M. Rahmani et al., "Machine learning (ML) in medicine: review, applications, and challenges," *Mathematics*, vol. 9, no. 22, p. 2970, Nov. 2021, doi: 10.3390/math9222970.
13. L.-T. Wang, C.-W. Wu, and X. Wen, *VLSI Test Principles and Architectures*. Elsevier, 2006.
14. S. M. Sze, *VLSI Technology*. McGraw-Hill Companies, 1983.
15. G. Rebala, A. Ravi, and S. Churiwala, *An Introduction to Machine Learning*. Springer, 2019.
16. R. Agalya and S. Saravanan, "Recent trends on Post-Silicon validation and debug: An overview," In *2017 International Conference on Networks & Advances in Computational Technologies (NetACT)*, Jul. 2017, doi: 10.1109/netact.2017.8076741.
17. E. Alpaydin, *Introduction to Machine Learning*. MIT Press, 2004.
18. J. D. Kelleher, B. Mac Namee, and A. D'Arcy, *Fundamentals of Machine Learning for Predictive Data Analytics,* 2nd edn. MIT Press, 2020.
19. V. Nasteski, "An overview of the supervised machine learning methods," *Horizons.B*, vol. 4, pp. 51–62, Dec. 2017, doi: 10.20544/horizons.b.04.1.17.p05.
20. T. G. Dietterich, "ensemble methods in machine learning," In *Multiple Classifier Systems*, pp. 1–15, 2000, doi: 10.1007/3-540-45014-9_1.
21. T. Hastie, R. Tibshirani, and J. Friedman, *The Elements of Statistical Learning*. Springer Science & Business Media, 2013.
22. G. Gan, C. Ma, and J. Wu, *Data Clustering: Theory, Algorithms, and Applications,* 2nd edn. SIAM, 2020.
23. R. S. Sutton and A. G. Barto, *Reinforcement Learning,* 2nd edn. MIT Press, 2018.
24. S. Bhattacharyya, V. Snasel, A. E. Hassanien, S. Saha, and B. K. Tripathy, *Deep Learning*. Walter de Gruyter GmbH & Co KG, 2020.
25. A. Zjajo, *Stochastic Process Variation in Deep-Submicron CMOS*. Springer Science & Business Media, 2013.
26. A. F. Roslan et al., "Comparative analysis of process parameter variations in DGFinFET device using statistical methods," *Journal of Physics: Conference Series*, vol. 1123, p. 012048, Nov. 2018, doi: 10.1088/1742-6596/1123/1/012048.
27. M. Olivieri and A. Mastrandrea, "Logic drivers: a propagation delay modeling paradigm for statistical simulation of standard cell designs," *IEEE Transactions on Very Large-Scale Integration (VLSI) Systems*, vol. 22, no. 6, pp. 1429–1440, Jun. 2014, doi: 10.1109/tvlsi.2013.2269838.
28. H. Carrillo-Nunez, N. Dimitrova, A. Asenov, and V. Georgiev, "Machine learning approach for predicting the effect of statistical variability in Si junction less nanowire transistors," *IEEE Electron Device Letters*, vol. 40, no. 9, pp. 1366–1369, Sep. 2019, doi: 10.1109/led.2019.2931839.
29. J.-S. Yoon, S. Lee, H. Yun, and R.-H. Baek, "Digital/analog performance optimization of vertical nanowire FETs using machine learning," *IEEE Access*, vol. 9, pp. 29071–29077, 2021, doi: 10.1109/access.2021.3059475.
30. C. Akbar, Y. Li, and N. Thoti, "Device-simulation-based machine learning technique for the characteristic of line tunnel field-effect transistors," *IEEE Access*, vol. 10, pp. 53098–53107, 2022, doi: 10.1109/access.2022.3174685.
31. A. Stillmaker and B. Baas, "Scaling equations for the accurate prediction of CMOS device performance from 180 nm to 7 nm," *Integration*, vol. 58, pp. 74–81, Jun. 2017, doi: 10.1016/j.vlsi.2017.02.002.

32. D. Amuru, A. Zahra, and Z. Abbas, "Statistical variation aware leakage and total power estimation of 16 nm VLSI digital circuits based on regression models," *Communications in Computer and Information Science*, pp. 565–578, 2019, doi: 10.1007/978-981-32-9767-8_47.
33. A. B. Kahng, M. Luo, and S. Nath, "SI for free: machine learning of interconnect coupling delay and transition effects," *2015 ACM/IEEE International Workshop on System Level Interconnect Prediction (SLIP)*, Jun. 2015, doi: 10.1109/slip.2015.7171706.
34. V. Govindaraj and B. Arunadevi, "Machine learning based power estimation for CMOS VLSI circuits," *Applied Artificial Intelligence*, vol. 35, no. 13, pp. 1043–1055, Aug. 2021, doi: 10.1080/08839514.2021.1966885.
35. K. Agarwal, A. Jain, D. Amuru, and Z. Abbas, "Fast and efficient ResNN and Genetic optimization for PVT aware performance enhancement in digital circuits," *2022 International Symposium on VLSI Design, Automation and Test (VLSI-DAT),* Apr. 2022, doi: 10.1109/vlsi-dat54769.2022.9768067.
36. S. Gupta, M. Imani, H. Kaur, and T. S. Rosing, "NNPIM: a processing in-memory architecture for neural network acceleration," *IEEE Transactions on Computers*, vol. 68, no. 9, pp. 1325–1337, Sep. 2019, doi: 10.1109/tc.2019.2903055.
37. J. Yang, X. Cui, S. Jia, and Y. Wang, "A configurable computing-in-memory structure based on convolutional neural network," In *2021 China Semiconductor Technology International Conference (CSTIC),* Mar. 2021, doi: 10.1109/cstic52283.2021.9461459.
38. B. Asgari, R. Hadidi, H. Kim, and S. Yalamanchili, "ERIDANUS: efficiently running inference of DNNs using systolic arrays," *IEEE Micro*, vol. 39, no. 5, pp. 46–54, Sep. 2019, doi: 10.1109/mm.2019.2930057.
39. Y. Parmar and K. Sridharan, "A resource-efficient multiplierless systolic array architecture for convolutions in deep networks," *IEEE Transactions on Circuits and Systems II: Express Briefs*, vol. 67, no. 2, pp. 370–374, Feb. 2020, doi: 10.1109/tcsii.2019.2907974.
40. K. Inayat, I. Ullah, and J. Chung, "Factored systolic arrays based on radix-8 multiplication for machine learning acceleration," In *IEEE Transactions on Very Large Scale Integration (VLSI) Systems*, pp. 1–11, 2024, doi: 10.1109/tvlsi.2024.3378197.
41. A. Mirhoseini et al., "A graph placement methodology for fast chip design," *Nature*, vol. 594, no. 7862, pp. 207–212, Jun. 2021, doi: 10.1038/s41586-021-03544-w.
42. X. Xie et al., "A transferable approach for partitioning machine learning models on multi-chip-modules," *arXiv (Cornell University)*, Jan. 2021, doi: 10.48550/arxiv.2112.04041.
43. T. -W. Huang, "Machine learning system-enabled GPU acceleration for EDA," In *2021 International Symposium on VLSI Design, Automation and Test (VLSI-DAT)*, Hsinchu, Taiwan, 2021, pp. 1–1, doi: 10.1109/VLSI-DAT52063.2021.9427323.
44. Y. -Y. Huang, C. -T. Lin, W. -L. Liang and H. -M. Chen, "Learning based placement refinement to reduce DRC short violations," In *2021 International Symposium on VLSI Design, Automation and Test (VLSI-DAT)*, Hsinchu, Taiwan, 2021, pp. 1–4, doi: 10.1109/VLSI-DAT52063.2021.9427321.
45. Y. -H. Yeh et al., "Substrate signal routing solution exploration for high-density packages with machine learning," In *2022 International Symposium on VLSI Design, Automation and Test (VLSI-DAT)*, Hsinchu, Taiwan, 2022, pp. 1–4, doi: 10.1109/VLSI-DAT54769.2022.9768081.
46. M. Phute, A. Sahastrabudhe, S. Pimparkhede, S. Potphode, K. Rengade and S. Shilaskar, "A survey on machine learning in lithography," In *2021 International Conference on Artificial Intelligence and Machine Vision (AIMV)*, Gandhinagar, India, 2021, pp. 1–6, doi: 10.1109/AIMV53313.2021.9670977.
47. Y. Meng, Y.-C. Kim, S. Guo, Z. Shu, Y. Zhang, and Q. Liu, "Machine learning models for edge placement error based ETCH bias," *IEEE Transactions on Semiconductor Manufacturing*, vol. 34, no. 1, pp. 42–48, Feb. 2021, doi: 10.1109/tsm.2020.3042803.

48. Y. Chen, Y. Lin, T. Gai, Y. Su, Y. Wei, and D. Z. Pan, "Semi-supervised hotspot detection with self-paced multi-task learning," *IEEE Transactions on Computer-Aided Design of Integrated Circuits and Systems,* Jan. 2019, doi: 10.1145/3287624.3287685.
49. S. Roy, S. K. Millican and V. D. Agrawal, "Multi-heuristic machine intelligence guidance in automatic test pattern generation," In *2022 IEEE 31st Microelectronics Design & Test Symposium (MDTS)*, Albany, NY, USA, 2022, pp. 1–6, doi: 10.1109/MDTS54894.2022.9826985.
50. C. Liu and J. Ou, "Smart sampling for efficient system level test: a robust machine learning approach," In *2021 IEEE International Test Conference (ITC)*, Anaheim, CA, USA, 2021, pp. 53–62, doi: 10.1109/ITC50571.2021.00013.

12 FinFET-Based 9T SRAM for Enhanced Performance in AI/ML Applications

Yagnadatt Sarangi, Kartik Singh, and Shelja Kaushal

12.1 INTRODUCTION

In the current world, the need for more efficient chip design has increased and therefore the need for more efficient circuits that consume less power with a fast response. In the realm of VLSI (very large-scale integration), the quest for a transistor which meets the requirement has increased. For a long time, CMOS (complementary metal oxide transistor) has significantly contributed to develop the electronic devices and attain our objectives. But by looking at the present scenario, there is a need for higher performance, less power consumption, and less delay. CMOS inabilities are becoming very evident to compete these needs. CMOS-based circuits face major problems like current leakage, more power consumption, and extreme short channel effects (SCEs) which limit its performance. To confront all these problems, the semiconductor industry has emerged with a revolutionary alternative: FinFET (fin field-effect transistor). The FinFET technology resists various SCEs very efficiently and hence provides calibre to reach new stages of efficiency in VLSI [1].

One of the important components in chip design is the SRAM (static random access memory) arrays. An essential component within SRAM arrays, the SRAM cell serves as the cornerstone for temporary data storage in computer systems and various electronic devices. Being a form of volatile memory, it necessitates a continuous power supply to uphold the stored data. Within the realm of VLSI design, SRAM cells hold significant importance by facilitating rapid data access and are frequently integrated as cache memory and internal registers within Central Processing Units (CPUs). The straightforwardness and swiftness of SRAM render it well-suited for tasks demanding swift data retrieval and the necessity for refreshing cycles. Moreover, modern microprocessors incorporate SRAM cells extensively across expensive buffers and caches, thereby enhancing their operational efficiency and overall performance. SRAM cells have a latch that stores data for as long as the power is on, eliminating the need for refresh operations. SRAM is commonly utilized as cache memory in microprocessors, mainframe computers, engineering

 DOI: 10.1201/9781003483038-12

workstations, and handheld devices due to its high speed and low power consumption. An SRAM stores each bit on four transistors, forming two cross-coupled inverters. The power consumption and speed of SRAM have led to several designs aimed at minimizing power usage during read and write operations. Modern computing systems rely heavily on high-performance on-chip caches to support their memory structure [2].

Emerging SRAM circuits signify a significant leap forward in memory technology, surpassing traditional 6T SRAM designs in several crucial aspects. Emerging SRAM circuits, incorporating elements from various configurations such as 8T, 7T, 9T, 10T, and 11T, present a compelling amalgamation of reliability, efficiency, and performance, positioning them as pivotal components in the landscape of future memory architectures. Moreover, these innovative circuits exhibit remarkable stability during low-power operation, rendering them exceptionally suitable for energy-efficient applications, such as Internet of Things (IoT) devices and mobile electronics. Among these architectures, the 8T and 10T SRAM configurations contribute their unique attributes; the former excels in balancing area efficiency and stability, while the latter amplifies robustness and performance, albeit at the cost of increased area overhead. Additionally, the 7T and 11T SRAM designs offer specialized solutions, with the former prioritizing low-power operation and the latter emphasizing enhanced reliability in harsh environments. But among the array of emerging designs, the 9T SRAM configuration emerges as a standout due to its heightened resilience against diverse forms of noise and disturbances, including process variations and radiation-induced errors. Furthermore, the 9T SRAM excels in low-power operation, making it an ideal choice for energy-efficient applications, which may not be as effectively addressed by the 7T or 11T designs. The 9T SRAM emerges as the preferred choice among its counterparts, offering a compelling combination of area efficiency, stability, reliability, and low-power operation, thus positioning it as a key component in advanced memory architectures.

In this chapter, we have proposed a 9T SRAM using 18 nm FinFET technology. To employ with the circuit design and various simulations, Cadence Virtuoso tool is used [3,4]. The various sections in this chapter are as follows: Section 12.2 explains the working of 9T SRAM and its various operations, Section 12.3 employs the optimization of FinFET-based 9T SRAM for low power and increased stability. While Section 12.4 shows comparative and environment variation analysis followed by conclusion in Section 12.5.

12.2 FINFET-BASED 9T SRAM

In this section, the proposed 9T SRAM using FinFET technology is explained. As the characteristics of FinFET technology is a substantial improvement over traditional CMOS technology, notably in terms of stability enhancement, by enabling better control over current flow, FinFET technology reduces the hazards of data corruption caused by leakage currents or noise in SRAM cells. This increased stability correlates to greater reliability and data integrity in a variety of computing settings [4]. The 9T FinFET-based SRAM cell has nine transistors placed in an optimal arrangement for stability, durability, and performance as shown in Figure 12.1.

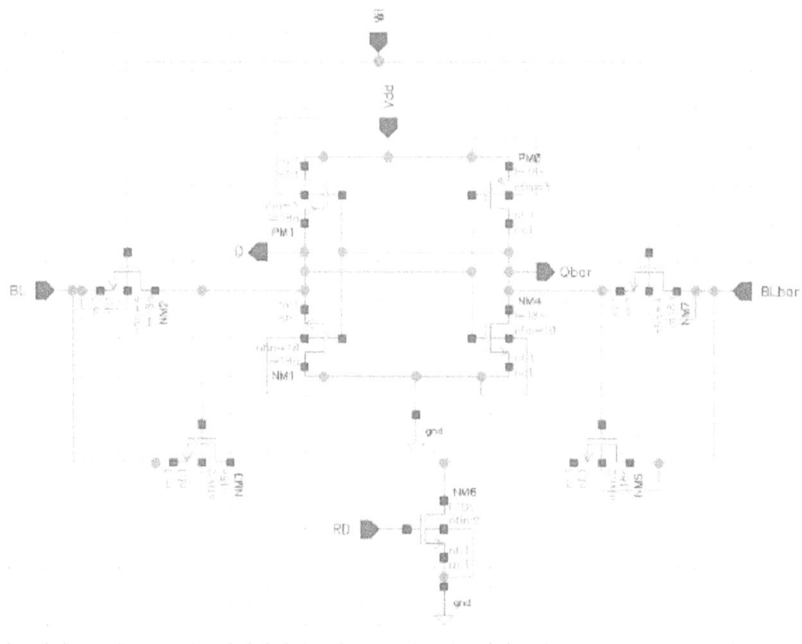

FIGURE 12.1 Structure of a 9T SRAM.

The cell's heart is made up of six storage transistors, which include two access transistors and two cross-coupled inverters. The read and write operations are controlled by access transistors, which provide access to the SRAM cell's storage nodes (bit lines). The cross-coupled inverters comprise the cell's latching mechanism, which stores data in a stable manner. Each inverter is made up of an n-type Metal-Oxide-Semiconductor (NMOS) and a p-type Metal-Oxide-Semiconductor (PMOS) transistor coupled in a feedback loop to ensure reliable data retention and rapid access times. In addition to the storage transistors, the 9T FinFET SRAM cell contains three auxiliary transistors that improve its stability and dependability. The pass transistors add extra routes for manipulating the states of the cross-coupled inverters, hence enhancing the cell's noise immunity and read stability. The assist transistor improves the read margin while reducing the effects of process fluctuations and external noise. Hence, the FinFET-based 9T SRAM cells may provide higher energy efficiency, particularly during standby or idle states, which contributes to overall power savings and longer battery life in mobile and battery-powered devices.

12.2.1 Operations of SRAM

SRAM circuits play a crucial role in modern electronic devices, providing fast and reliable data storage. Understanding the operational modes of SRAM cells, such as the hold, write, and read modes, is essential for comprehending their functionality.

Hold Operation: During the hold operation, the SRAM cell remains in an idle state, with data held in its latch without any alterations until the next read or write command is initiated. To maintain the stored data during this phase, the access transistors within the cell are disconnected by setting the word line (WL) to a low voltage or ground level (WL = 0). As a result, access transistors NM2 and NM7 are turned off, effectively isolating the storage nodes, which are composed of cross-coupled inverter FinFETs (PM0, PM1, NM1, and NM4) from the bit lines. This isolation ensures that the stored data remains intact and unaffected by external disturbances.

Write Operation: In the write mode, the objective is to modify the data stored in the SRAM cell. This process involves placing the desired data values, either logic 0 or logic 1, on the Bit Lines (BL) and BL-bar. Typically, logic 0 is stored at node 1, while logic 1 is stored at node 2, leading to specific configurations of FinFETs within the cell, i.e. PM3 and NM1 being turned ON while PM1 and NM4 are turned OFF. To perform the write operation, the WL is activated, setting it to a high voltage (WL = 1), thereby turning on the access FinFETs. Subsequently, the bit lines BL and BL-bar are set to the desired data values intended to be written into the cell. This action triggers a voltage difference across the cell, causing certain FinFETs to turn on or off selectively, e.g. if BL = 1 and BL-bar = 0, then PM1 and NM4 will turn ON while NM1 and PM3 are forced to turn OFF. Consequently, the stored data is overwritten, with node 1 being updated to logic 1 and node 2 being updated to logic 0.

Read Operation: In the read mode, the SRAM cell retrieves the stored data for further processing. Assuming logic 0 is stored at node 1, M4 and M1 are turned on, leading to node 2 charging up to the supply voltage (V_{DD}) while PM1 and NM4 remain off. Enabling the access transistors connecting the bit lines to the storage nodes by activating the WL (WL = 1), the bit lines BL and BL-bar are precharged to V_{DD}. This precharging initiates a potential voltage difference, causing a current to flow through NM2. Consequently, the sensed voltage difference between BL and BL-bar determines the stored data value, where a higher voltage on BL-bar indicates a stored logic 1, while a higher voltage on BL indicates a stored logic 0. The read data is then made available on the bit lines for subsequent processing, completing the read operation of the SRAM cell.

The various control signals to perform these operations in proposed FinFET-based 9T SRAM cell are well tabulated in Table 12.1.

SRAM cells are characterized by several key parameters that define their performance and reliability. These parameters include:

TABLE 12.1
Control Signals for the Proposed 9T FinFET SRAM Cell

Signals	Hold	Read	Write 0	Write 1
WL	0	1	1	1
BL	0	1	1	0
BL-bar	0	1	0	0
RL	0	1	0	0

Static Noise Margin: Static noise margin (SNM) refers to the stability or robustness of a FinFET 9T SRAM cell against noise-induced errors in maintaining stored data. SNM serves as a critical metric in assessing the reliability of the memory cell. In the context of a 9T SRAM cell, SNM is evaluated using various operational parameters, including voltage levels and transistor characteristics. It quantifies the voltage difference at which the cell transitions from one stable state to another, whether from '0' to '1' or vice versa, in the presence of noise. A higher SNM value indicates superior noise immunity and, consequently, enhances the overall reliability of the SRAM cell [5,6].

Access Time: The access time for both read and write operations is a critical parameter that directly influences the overall performance of the memory. Read access time refers to the duration required to retrieve data from an SRAM cell. It involves activating the appropriate WL and sensing the data stored in the cell, typically by detecting voltage differences on the bit lines. SRAMs are renowned for their fast read access times, often measured in nanoseconds, making them suitable for applications requiring rapid data retrieval, such as cache memory in processors. On the other hand, write access time denotes the time taken to modify or store data into an SRAM cell. This process involves setting the desired data values on the bit lines, activating the WL, and updating the content of the cell accordingly. Write access times in SRAMs are generally comparable to read access times, although they may vary depending on factors such as cell configuration and circuit design. Ensuring balanced read and write access times is crucial for maintaining efficient memory operations and achieving optimal system performance in various computing applications.

Power Consumption: Power stability is crucial for SRAM design, ensuring reliable operation and data integrity. Techniques like voltage regulation, power gating, and static power management are employed to minimize fluctuations in power supply and reduce both dynamic and static power consumption. By prioritizing power stability, SRAMs deliver consistent performance across various operating conditions, making them essential components in a wide range of computing systems.

12.3 OPTIMIZATION OF 9T SRAM

In the realm of digital circuit design, achieving efficient and reliable operation is paramount. A key factor in this pursuit is optimizing transistor sizing, especially in FinFET-based SRAM. Transistor sizing in FinFET SRAM is intricately linked to the number of fins per finger, directly influencing both the SNM and power consumption. This section undertakes a systematic exploration of transistor size, focusing particularly on manipulating the fins per finger within a FinFET design. Our research aims to unravel the intricate relationship between number of fins per finger and key circuit performance parameters, such as SNM and power consumption. Through meticulous examination and analysis of the ramifications of various fin topologies, we aspire to offer valuable insights into the design landscape of circuits, ultimately facilitating the advancement of more efficient and resilient digital systems [7]. In Figures 12.2a and b, the number of fins for access NMOS transistor, i.e. NM2 and NM7, is varied simultaneously and analysed for SNM and power stability, respectively. It is observed that with increase in number of fins in access transistor, the drive

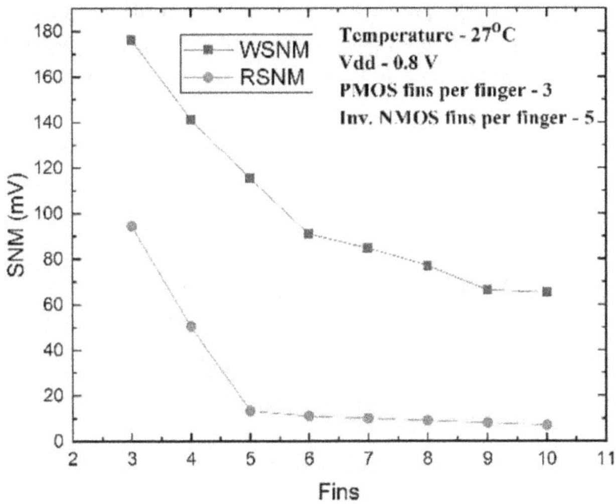

FIGURE 12.2a Variation of static noise margin with respect to change in fins per finger of access NMOS.

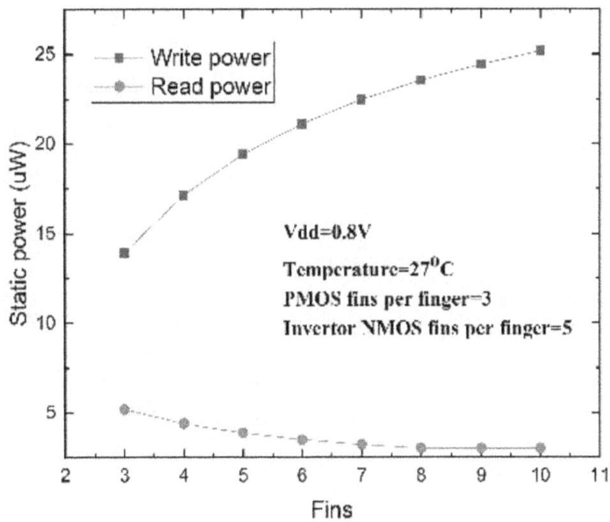

FIGURE 12.2b Variation of static power with respect to change in fins per finger of access NMOS.

strength of the access transistors also increases. This heightened drive strength leads to faster switching speeds and higher current flow, resulting in a smaller voltage difference between the stable states ('0' and '1') of the SRAM cell. Consequently, the decreased voltage difference makes the cell more susceptible to noise-induced errors, ultimately reducing the SNM. Due to this, the power in read operation is also increasing with the upsurge in number of fins.

In Figures 12.3a and b, the number of fins for inverter NMOS transistor, i.e. NM1 and NM4, is varied simultaneously and analysed for SNM and power stability, respectively. It clearly seems that the Write Static Noise Margin (WSNM) is reduced because inverter NMOS transistor may enhance certain performance metrics, such as speed; it can also compromise the noise immunity and stability of the SRAM cell, leading to a

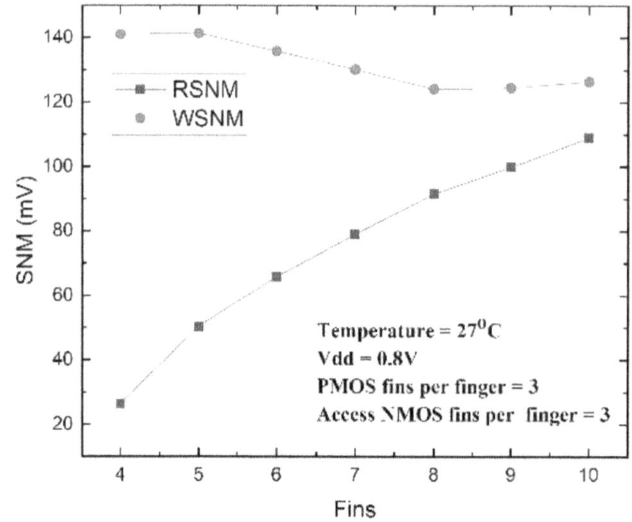

FIGURE 12.3a Variation of static noise margin with respect to change in fins per finger of inverter NMOS.

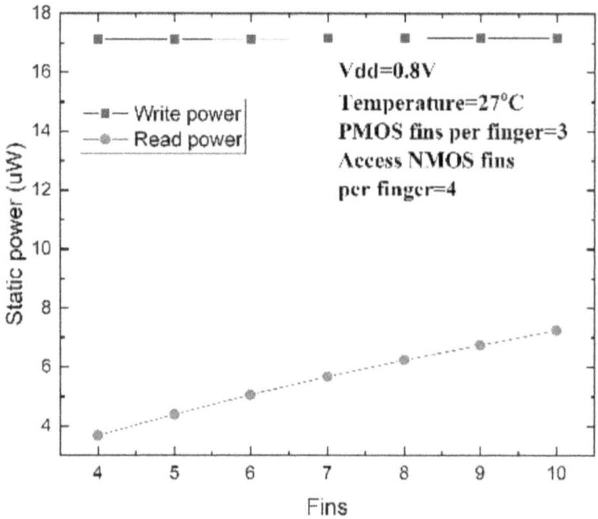

FIGURE 12.3b Variation of static power with respect to change in fins per finger of inverter NMOS.

decrease in the SNM. Due to this, the static power is also increased during the write operation. As the number of fins in the NMOS inverter transistor increases, the Read Static Noise Margin (RSNM) also improves. This increase in the number of fins alters the effective channel doping concentration, which in turn affects the threshold voltage. As a result, the improved subthreshold slope, which measures the rate of change of drain current concerning gate voltage, with more fins enhances the cell's ability to retain data reliably, further contributing to the increased SNM.

Finally, Figures 12.4a and b show the variation in SNM and power stability, with the variation in number of fins per finger of inverter PMOS transistor, PM0 and

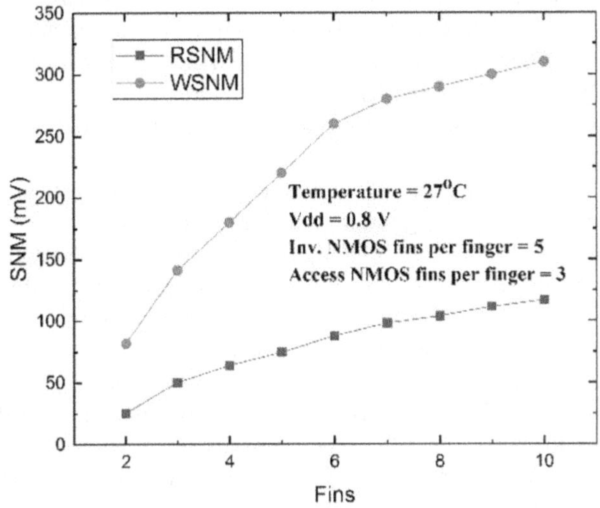

FIGURE 12.4a Variation of static noise margin with respect to change in fins per finger of inverter PMOS.

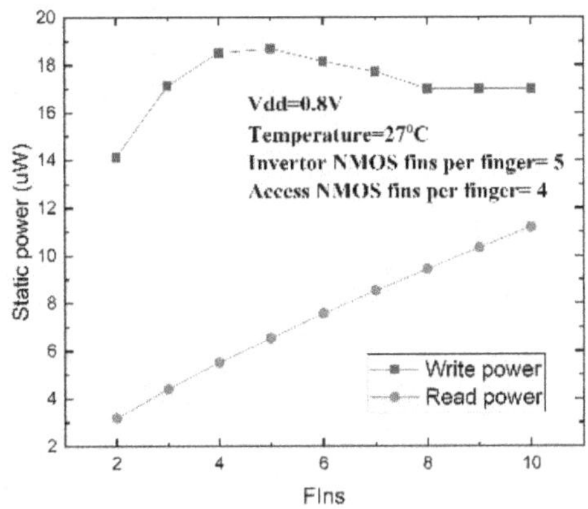

FIGURE 12.4b Variation of static power with respect to change in fins per finger of inverter PMOS.

TABLE 12.2
Optimized Number of Fins per Finger for the Proposed 9T SRAM Cell

Transistor	Number of Fins per Finger
Fins per finger for access NMOS, i.e. NM2 and NM7	4
Fins per finger for inverter NMOS, i.e. NM1 and NM4	10
Fins per finger for PMOS, i.e. PM0 and PM1	3

PM1. Increased drive strength and potential threshold voltage adjustment with more fins in the inverter PMOS transistor can lead to an overall increase in the SNMs of the SRAM cell. The increase in drive strength will increase collision and hence the power dissipation. Therefore, the read power is also increasing with increase in number of fins of PMOS. Whereas the write power is increasing with increase in number of fins till number 5, and after that, it started reducing because of better current flow and less leakage current.

The results for SNM and static power were analyzed with varying numbers of fins. The standard hierarchical sizing order of the transistors—starting with the inverter PMOS, followed by the inverter NMOS, and then the access NMOS—was maintained, based on their specific roles within the cell [8]. According to this hierarchical pattern of sizing, the PMOS transistor, responsible for the pull-up functionality during write operations and the standby mode, tends to be the smallest in size to ensure adequate drive strength for stable operation. The access NMOS transistor, facilitating data read and write access to the cell, is usually larger than inverter PMOS in size and smaller than the inverter NMOS, as it primarily serves to enable efficient access to the storage node without requiring excessive area. The inverter NMOS essential for data sensing and amplification during read operations is typically sized largest. Finally, this sizing hierarchy ensures balanced performance, power efficiency, and area utilization in the 9T SRAM cell design. Also, the SNM and static power consumption plays an important role in determining the overall efficiency of the SRAM cell. Thus, the number of fins is optimised by taking the simulated results as well as the hierarchical pattern of sizing into consideration to increase stability with low power operations. The optimised number of fins per finger for 9T SRAM cell is tabulated in Table 12.2.

12.4 RESULT ANALYSIS

12.4.1 Comparative Analysis

When evaluating the proposed 9T SRAM design, it becomes essential to conduct thorough comparisons with existing SRAM models. This approach is crucial for gaining a comprehensive understanding of its performance and benefits. By juxtaposing the 9T SRAM with other designs like the conventional 6T SRAM and possibly more recent advancements such as 10T SRAM, these comparative analyses enable us to pinpoint specific strengths and weaknesses of the proposed 9T SRAM design in comparison to its counterparts. Furthermore, benchmarking against established

TABLE 12.3a
Comparative Analysis of the Proposed 9T SRAM SNM with Other Existing SRAM

No. of Transistors	Ref. No.	Technology (nm)	Parameter	Value (mV)
8T	[9]	16	WSNM	504.31
8T	[9]	16	HSNM	517.73
10T	[10]	18	RSNM	198
10T	[10]	18	WSNM	280
Proposed 9T		18	HSNM	380
Proposed 9T		18	RSNM	115
Proposed 9T		18	WSNM	146

TABLE 12.3b
Comparative Analysis of the Proposed 9T SRAM SNM with Other Existing SRAM

			Access time (pS)	
No. of Transistors	Ref. No.	Technology (nm)	Read	Write
8T	[9]	16	708	439
10T	[8]	18	28.54	92.4
Proposed 9T		18	13.335	22.795

SRAM architectures provides valuable insights into potential trade-offs and areas for refinement, facilitating the optimization of the proposed 9T SRAM for superior performance in real-world applications. Table 12.3a shows the comparative analysis of stability margins with already published results for various SRAM designs. While Table 12.3b shows the comparative analysis of read and write access time with already published results, it has been found that the proposed 9T SRAM is showing 2.14- and 4.05- times fast read and write operation, respectively, as compared to already published 10T SRAM [8]. While stability is reduced a little by 1.7- and 1.9- times for read and write operation, respectively, the proposed SRAM is also compared with an already published 8T SRAM cell [9]. It has been found that the proposed 9T SRAM is almost 42- and 20- times faster as compared to already published 8T SRAM cell. The stability is reduced enough as compared to 8T SRAM cell. Hence, it has been shown that the proposed 9T SRAM cell provides fast read and write operations and is very beneficial for upcoming AI and ML applications. It still provides the stability within the required range and further can be improved by varying the number of fins as per requirement.

12.4.2 Environmental Variation

In this section, the performance of the proposed 9T SRAM is analysed with varying power supply as well as temperature. Figures 12.5a–c show the variation of SNMs,

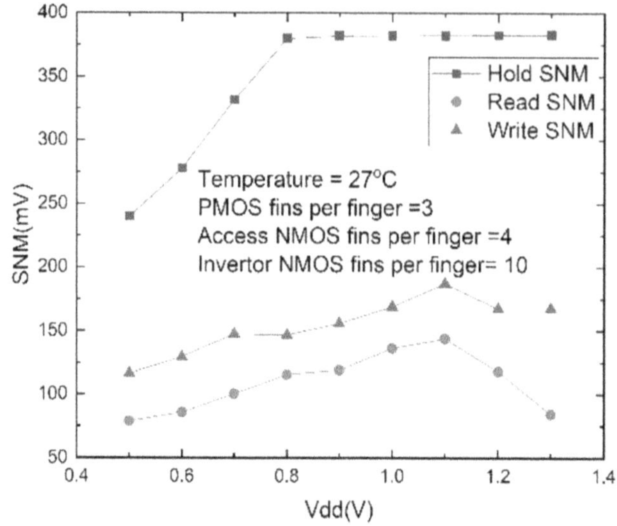

FIGURE 12.5a Variation in the hold SNM, read SNM, and write SNM of the proposed 9T SRAM with respect to V_{DD}.

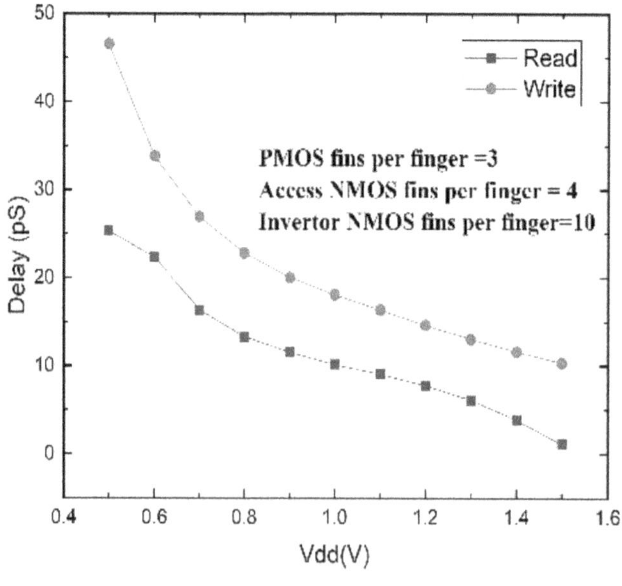

FIGURE 12.5b Variation in the read and write access time of the proposed 9T SRAM with respect to V_{DD}.

access time, and power consumption, respectively, of the proposed 9T SRAM with power supply. Figure 12.5a shows that as the supply voltage (V_{DD}) levels increase, a noticeable trend emerges in the proposed 9T SRAM performance. As the high power mitigates the leakage currents in the SRAM cells, leakage currents can corrupt

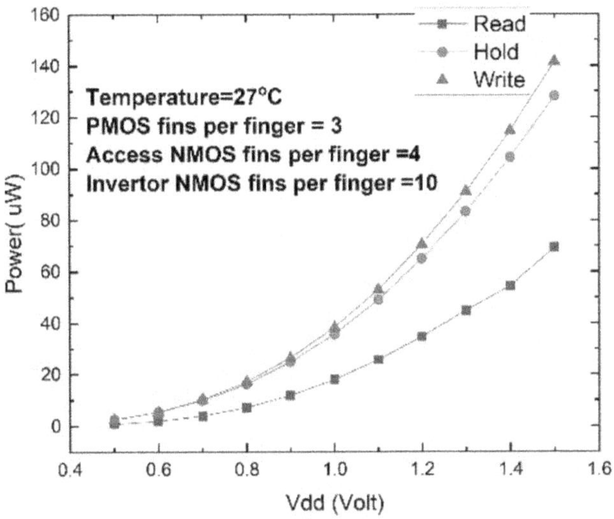

FIGURE 12.5c Variation in power consumption of the proposed 9T SRAM with respect to V_{DD}.

stored data over time, especially in low-power scenarios. By providing ample power, these leakage currents can be minimized, promoting stability. But it is noticed that V_{DD} above 1.1 V will start reduction due to higher voltage levels exacerbating various forms of noise, such as process variation and transistor mismatch, leading to increased susceptibility to errors and reduced stability. Additionally, in Figure 12.5b, the speed of SRAM operations tends to decrease at higher V_{DD} levels due to longer signal propagation times caused by larger voltage swings and increased parasitic capacitance. While Figure 12.6c shows that with increased V_{DD} levels in SRAM, there is a corresponding rise in power consumption, this escalation occurs due to the direct relationship between supply voltage and power dissipation in integrated circuits. When V_{DD} is elevated, each memory cell within the SRAM array requires more energy to store and retain data, contributing to higher overall power consumption. Hence, it is observed that the increased V_{DD} will fasten up the proposed 9T SRAM with minute reduction in stability. But it will lead to increased power consumption. Hence, the operation V_{DD} can be optimized to achieve a balance between minimizing static power and maintaining adequate performance and reliability in 9T SRAM cell design.

Figures 12.5d–i show the variation of stability, access time, and power consumption of the proposed 9T SRAM with temperature. Figures 12.5d–g show that the SNM of SRAM typically decreases as temperature rises. As the elevated temperatures lead to increased leakage currents in transistors, it affects the stability of stored data. Additionally, higher temperatures can exacerbate process variations, leading to inconsistencies in transistor characteristics and reducing SNM. While the read access time of SRAM tends to increase as temperature rises as shown in Figure 12.5h, this phenomenon occurs due to increased thermal agitation of carriers which results in higher parasitic and increased delay. The write access time is reduced up to some

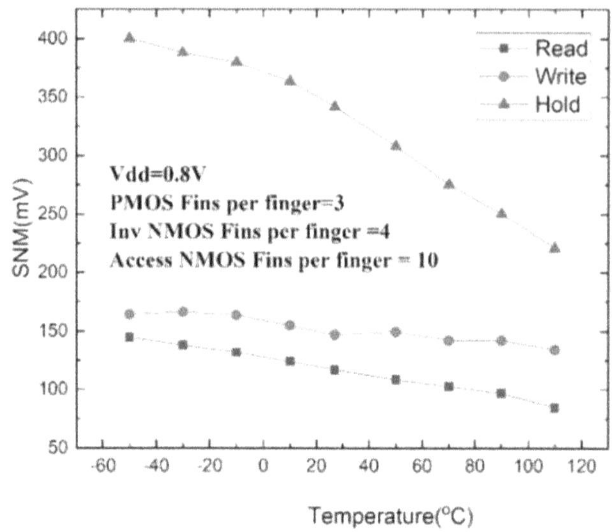

FIGURE 12.5d Variation of SNM of the proposed 9T SRAM with respect to temperature.

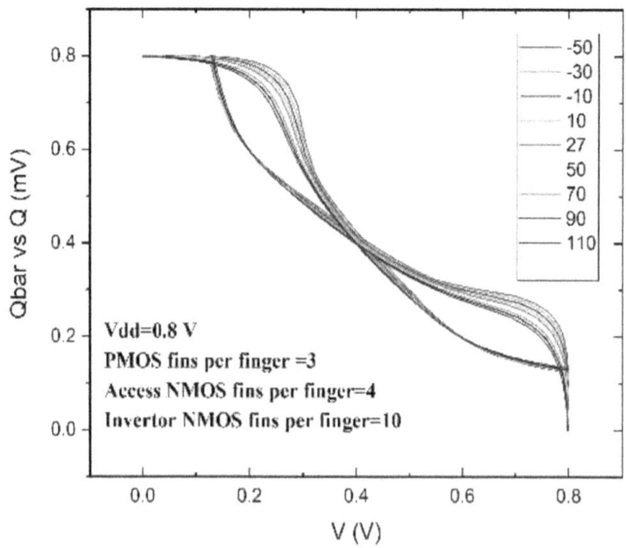

FIGURE 12.5e Variation of read SNM with respect to temperature.

extent as mobility increases but it's almost similar at different temperatures. Finally, Figure 12.5i shows the variation in power consumption of the proposed 9T SRAM with regard to temperature. It is observed that during the hold and write operation the mobility increases with rise in temperature and hence the leakage current. Because of this, the power consumption during the hold and write operation is increased with temperature. While it seems counterintuitive, power consumption in the proposed 9T

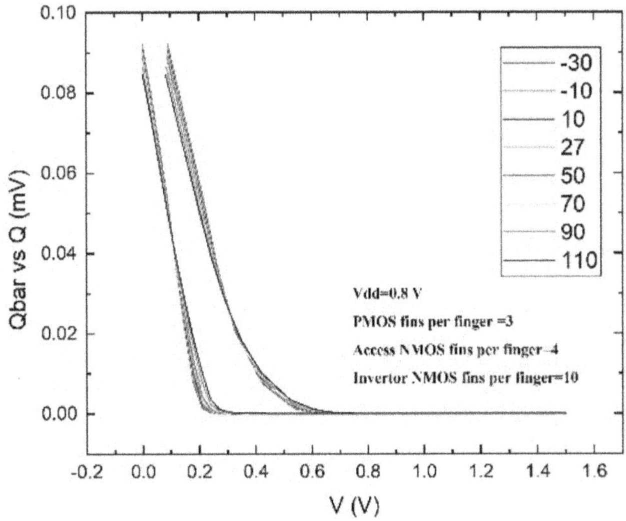

FIGURE 12.5f Variation of write SNM with respect to temperature.

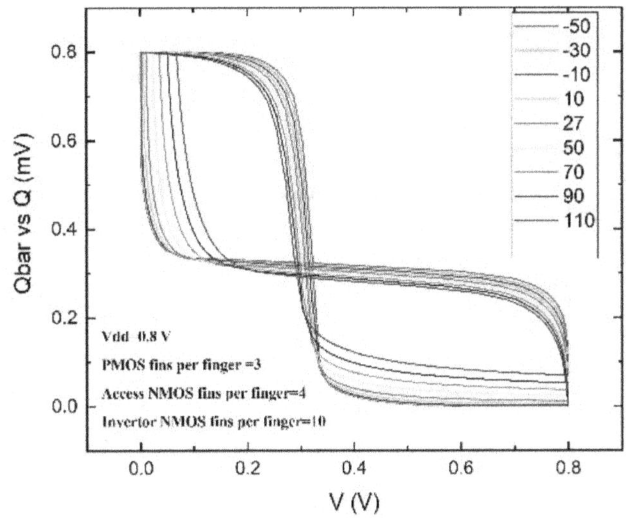

FIGURE 12.5g Variation of hold SNM with respect to temperature.

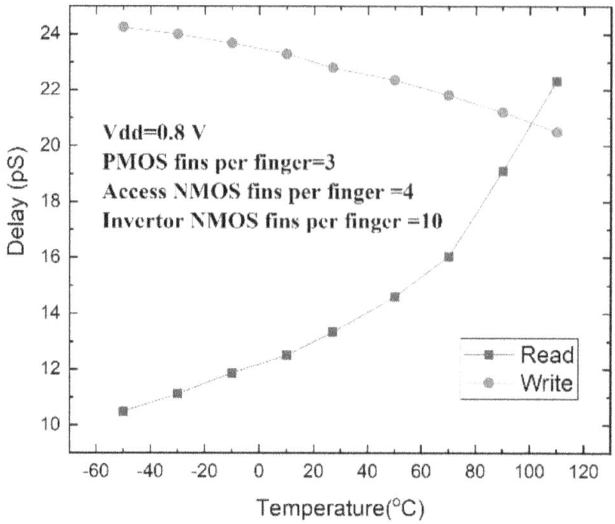

FIGURE 12.5h Variation of read and write access time of the proposed 9T SRAM with respect to temperature.

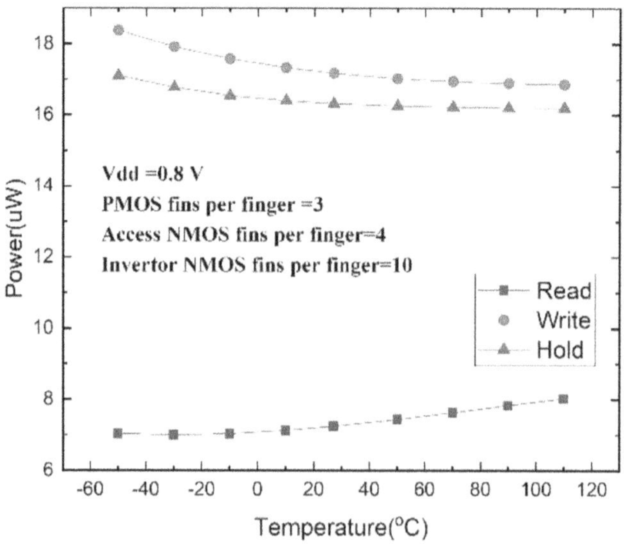

FIGURE 12.5i Variation of power consumption in the proposed 9T SRAM with respect to temperature.

SRAM during the read operation decreases with a rise in temperature. Because the increase in temperature increases the thermal energy that can activate carriers in the semiconductor material, it allows them to more effectively fill the traps and states that would otherwise contribute to leakage currents. Consequently, the leakage current decreases, leading to lower overall power consumption. Therefore, it can be said that the proposed 9T SRAM will lead to overall constant power consumption with reduced stability and slower read operation with the rise in temperature.

12.5 CONCLUSION

In this chapter, a comprehensive study on the design and performance evaluation of FinFET-based 9T SRAM for advanced AI/ML applications is presented. Through exhaustive analysis and simulation, this chapter has showcased the exceptional attributes of the proposed architecture, highlighting its heightened stability, diminished leakage power, enhanced read and write performance, and fortified resilience against various process variations and environmental factors. The result shows that the innovative integration of FinFET technology alongside the novel 9T SRAM cell design presents a promising avenue to fulfil the demanding prerequisites of forthcoming AI/ML systems. The optimized results and trade-off between various parameters for crafting efficient and dependable memory subsystems essential for realizing high-performance computing platforms tailored for AI/ML workloads is well explained in this chapter. It has shown that the FinFET-based 9T SRAM leads to fast read and write operation with some trade-off with stability and power. While the FinFET-based 9T SRAM is very less prone to environmental changes such as temperature, it increases power consumption with increase in supply voltage. By addressing the proposed SRAM, forthcoming research can further analyse it for various process variations, multi cell design, and memory arrays, etc. for next-generation AI/ML hardware, meeting stringent performance and power requisites.

12.6 ACKNOWLEDGEMENTS

The authors would like to thank the Department of Electronics and Communication Engineering, Vellore Institute of Technology, Vellore, Tamil Nadu, India, for providing valuable support to carry out this study in VLSI & Nano Laboratory.

REFERENCES

1. Kajal and V. K. Sharma, "An efficient low power method for FinFET Domino or Logic Circuit," *Microprocessors and Microsystems*, vol. 95, p. 104719, 2022. doi: 10.1016/j.micpro.2022.104719
2. R. H. Vanlalchaka, R. Maity, and N. Pratap Maity, "A low power design using FinFET based adiabatic switching principle: Application to 16-bit arithmetic logic unit," *Ain Shams Engineering Journal*, vol. 14, no. 4, p. 101948, 2023. doi: 10.1016/j.asej.2022.101948
3. R. R. Vallabhuni, D. V. L. Sravya, M. S. Shalini, and G. U. Maheshwararao, "Design of comparator using 18nm FinFET technology for analog to digital converters," *2020 7th International Conference on Smart Structures and Systems (ICSSS)*, 2020. doi:10.1109/icsss49621.2020.9202164

4. S. Birla, N. Singh, N. K. Shukla, and S. Sharma, "A 9t finfet SRAM cell for ultra-low power application in the subthreshold regime," *Bulletin of Electrical Engineering and Informatics*, vol. 10, no. 6, pp. 3094–3101, 2021. doi:10.11591/eei.v10i6.3175
5. A. J. Bhavnagarwala, Xinghai Tang, and J. D. Meindl, "The impact of intrinsic device fluctuations on CMOS SRAM cell stability," *IEEE Journal of Solid-State Circuits*, vol. 36, no. 4, pp. 658–665, 2001. doi:10.1109/4.913744
6. C. Roy and A. Islam, "Design of low power, variation tolerant single bit line 9t SRAM cell in 16-NM technology in subthreshold region," *Microelectronics Reliability*, vol. 120, p. 114126, 2021. doi: 10.1016/j.microrel.2021.114126
7. R. R. Vallabhuni, D. V. L. Sravya, M. S. Shalini, and G. U. Maheshwararao, "Design of comparator using 18nm finfet technology for analog to digital converters," *2020 7th International Conference on Smart Structures and Systems (ICSSS)*, 2020. doi:10.1109/icsss49621.2020.9202164
8. P. U. Jain and V. Tomar, "Expanded noise margin 10T SRAM cell using Finfet device," *International Journal on Recent and Innovation Trends in Computing and Communication*, vol. 11, no. 9s, pp. 767–776, 2023. doi:10.17762/ijritcc.v11i9s.7959
9. R.P Rao, Design of static RAM cell through 16 nm FinFET technology", *International Journal of Management, Technology and Engineering*, vol. 8, pp. 2880.
10. R. R. Vallabhuni, K. C. Koteswaramma, B. Sadgurbabu, and G. A, "Comparative validation of SRAM cells designed using 18nm FinFET for memory storing applications," *SSRN Electronic Journal*, 2020. doi:10.2139/ssrn.3733530

13 Power Consumption and SNM Analysis of 6T and 7T SRAM Using 90 nm Technology

Devanshi Tripathi, Vedha Tripathi, Siddhant Bhushan, and Prachi Sharma

13.1 INTRODUCTION

Static random access memory (SRAM) is one of the most commonly used components in today's Very Large Scale Integrated Circuit (VLSI)-based circuits for their small size, low power consumption, and high-speed ability [1]. They also take a large portion in VLSI circuit designs because of the number of transistors involved in making one cell [2,3]. As the technology advances and the size of the SRAM cell reduces to micron and even sub-micron levels, the density of these chips increases significantly [3]. The positive part is that the speed of these chips increase significantly, but at the same time, the power consumption increases also.

The topologies of SRAMs can vary based on the number of transistors, for example, 6T SRAMs use six transistors to realize memory operations and 7T SRAMs use seven transistors to realize memory operations and so on; various types of SRAMs with different transistor numbers can be made depending on the application [4]. Another way SRAMs can differ is based on the size used, i.e. 45, 65, 90, 180 nm, etc. Based on the usage of the chips in various applications, the appropriate SRAM is used.

Low power consumption and minimal delay, resulting in high speed, are essential characteristics of SRAM circuits. These attributes have driven numerous researchers and scientists to explore methods to enhance or maximize their potential [5]. But many issues are faced while exploring; due to the size, lesser voltage has to be provided. This increases the delay of the SRAM, and in addition, the threshold voltage of the chip also has to be reduced.

There are many key factors to designing an optimized SRAM. An SRAM is chosen for a particular task depending on its architecture and purpose fitting the performance required. When designed while keeping in mind the architecture, algorithm, logic, layout, etc., there are many possibilities to create various types of SRAMs [4].

The next section discusses the model and working of 6T and 7T SRAM cells. The design criteria which we have adopted for power dissipation are discussed.

DOI: 10.1201/9781003483038-13

Then, with the help of the butterfly curve of each SRAM, we can determine the static noise margin (SNM), and it is used to make a stability analysis of the cells.

13.2 SIX TRANSISTOR SRAM OR 6T SRAM

The 6T SRAM consists of six transistors which are interconnected to serve as memory holders for the information we intend to supply. Two CMOSs are cross coupled to each other. The data stored within these inverters is accessed via the bit and bit line bar, using access inverters. The bit and bit line bar function as input lines during memory writing, while the (BN) Bit-line Bar serves as an output line for reading from the memory [6] (Figure 13.1).

When the word line (WL) is activated alongside the appropriate bit line and its complement (bit line bar), the system transitions into a state primed for read and write functionalities; if not, then the system transitions to a hold state [7].

Read Operation:

When the WL is activated, and the accessed memory cell is in a high state, the pre-charged capacitors maintain a stable voltage level of V_{DD}, as no voltage difference between Q and N3 (access transistor) occurs, thus no changes happen. However,

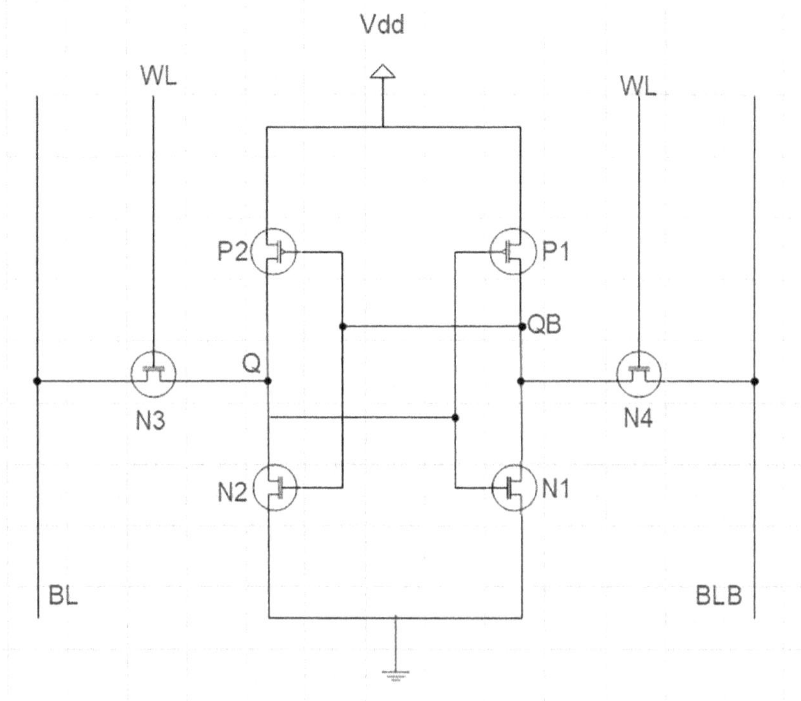

FIGURE 13.1 Circuit schematic of 6T SRAM.

in the case of Q bar and N4, a voltage disparity exists due to Q being at zero, and the access transition near Q is at V_{DD}, leading to capacitor discharge and a subsequent decrease in the bit bar voltage, hence providing an output of 1 [8].

Conversely, when the accessed memory cell holds a low state, the presence of a voltage differential between the bit and N3 nodes triggers an output of 0.

Write Operation:

Assuming low state, the bit line bar is grounded to create a voltage disparity, which is affected by the threshold voltage of N2: If V is less than Vth of N2, then P2 is turned on while the N2 is at off state, getting us an output as 1; hence, Q is successfully written to the memory from 0 to 1 [6].

Conversely, when writing to a memory cell set to a high state, and bit is grounded, if V is less than Vth of N1, then P1 is turned on, shutting off N1, respectively, giving Q as zero; hence, writing over is successful.

13.3 SEVEN TRANSISTOR SRAM OR 7T SRAM

The 7T SRAM circuit too consists of two CMOS cross coupled to each other; in this circuit, we additionally connect an N-type Metal-Oxide-Semiconductor (NMOS) transistor to the write line. And it too has to pass NMOS transistors connected to the bit and bit bar line. The transistors, NMOS 3 and NMOS 4, are connected to the read and write line to perform the read and write operations [5] (Figure 13.2).

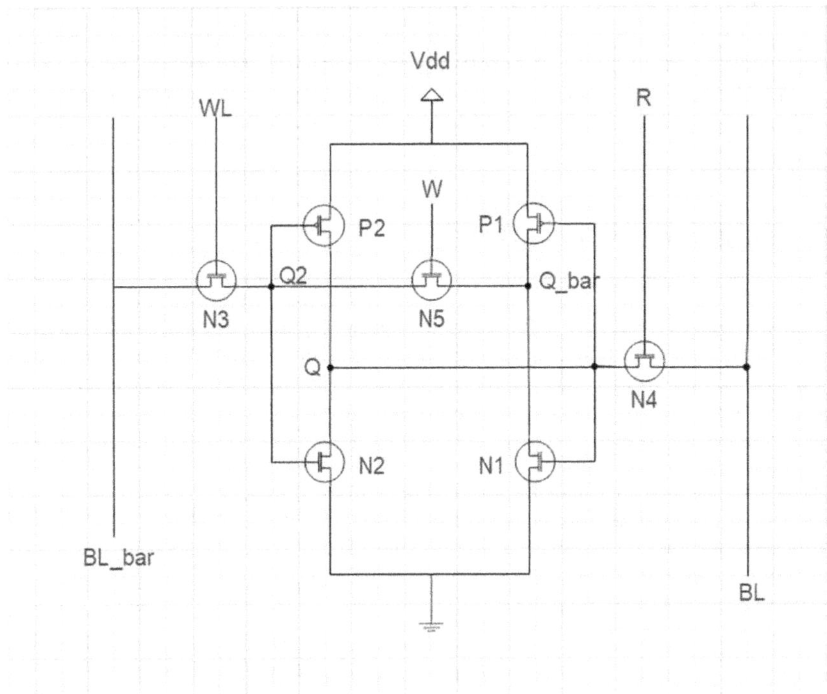

FIGURE 13.2 Circuit schematic of 7T SRAM.

13.3.1 Read Operation

In a given scenario, when activating the write line and read signals at a value of 1, while keeping NMOS5 active, the functionality of the SRAM cell hinges on the data stored in the memory bit Q. If Q assumes a value of "0", the read path involves just NMOS2 and NMOS4 transistors, enabling smooth data retrieval. However, if Q is "1", the read path becomes more intricate and sensitive. Now, it comprises a series connection of transistors NMOS1, NMOS5, and NMOS3. This setup presents a challenge as the series configuration reduces the cell's driving capability, potentially impacting its performance. To address this challenge, precise adjustments to the sizes of these transistors are necessary. This ensures the SRAM cell maintains reliability and operates optimally even in these demanding read scenarios [9]. This highlights the significance of meticulous transistor sizing in SRAM cell design, ultimately contributing to improved memory system performance.

13.3.2 Write Operation

The write operation of the 7T SRAM cell initiates by deactivating NMOS 5 to sever the feedback connection. BL_Bar, carrying the complement of the input data, activates NMOS 3 while keeping NMOS 4 inactive. Structurally, the 7T SRAM cell resembles two consecutive inverters, with inverter 2 preceding inverter 1. Through NMOS3 transistor, data is transferred from BL_Bar to Q2, subsequently driving inverter 2 along with PMOS 2 and NMOS 2 to generate Q, representing the cell's data. Similarly, Q drives inverter 1, comprising PMOS 1 and NMOS 1, to generate Q_Bar, which mirrors Q2 when the data is "0" and slightly exceeds Q2 when the data is "1". Following this, the write line is deactivated, and NMOS5 is activated to reestablish the feedback link between the two inverters, ensuring stable storage of the updated data [9,10].

13.4 RESULTS AND CONCLUSION

This chapter presents the design and implementation of SRAM memory and its power consumption and SNM analysis. We designed 6T and 7T SRAM memory cells for this research and have simulated the circuits in Cadence Virtuoso Software. In Figure 13.3a, we observe the circuit diagram of 6T SRAM in Cadence Virtuoso in 90nm technology. We have added a small circuit in the bottom containing pin writedata and a CMOS inverter in order to increase the efficiency of the SRAM circuit model. We are feeding in values to pins V_{DD}, writedata, WL, and write so as to observe the dc and transient analysis of the circuit accordingly. Similarly, in Figure 13.4a, we have designed the 7T SRAM cell in 90nm technology with same pin parameters so that we can observe the comparison of both based on same input values. The delay of 6T and 7T can subsequently be observed through transient analysis of 6T (Figures 13.3b and c) and transient analysis of 7T (Figures 13.4b and c). By performing dc analysis, we can observe the graph of qb vs q for both the circuits. On plotting q and qb of each graph with each other, we can obtain the butterfly curve of 6T SRAM (Figure 1.3d) and butterfly curve of 7T SRAM (Figure 13.4d).

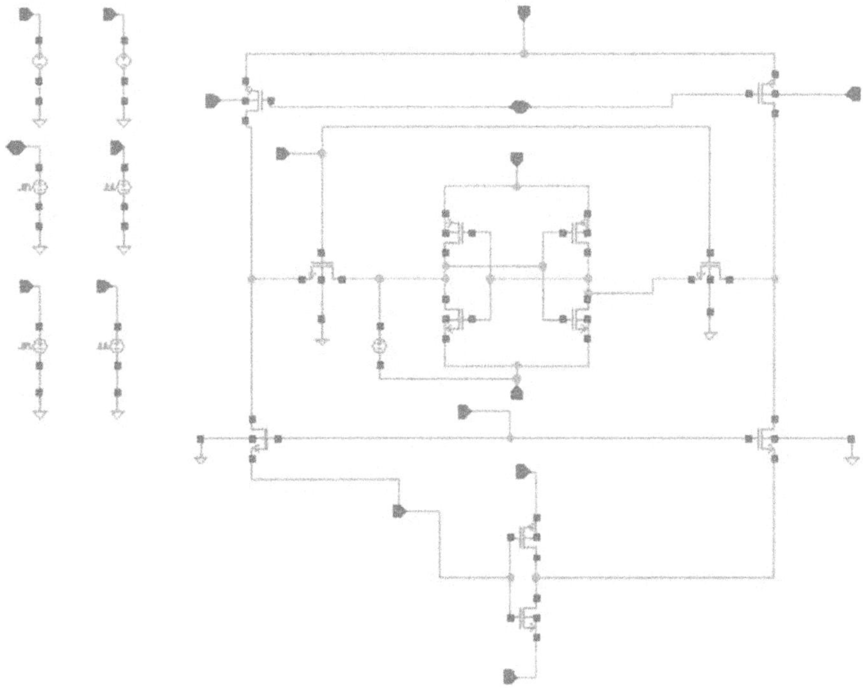

FIGURE 13.3a 6T SRAM circuit.

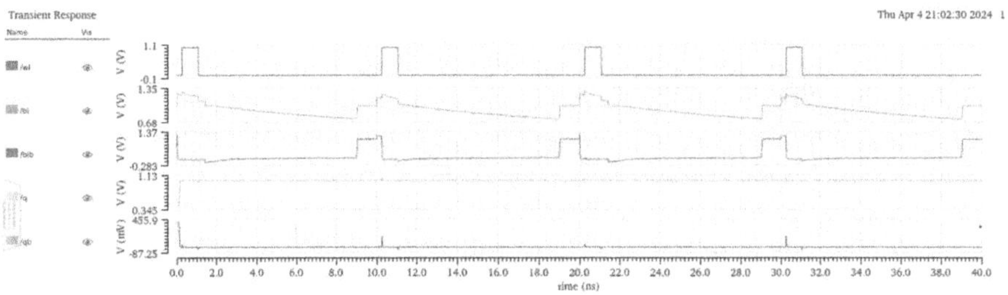

FIGURE 13.3b Transient analysis 1 of 6T SRAM.

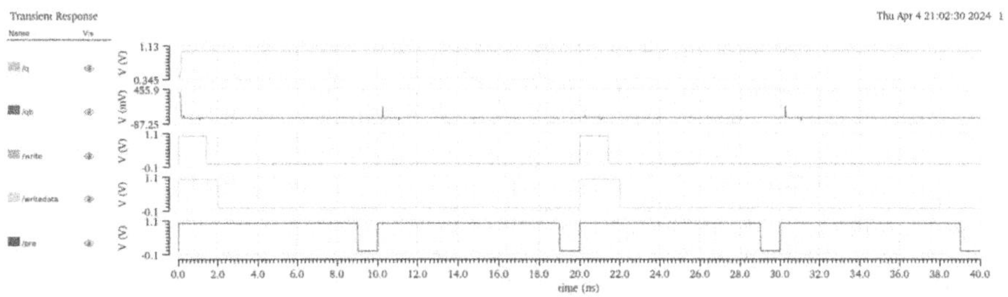

FIGURE 13.3c Transient analysis 2 of 6T SRAM.

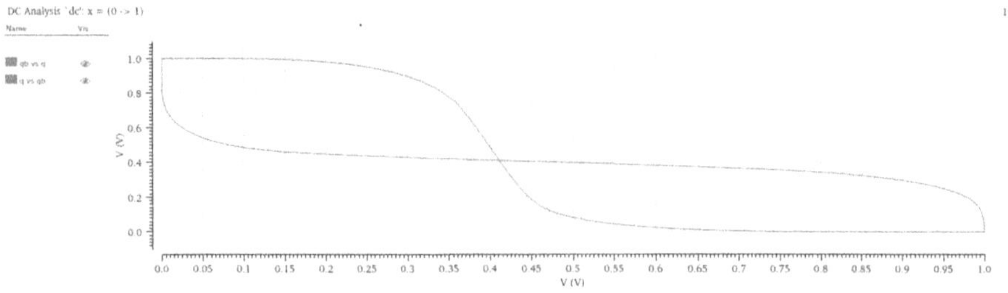

FIGURE 13.3d Butterfly curve of 6T SRAM.

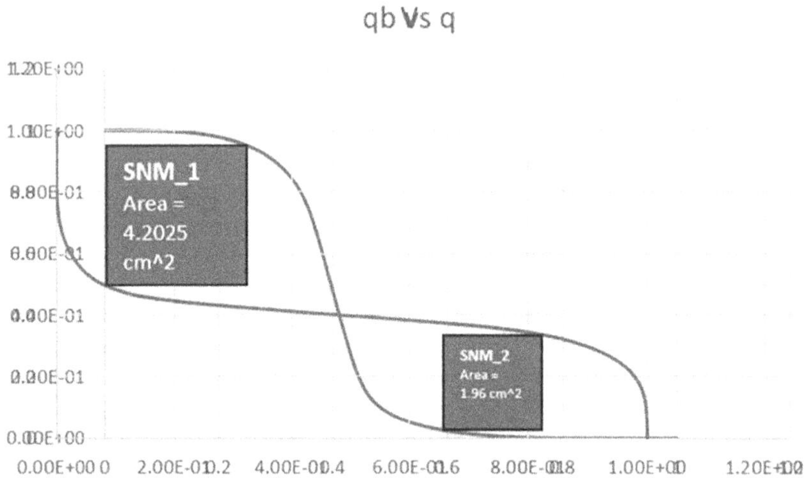

FIGURE 13.3e SNM of 6T SRAM.

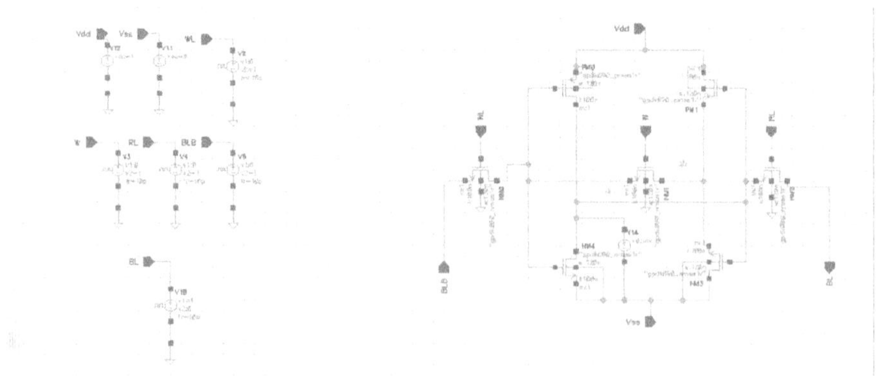

FIGURE 13.4a Circuit diagram of 7T SRAM.

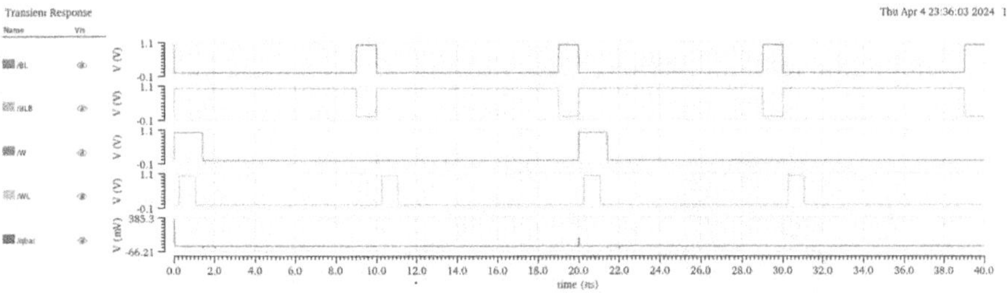

FIGURE 13.4b Transient analysis 1 of 7T SRAM.

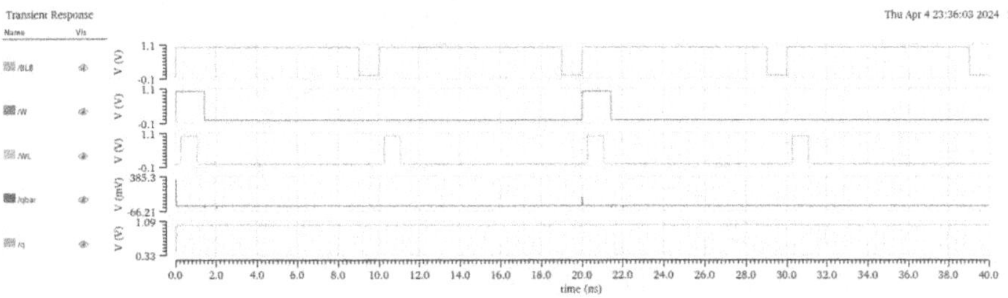

FIGURE 13.4c Transient analysis 2 of 7T SRAM.

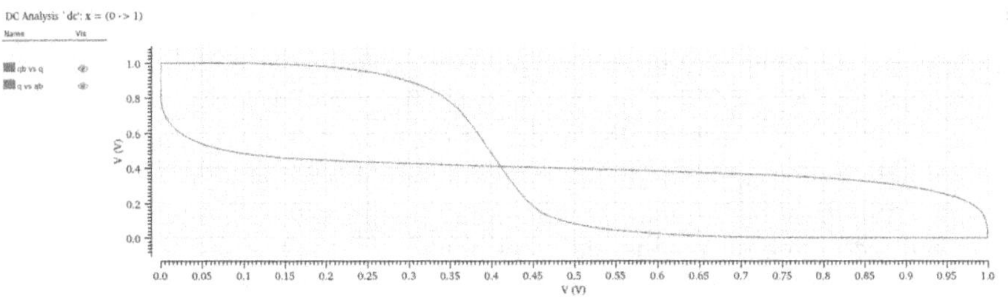

FIGURE 13.4d Butterfly curve of 7T SRAM.

The butterfly curve is a very important component through which we calculate the values of SNM (static noise margin) of both 6T and 7T SRAM (Figures 13.3e and 13.4e).

Power consumption of a component depends on various factors such as load capacitance and voltage supply [11]. Therefore, there is a significant focus on reducing power consumption as it offers numerous benefits, especially in energy-constrained applications such as portable devices and IoT-based sensors. This reduction in power consumption is instrumental in extending battery life and enhancing the overall energy efficiency of a device [12]. Our findings in Table 13.1 indicate that the power consumption of 6T SRAM is lower than that of 7T SRAM.

SNM is another crucial factor to consider when integrating SRAM into circuits. SNM evaluates the memory cell's ability to withstand errors induced by noise [12].

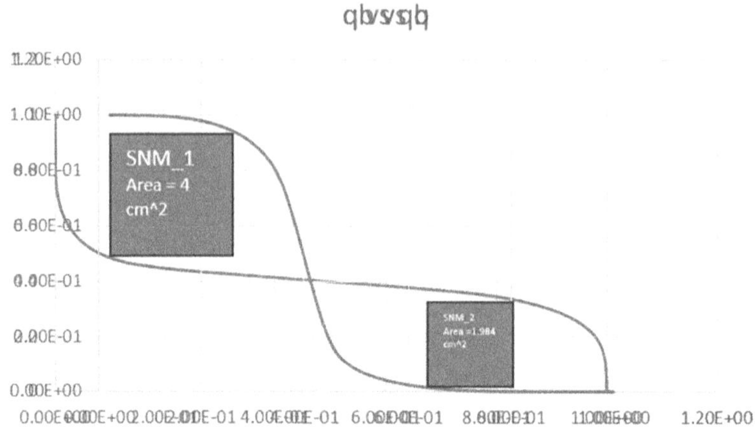

FIGURE 13.4e SNM of 7T SRAM.

TABLE 13.1
Power consumption and SNM comparison

SRAM	Supply Voltage (V)	Power (W)	SNM Value 1	SNM Value 2
6T	1	1.00854×10^{-7}	4.2025	1.96
7T	1	2.088854×10^{-5}	4	1.9044

Through analysis using the butterfly curve, we observe that the SNM of 7T SRAM surpasses that of 6T SRAM. This highlights the better performance of 7T SRAM, indicating its reduced susceptibility to data errors caused by noise. Such resilience to noise-induced errors makes 7T SRAM particularly beneficial for communication systems applications [13].

To conclude, our observations and analysis lead us to the conclusion that the performance of 7T SRAM, in terms of power consumption and resistance to data errors caused by noise, exceeds that of 6T SRAM.

ACKNOWLEDGMENTS

We would like to give our warmest thanks to our faculty supervisor, Dr. Prachi Sharma who made this work possible. Her guidance and advice carried us through all stages of writing this research. We would also like to thank the support staff/teachers (Department's staff and teachers) who helped us through all the doubts we faced using the software.

We give our special regards to our college, Vellore Institute of Technology, for providing us the resources and platform to finish this research paper without any hurdles and providing us ample amount of materials and mediums to access them.

Finally, we want to thank our parents and siblings for their wholehearted support and understanding when undertaking our research and writing the paper. Their prayer for us is what sustained us this far.

REFERENCES

1. Jennifer Eunice, R., S. AlimaAashika, and T. Bharathisankari. "A Study on 6T SRAM and 7T SRAM Cells." *Saun*, 2018.
2. Saun, S., and H. Kumar. "Design and Performance Analysis of 6T SRAM Cells on Different CMOS Technologies with Stability Characterization." In *IOP Conference Series: Materials Science and Engineering*, 561, no. 1 (2019): 012093. IOP Publishing.
3. Kaur, B., S. Kaur, and J. S. Kalsi. "Analysis of 64-Bit 6T, 7T & 8T SRAM Cell - 90 and 180 nm Technology." In *2 Days International Conference on Recent Innovations in Engineering, Science, Humanities and Management*, 2017.
4. Satheesh, B., and P. Benakop. "Comparison of Performance Analysis of 6T, 7T, 8T, 9T, and 10T-SRAM Cells in Terms of Low Power Dissipation and Delay." *Think India Journal*, 22, no. 14 (2019): 9732–9739.
5. Sujata. "A Comparative Performance Analysis of 6T, 7T, and 8T SRAM." *Impact Factor*, 11, no. 9 (2022).
6. Sah, R. K., I. Hussain, and M. Kumar. "Performance Analysis of a 6T SRAM Cell in 180nm CMOS Technology." *IOSR Journal of VLSI and Signal Processing (IOSR-JVSP)*, 5, no. 2 (2015): 20–22.
7. Koundinya, V., J. Pulivarthi, M. K. Chand, and D. Jayanth Sai Kumar. "Performance Evaluation of 6T SRAM Cell Using 90 nm Technology." *International Journal of Engineering Research & Technology (IJERT)*, 10, no. 9 (2021).
8. Rath, S., and S. Panda. "Analysis of 6T SRAM Cell in Different Technologies." *Circulation in Computer Science* (2017): 7–10.
9. Poojasree, N., and A. Pulla Reddy. "Design of 7T SRAM for Low Power Applications." 2020.
10. Mehrabi, K., B. Ebrahimi, and A. Afzali-Kusha. "A Robust and Low Power 7T SRAM Cell Design." In *2015 18th CSI International Symposium on Computer Architecture and Digital Systems (CADS)*, 1–6. IEEE, 2015.
11. Maravi, S. K., and P. S. Chouhan. "Comparative Analysis of 7T and 6T SRAM Using 0.18 μm Technology." 2012.
12. Singhal, V. K., and B. Singh. "Comparative Study of Power Reduction Techniques for Static Random Access Memory." *International Journal of VLSI and Signal Processing Applications*, 1, no. 2 (2011): 80–88.
13. Mittal, D., and V. K. Tomar. "Performance Evaluation of 6T, 7T, 8T, and 9T SRAM Cell Topologies at 90 nm Technology Node." In *2020 11th International Conference on Computing, Communication and Networking Technologies (ICCCNT)*, 1–4. IEEE, 2020.

14 Transforming Electronics
An Extensive Analysis of Hyper-FET Technological Developments and Utilisation

Debashish Dash and Manwendra Chauhan

14.1 INTRODUCTION

The development of transistors, the essential components of contemporary electronics, has been fuelled by the semiconductor industry's unrelenting march towards miniaturisation. Every generation of transistors, starting from the modest bipolar junction transistor (BJT) and ending with the widely used metal-oxide-semiconductor field-effect transistor (MOSFET), has improved performance and efficiency, accelerating technological progress in a variety of disciplines. However, as conventional MOSFET scaling approaches its limits, the search for a new transistor architecture that can maintain Moore's Law has been more intense. The hyper-FET is a cutting-edge competitor that has the potential to completely change the transistor market. Unlike the planar construction of MOSFETs, hyper-FETs claim a gate-all-around (GAA) architecture, where the channel is fully encompassed by the gate electrode [1–4]. This creative design provides several significant benefits.

Decreased Gate Tunnelling: Leakage currents are reduced by the close contact between the gate and channel, which lowers power consumption and boosts energy efficiency.

Improved Mobility: Higher carrier mobility and quicker device operation are made possible by the GAA structure's greater electrostatic control over the channel.

Better Scalability: By reducing the channel length without sacrificing gate control, the GAA shape pushes the limits of transistor density [5,6].

These inherent advantages have propelled hyper-FETs to the forefront of transistor research, attracting significant attention from both academia and industry [7–9]. Numerous research groups worldwide are actively exploring the potential of this technology, and the field is witnessing rapid advancements in material engineering, device fabrication, and circuit design.

This chapter explores the fascinating field of hyper-FETs, carefully analysing their basic ideas, features, and state-of-the-art at the moment [10]. We set out on a quest through the complex field of study, examining significant discoveries, recognising

DOI: 10.1201/9781003483038-14

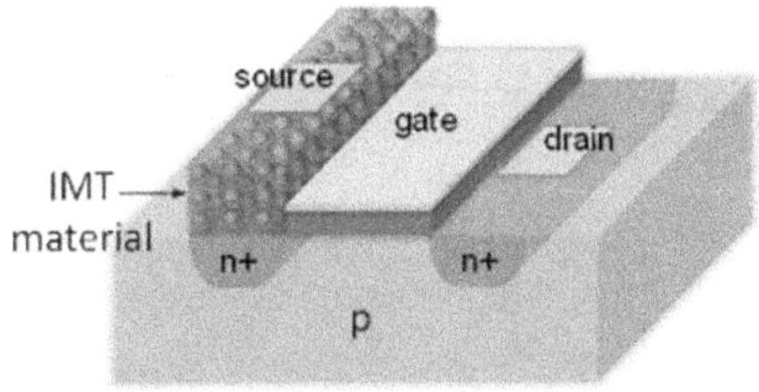

FIGURE 14.1 Hyper-FET structure.

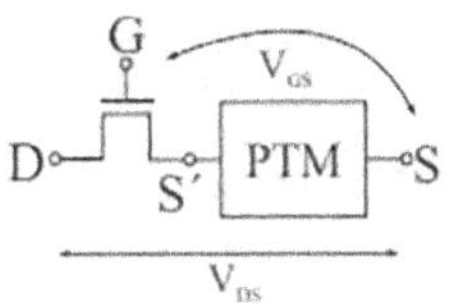

FIGURE 14.2 Schematic of the hyper-FET [40].

pressing issues, and considering viable remedies. We also illuminated how new device designs and developing materials will play a critical role in determining the direction of hyper-FET technology in the future (Figure 14.1).

14.2 HYPER-FET DEVICES AND APPLICATIONS

The search for efficient, high-performance transistors for next-generation electronics has prompted the investigation of novel materials and structures. Among these fascinating competitors is the hyper-FET, a new transistor with a GAA design and the tantalising prospect of vanadium dioxide (VO2) integration [11–15].

Of the various ways to improve the performance of hyper-FETs, one intriguing path involves the use of vanadium dioxide (VO2) [16]. This material is interesting because it shows an amazing insulator-to-metal phase transition that can be induced by electrical or thermal stimulation. Through the intentional integration of VO2 into hyper-FET architectures, scientists are opening new and fascinating avenues for ultra-low-power electronics.

In its insulating condition, VO2 functions as a dynamic gate, essentially turning off current via the hyper-FET channel while in standby mode. This significantly reduces leakage currents and static power consumption, which are a major bottleneck in conventional transistors. By introducing a voltage or heat pulse, VO2 quickly transitions to its metallic phase, restoring conductivity and allowing active device functioning. This breakthrough method provides the path for new circuit designs with near-zero standby power, significantly increasing overall energy efficiency.

The integration of VO2 presents its own set of challenges, requiring precise control over material properties and device fabrication techniques. However, the potential rewards are substantial, pushing the boundaries of low-power electronics and opening doors for applications in battery-powered devices, mobile computing, and the ever-expanding internet of things (IoT) (Figure 14.2).

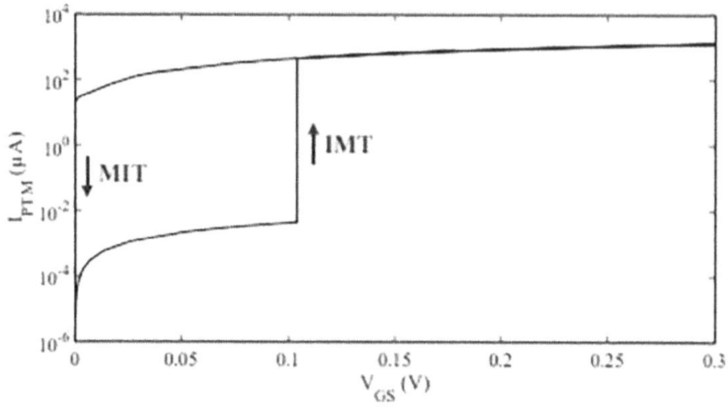

FIGURE 14.3 I-V curve of the PTM device [40].

Near 68°C, this unusual material transitions from an insulator to a metallic state. Below this temperature, its I–V curve mimics that of an insulator, with very little current flow. Above the transition, the curve becomes linear, indicating metallic behaviour and higher conductivity.

The distinct I–V properties of phase transition materials have applications in a variety of sectors, including memory devices: Hysteretic behaviour can be utilised to produce non-volatile memories, such as memristors, whose resistance state is determined by the voltage history [17].

Sensors: The sensitivity of conductivity to temperature, pressure, or magnetic field can be used for sensing purposes.

Electronic Switches: The rapid change in conductivity during the phase transition can be used to switch and control devices. Understanding the I–V properties of phase transition materials allows us to realise their enormous promise for novel technology and intriguing future applications (Figure 14.3).

This complicated dance of electrons and phases opens the way for revolutionary circuit designs, promising ultra-low-power operation is ideal for wearables, IoT devices, and mobile computing [18–20]. Hyper-FETs help to make devices more environmentally friendly by reducing energy usage.

Faster Data Processing: Increased transistor speed enables powerful, efficient computing and communication systems.

14.3 POWER AND SPEED EVALUATION OF HYPER-FETs

Hyper-FETs, with their GAA architecture and improved electron control, are promising candidates for the next generation of transistors [21–23]. However, the pursuit of quicker, more efficient electronics necessitates a careful examination of their power and speed capabilities. This section digs deeply into the methodology and criteria used to assess these essential features of hyper-FETs, providing a thorough analysis.

However, before they can dominate the circuits, we must first grasp their power consumption. This requires analysing three essential metrics:

DC Power Consumption: This shows how much power the hyper-FET consumes while idle (leakage) and during switching. Minimising both is critical to efficiency.

Leakage Current: This cunning culprit keeps the current flowing even while the hyper-FET is cooling. Hyper-FETs, with their improved gate control, significantly minimise this when compared to ordinary MOSFETs.

Energy Efficiency: This indicates how much energy is consumed by each flip of the hyper-FET switch, measured in Joules per operation. The lower the number, the more effective the device (Figure 14.4).

To accurately assess these variables, we use a three-pronged attack:

Device Level: We directly measure voltage and current on the hyper-FET, peeking inside like a microscopic detective.

Circuit Level: We place the hyper-FET in real-world circuit simulations, observing how it reacts under various workloads and determining its practical power consumption.

Benchmarking: We compare the hyper-FET to known champions such as FinFETs, using power measurements to objectively analyse efficiency advances. By using these methods, we can unravel the secrets of hyper-FET power, paving the path

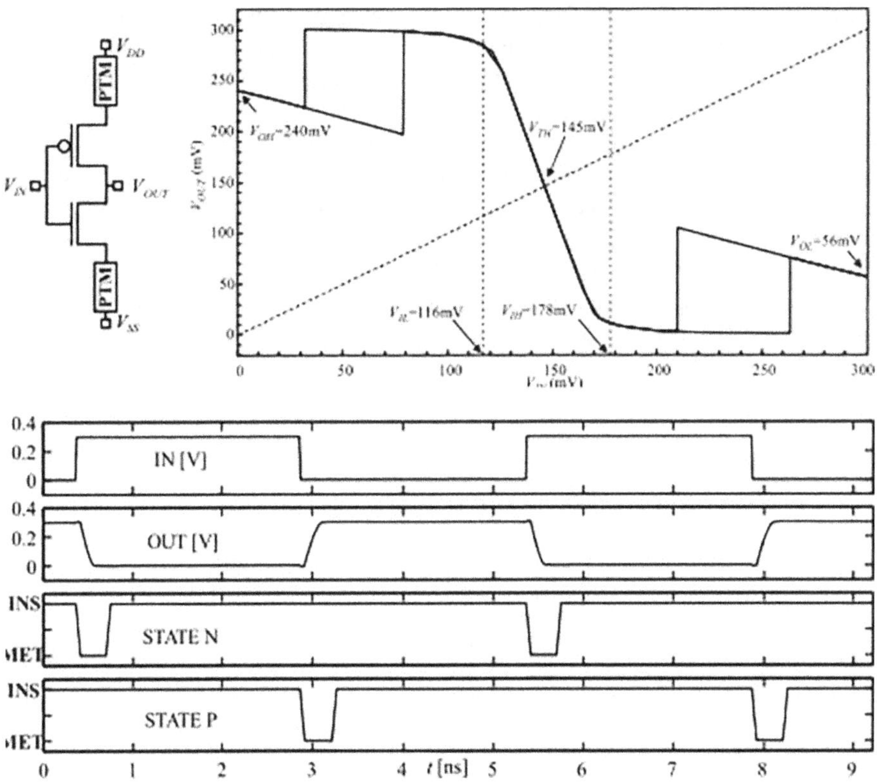

FIGURE 14.4 (a) Voltage transfer characteristic of a hyper-FET inverter (schematic also shown) [41]. (b) Waveforms corresponding to a hyper-FET inverter with the internal states of the n-type and p-type hyper-FETs [41].

for a future in which transistors operate harder but not on our batteries. To determine the full speed potential of hyper-FETs, researchers use a set of critical metrics and evaluation methods. Switching delay [24] is an important metric that reveals how long it takes for the hyper-FET to gracefully flip between logic states, much like a dancer changing positions. Minimising this delay is critical for increasing circuit speed and facilitating faster processing. Cut-off frequency (fT), a metric of the hyper-FET's high-frequency capability, denotes the frequency at which its current gain gracefully drops to unity. A higher fT, like a musician's broader vocal range, indicates faster signal amplification and opens possibilities for high-speed applications. Propagation delay estimates how long it takes for a signal to travel through a single stage of a multi-stage hyper-FET circuit, like a relay race in which each stage must pass the baton quickly. Reducing this delay increases the overall circuit speed and ensures a smooth flow of information. To evaluate these metrics, researchers use a variety of techniques. Electrical measurements require time-domain characterisation via pulsed pulses, like a stopwatch measuring a sprinter's split timings [25,26]. This directly evaluates switching delays and other transient responses, demonstrating the hyper-FET's agility. Frequency-domain analysis examines the hyper-FET's gain and phase response, like mapping a dancer's choreography onto a musical score. This examination reveals its high-frequency performance and cut-off frequency, demonstrating its ability to execute graceful manoeuvres at high speeds. Circuit simulations provide a virtual stage for testing hyper-FET-based circuits under a variety of conditions, much like a dress rehearsal before the big show. Observing propagation delays and total speed across diverse applications demonstrates their versatility and adaptation to different choreographies. Understanding hyper-FET performance necessitates recognising the complex link between power and speed. Optimising one frequently impacts the other, creating a fascinating dance of trade-offs: high speed, high power: aggressive voltage scaling for faster switching results in increased dynamic power usage. Balancing speed needs with adequate voltage scaling is critical for optimal performance (Figures 14.5 and 14.6).

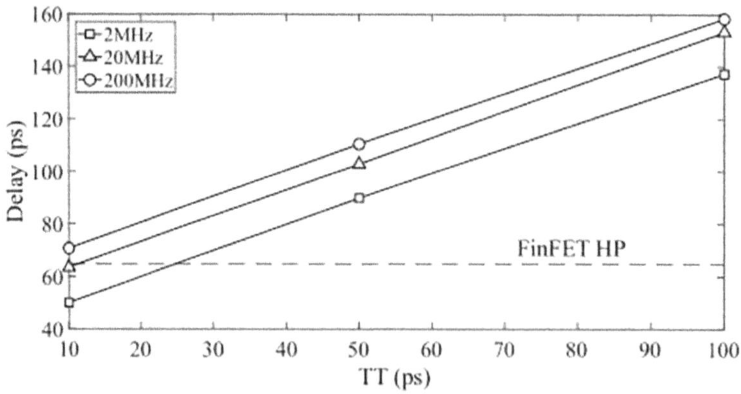

FIGURE 14.5 Delay versus PTM transition time for hyper-FET inverter [41].

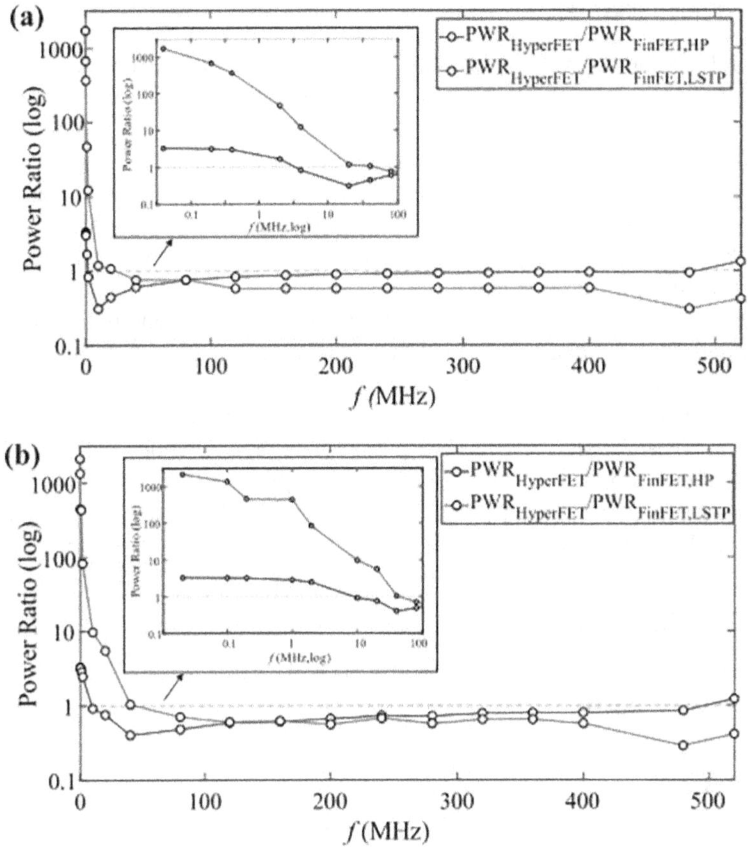

FIGURE 14.6 Power ratio versus frequency for inverter chain (a) periodic input and (b) random input [41].

Material Selection: Materials with higher mobility have lower leakage currents, perhaps resulting in a better power-speed trade-off. Exploring innovative materials is critical to achieving more optimisation [27–29].

Circuit Design: Tailoring circuit design to individual application requirements can improve power-speed balance. Techniques like logic families and power gating can provide considerable advantages. Evaluating hyper-FETs primarily based on power and speed measurements provides an incomplete picture [30]. Real-world performance requires a more holistic approach, considering additional factors.

Fabrication Complexity [31,32]: The manufacturing issues connected with GAA [33] structures might have an impact on scalability and cost, affecting their overall viability.

Integration Challenges: To ensure seamless integration into real-world systems, compatibility with existing circuits and design processes must be carefully considered.

Emerging Technologies: Integrating innovative materials such as vanadium dioxide for dynamic threshold control has the potential to boost power efficiency even more [34–39].

14.4 CONCLUSION

Hyper-FETs are on the verge of revolutionising the world of electronics. Their GAA architecture and improved electrostatics point to a future of faster, more efficient transistors capable of catapulting computers to new heights and altering areas ranging from mobile devices to renewable energy technology.

This chapter has explored the complex environment of hyper-FET research, focusing on power and speed characteristics, integration issues, and prospective applications. While various challenges remain, such as fabrication complexity and material restrictions, the research community is persistently seeking novel solutions. Novel materials, such as vanadium dioxide, hold great promise for additional efficiency improvements, while advances in circuit design and manufacturing techniques pave the way for more seamless integration with existing systems.

As we conclude this chapter, the enthusiasm for hyper-FETs is obvious. Their dance of force and speed is rife with possibilities, and the potential rewards are apparent. While complete acceptance of this disruptive technology may take time, the seeds of change have been planted, and the future of electronics is undeniably bright with the promise of hyper-FETs.

ACKNOWLEDGEMENT

This work was supported by the Department of Electronics and Communication Engineering, Vellore Institute of Technology.

REFERENCES

1. Aziz, Ahmedullah, et al. "Steep switching hybrid phase transition FETs (hyper-FET) for low power applications: A device-circuit co-design perspective: Part II." *IEEE Transactions on Electron Devices* 64(3) (2017): 1358–1365.
2. Ladd, Larry A., and William Paul. "Optical and transport properties of high quality crystals of V2O4 near the metallic transition temperature." *Solid State Communications* 7(4) (1969): 425–428.
3. Allam, Mohamed W., and Mohamed I. Elmasry. "Dynamic current mode logic (DyCML): A new low-power high-performance logic style." *IEEE Journal of Solid-State Circuits* 36(3) (2001): 550–558.
4. Alvandpour, Atila, et al. "A sub-130-nm conditional keeper technique." *IEEE Journal of Solid-State Circuits* 37(5) (2002): 633–638.
5. Anis, Mohab H., Mohamed W. Allam, and Mohamed I. Elmasry. "Energy-efficient noise-tolerant dynamic styles for scaled-down CMOS and MTCMOS technologies." *IEEE Transactions on Very Large Scale Integration (VLSI) Systems* 10(2) (2002): 71–78.
6. Tiri, Kris, Moonmoon Akmal, and Ingrid Verbauwhede. "A dynamic and differential CMOS logic with signal independent power consumption to withstand differential power analysis on smart cards." *Proceedings of the 28th European Solid-State Circuits Conference*, Manchester, United Kingdom. IEEE (2002).
7. Kursun, Volkan, and Eby G. Friedman. "Domino logic with variable threshold voltage keeper." *IEEE Transactions on Very Large-Scale Integration (VLSI) Systems* 11(6) (2003): 1080–1093.

8. Tiri, Kris, and Ingrid Verbauwhede. "A logic level design methodology for a secure DPA resistant ASIC or FPGA implementation." *Proceedings Design, Automation and Test in Europe Conference and Exhibition*, Vol. 1, Paris, France. IEEE (2004).
9. Tiri, Kris, and Ingrid Verbauwhede. "Design method for constant power consumption of differential logic circuits." *Design, Automation and Test in Europe*, Munich, Germany. IEEE (2005).
10. Tiri, Kris, and Ingrid Verbauwhede. "A VLSI design flow for secure side-channel attack resistant ICs." *Design, Automation and Test in Europe*, Munich, Germany. IEEE (2005).
11. Hua, Chung-Hsien, Wei Hwang, and Chih-Kai Chen. "Noise-tolerant XOR-based conditional keeper for high fan-in dynamic circuits." *2005 IEEE International Symposium on Circuits and Systems (ISCAS)*, Kobe, Japan. IEEE (2005).
12. Bucci, Marco, et al. "Three-phase dual-rail pre-charge logic." *Cryptographic Hardware and Embedded Systems-CHES 2006: 8th International Workshop*, Yokohama, Japan, October 10–13, 2006. Springer: Berlin Heidelberg (2006).
13. Wright, Peter, and Minghui Fan. "A dfm methodology to evaluate the impact of lithography conditions on the speed of critical paths in a VLSI circuit." *7th International Symposium on Quality Electronic Design (ISQED'06)*, San Jose, California, USA. IEEE (2006).
14. Hassoune, Ilham, et al. "Low-swing current mode logic (LSCML): A new logic style for secure and robust smart cards against power analysis attacks." *Microelectronics Journal* 37(9) (2006): 997–1006.
15. Zhao, Wei, and Yu Cao. "New generation of predictive technology model for sub-45 nm early design exploration." *IEEE Transactions on Electron Devices* 53(11) (2006): 2816–2823.
16. Ruzmetov, Dmitry, et al. "Correlation between metal-insulator transition characteristics and electronic structure changes in vanadium oxide thin films." *Physical Review B* 77(19) (2008): 195442.
17. Gopalakrishnan, Gokul, Dmitry Ruzmetov, and Shriram Ramanathan. "On the triggering mechanism for the metal–insulator transition in thin film VO_2 devices: Electric field versus thermal effects." *Journal of Materials Science* 44 (2009): 5345–5353.
18. Ruzmetov, Dmitry, et al. "Electrical triggering of metal-insulator transition in nanoscale vanadium oxide junctions." *Journal of Applied Physics* 106(8) (2009).
19. Bernstein, Kerry, et al. "Device and architecture outlook for beyond CMOS switches." *Proceedings of the IEEE* 98(12) (2010): 2169–2184.
20. Dadgour, Hamed F., and Kaustav Banerjee. "A novel variation-tolerant keeper architecture for high-performance low-power wide fan-in dynamic OR gates." *IEEE Transactions on Very Large Scale Integration (VLSI) Systems* 18(11) (2009): 1567–1577.
21. Theis, Thomas N., and Paul M. Solomon. "In quest of the "next switch": Prospects for greatly reduced power dissipation in a successor to the silicon field-effect transistor." *Proceedings of the IEEE* 98(12) (2010): 2005–2014.
22. Ionescu, Adrian M., and Heike Riel. "Tunnel field-effect transistors as energy-efficient electronic switches." *Nature* 479(7373) (2011): 329–337.
23. Jeyasingh, Rakesh Gnana David, Navakanta Bhat, and Bharadwaj Amrutur. "Adaptive keeper design for dynamic logic circuits using rate sensing technique." *IEEE Transactions on Very Large Scale Integration (VLSI) Systems* 19(2) (2009): 295–304.
24. Zhang, Meng, and Niraj K. Jha. "FinFET-based power management for improved DPA resistance with low overhead." *ACM Journal on Emerging Technologies in Computing Systems (JETC)* 7(3) (2011): 1–16.
25. Cavin, Ralph K., Paolo Lugli, and Victor V. Zhirnov. "Science and engineering beyond Moore's law." *Proceedings of the IEEE* 100. Special Centennial Issue (2012): 1720–1749.

26. Nikonov, Dmitri E., and Ian A. Young. "Uniform methodology for benchmarking beyond-CMOS logic devices." *2012 International Electron Devices Meeting*, San Francisco, California, USA. IEEE (2012).
27. Salvatore, Giovanni A., Alexandru Rusu, and Adrian M. Ionescu. "Experimental confirmation of temperature dependent negative capacitance in ferroelectric field effect transistor." *Applied Physics Letters* 100(16) (2012).
28. Peiravi, Ali, and Mohammad Asyaei. "Current-comparison-based domino: New low-leakage high-speed domino circuit for wide fan-in gates." *IEEE Transactions on Very Large Scale Integration (VLSI) Systems* 21(5) (2012): 934–943.
29. Markov, Igor L. "Limits on fundamental limits to computation." *Nature* 512(7513) (2014): 147–154.
30. Alioto, Massimo, and David Esseni. "Tunnel FETs for ultra-low voltage digital VLSI circuits: Part II–Evaluation at circuit level and design perspectives." *IEEE Transactions on Very Large Scale Integration (VLSI) Systems* 22(12) (2014): 2499–2512.
31. Wang, Tianshi, and Jaijeet Roychowdhury. "PHLOGON: Phase-based logic using oscillatory nano-systems." *Unconventional Computation and Natural Computation: 13th International Conference, UCNC 2014*, London, ON, Canada, July 14–18 2014. Springer International Publishing (2014).
32. Tena-Sánchez, Erica, Javier Castro, and Antonio J. Acosta. "A methodology for optimized design of secure differential logic gates for DPA resistant circuits." *IEEE Journal on Emerging and Selected Topics in Circuits and Systems* 4(2) (2014): 203–215.
33. Tsai, Wei-Yu, et al. "Low-power high-speed current mode logic using tunnel-fets." *2014 22nd International Conference on Very Large Scale Integration (VLSI-SoC)*, Kyoto, Japan. IEEE (2014).
34. Lu, Hao, and Alan Seabaugh. "Tunnel field-effect transistors: State-of-the-art." *IEEE Journal of the Electron Devices Society* 2(4) (2014): 44–49.
35. Maffezzoni, Paolo, et al. "Modeling and simulation of vanadium dioxide relaxation oscillators." *IEEE Transactions on Circuits and Systems I: Regular Papers* 62(9) (2015): 2207–2215.
36. Dagtekin, Nilay, and Adrian Mihai Ionescu. "Impact of super-linear onset, off-region due to uni-directional conductance and dominant C_{GD} on performance of TFET-based circuits." *IEEE Journal of the Electron Devices Society* 3(3) (2014): 233–239.
37. Jo, Jaesung, and Changhwan Shin. "Impact of temperature on negative capacitance field-effect transistor." *Electronics Letters* 51(1) (2015): 106–108.
38. Shukla, Nikhil, et al. "A steep-slope transistor based on abrupt electronic phase transition." *Nature Communications* 6(1) (2015): 7812.
39. Nikonov, Dmitri E., and Ian A. Young. "Benchmarking of beyond-CMOS exploratory devices for logic integrated circuits." *IEEE Journal on Exploratory Solid-State Computational Devices and Circuits* 1 (2015): 3–11.
40. Avedillo, María J., and Juan Núñez. "Insights into the operation of hyper-FET-based circuits." *IEEE Transactions on Electron Devices* 64(.9) (2017): 3912–3918.
41. Núñez, Juan, and María J. Avedillo. "Power and speed evaluation of hyper-FET circuits." *IEEE Access* 7 (2018): 6724–6732.

15 VLSI Realization of Smart Systems Using Blockchain and Fog Computing

R. Hanumantharaju, K. N. Shreenath, Sagarika Choudhury, Swagata Devi, and Naushad Manzoor Laskar

15.1 INTRODUCTION

The recent development of technology is having an impact on every aspect of human existence and altering how we utilize and view the world. Similar to the improvements the technology has brought about in other spheres of life, it is likewise discovering new avenues for advancement in the healthcare industry.

The progress of embedded systems has been upward-sloping during the last few decades. Indeed, a variety of industries have seen the use of embedded systems. The significance of innovative technology for human health and well-being has led to an expansion in its application across several domains in recent times. The use of artificial intelligence algorithms to evaluate data from multiple sensors is actually a useful new technique for detecting anomalies, diseases, problems, or alerts. Actually, it paves the way for the development of an effective medical diagnostic assistance system based on embedded systems.

The enhancement of security, the user experience, and the other aspects of the healthcare industry are the primary advantages that come with technological innovation. Electronic medical record (EMR) and electronic health record (EHR) systems provide these advantages. They still have certain problems with data integrity, user ownership, security of medical information, and so on. Blockchain technology could be the new technology that solves these problems. This technology promises to give medical records and other healthcare-related data a safe, uncheckable platform to be stored on. Before the development of contemporary technology, the medical field stored medical records utilizing handwritten mechanisms on paper-based systems. This system of medical records on paper was unorganized, insecure, inefficient, and susceptible to tampering. Because the patient's medical records were duplicated and redundant in every institution the patient visited, it also had to deal with this problem. There was a shift in the healthcare industry toward EHR systems, which integrated paper-based records and EMRs. Clinical notes and laboratory results were stored in these systems' various components. It was suggested that they improve patient safety by reducing errors and facilitating easier access to information. EHR systems

DOI: 10.1201/9781003483038-15

were designed to address the issues with paper-based medical records and offer a productive solution that would revolutionize the healthcare industry. Many hospitals throughout the world have adopted EHR systems because of their many advantages, chief among them being increased security and cost-effectiveness. Given that it gives healthcare a lot of functionality, they are regarded as an essential component of the healthcare industry. These features include lab test results, patient appointment scheduling, billing and accounts, and EMR storage. Many of the EHR systems in use in the healthcare industry have them available. Providing safe, tempered-proof, and scalable medical records across various platforms is the main goal. The proposed work aims to clarify data privacy and security issues in smart healthcare systems, analyze patient health records in fog nodes, protect data privacy through blockchain technology, allow authorized access to records, assist businesses in processing registered claims, and help files transfer securely using InterPlanetary File System (IPFS). In order to safely satisfy the needs of medical privacy and secrecy, the blockchain is used. In this work, the heart rate of patients is collected using smart figure sensors. Gathered by an embedded platform, this encrypted data will be processed on a back-end server before being sent on, functioning as a clever data relay.

15.2 LITERATURE SURVEY

The survey primarily focuses on the advantages of blockchain and its applications for improving the security and privacy of EHRs. It also examines problems with the current EHR system and how to fix them with cutting-edge technologies like blockchain. Shahnaz et al. [1] elucidate a system that could be used to implement blockchain innovation for EHRs in the medical services sector. Our proposed system's primary goal is to implement blockchain technology for EHRs. It also aims to provide secure electronic record storage by defining granular access rules that are specific to the system's users. According to Jetley and Zhang [2], the growing adoption of EHR as a supplemental information source in both Information Systems (IS) and supporting fields is a result of the medical services associations' increased acceptance of EHR frameworks. According to Cowie et al. [3], EHR frameworks are gradually storing imaging data that provide a more accurate picture of infection than is currently available in the majority of observational data sets. In [4] some countries, the EHR can be connected, on an individual basis, to various informational indices, such as general public well-being and lifestyle studies, disease vaults, and data collected in various areas. According to Mayer et al. [4], a comprehensive literature review on EHRs within a blockchain was carried out to identify and debate the primary problems, obstacles, and potential advantages of blockchain adoption in the healthcare industry. Zheng et al. [5] describe an outline of blockchain design and some commonplace agreement calculations utilized in various blockchain. Moreover, specialized difficulties and late advances are briefly recorded, additionally spreading out conceivable future patterns for blockchain. This chapter centers on blockchain innovation rather than advanced monetary forms. The importance of interoperability in EHRs for the smooth exchange of data among various healthcare stakeholders is discussed in the work done by Sonkamble et al.. [6]. According to Milojkovic [7], the blockchain functions as an appropriate record and is an attached information structure. To do this, all of the data on the blockchain must be recreated for each hub in the framework [8]. In comparison to current systems, a decentralized EHR system was proposed in this chapter, offering improved security and effective data sharing. The systematic analysis of existing blockchain-based solutions for safeguarding clinical data independent of distributed computing is described by Mahajan et al. [9]. The review focused on incomplete studies published in the last few years regarding blockchain-based security and identified various gaps and difficulties in the analysis. An architecture with the user, fog, and cloud tiers is proposed by Hanumantharaju et al. [10]. According to Zīle and Strazdiņa [11] who reviewed the literature, this chapter's goal is to provide a vague synopsis of the blockchain use cases that are currently being used in the data innovation sector. In addition, this chapter provides an overview of the core innovative perspectives and operational principles of blockchain technology, enabling an assessment of the presented use cases. An architecture with the user, fog, and cloud tiers is proposed by Hanumantharaju et al. [12]. Lakhan et al. [13] elucidate how each medical clinic haze hub is directly connected to the remote cloud. Fatokun et al. [14] describe the demonstration of how blockchain technology can be used to improve EHR systems and how the problems with privacy, security, and data exchange can be resolved by their suggested model.

According to a recent projection [15], the blockchain is recognized in the healthcare sector as a helpful tool to save costs, enhance the quality of medical records, and avoid data breaches. Healthcare practitioners are adopting internet of things (IoT) and modernizing the analysis and care process. The internet has made it possible to connect billions of sensors and devices. In patient care and therapy, the technology for remote patient monitoring is now commonly employed. These tools do, however, also pose serious privacy and security risks when it comes to data transfer and transaction logging. Security and confidentiality problems pertaining to medical information could potentially result in treatment delays or even endanger the patient's life. Blockchain is used to securely manage and evaluate sensitive medical data [16]. IoT architecture is discussed in the article by Dammak et al. [17] in order to monitor personal healthcare data. For the purpose of gathering Electrocardiogram (ECG) data, they employed an Arduino board equipped with a sensor. In order to expedite proof-of-work consensus, it was proposed in the article by Frikha et al. [18] to leverage an embedded architecture to generate a Hardware Intellectual Property (HW IP) in order to improve system efficiency, particularly with regard to real-time and low-power limits. A Verilog IP is employed by Tien et al. [19] with field programmable gate arrays (FPGAs).

15.3 DESIGN AND IMPLEMENTATION

This section provides a thorough overview and insight into the potential design of our suggested solution by discussing the architecture that was employed and the initial user interface that was conceived. Figure 15.1 shows the overall architectural design of the system. Admin can be a government entity or an authorized organization. Only the administrator can add the participants (hospital, doctor, insurance company, and patient) to the blockchain. Hospitals further can add doctors to their

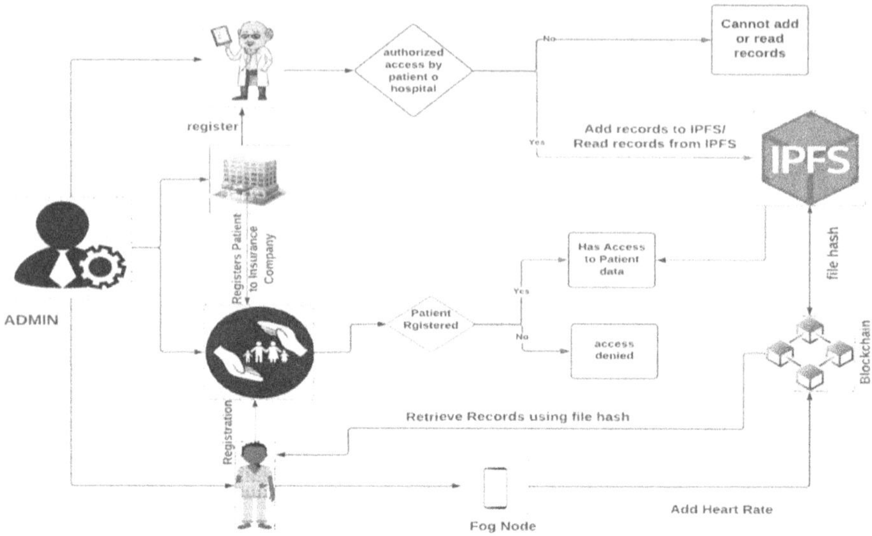

FIGURE 15.1 Architecture design.

network. Doctors can view the previous records of patients only when a patient has given access to a particular doctor; the doctors cannot make any changes to existing data in the records but can add information to it. Patients can register themselves with insurance companies or the hospitals can also register patients with insurance companies. Insurance companies have access to patients' data (view only). Patients can also add their records directly or using the fog node (block) and add them to IPFS storage. The patient can only view his/her records.

15.3.1 VLSI Hardware-Based Design

A heterogeneous embedded platform is used to create a multi-input EHR that takes advantage of numerous sensors connected via the IoT. The permissioned Ethereum blockchain, which this application employs, double-secures the data. In this endeavor, the goal is to apply a secure algorithm which ensures health information protection. It is imperative that confidential information will be difficult to obtain or readily compromised. The literature illustrates the energy-intensive efficiency of blockchain technology. Strong Personal Computer (PCs), servers, and Graphics Processing Unit (GPUs) are therefore used. We provide a system that is faster, more secure, energy-efficient, and uses fewer resources than a conventional PC. Energy conservation should also be handled by the hardware and software working together. With the use of FPGA, a fast system is the one that is safe and low power. Various types of sensors with a variety of data are employed. These medical data are obtained from a variety of sources, such as healthcare facilities and paramedical patient data (heart rate). These specifics are either immediately used in transactions or stored in a database. The database link is used to protect transactions after it has been encrypted.

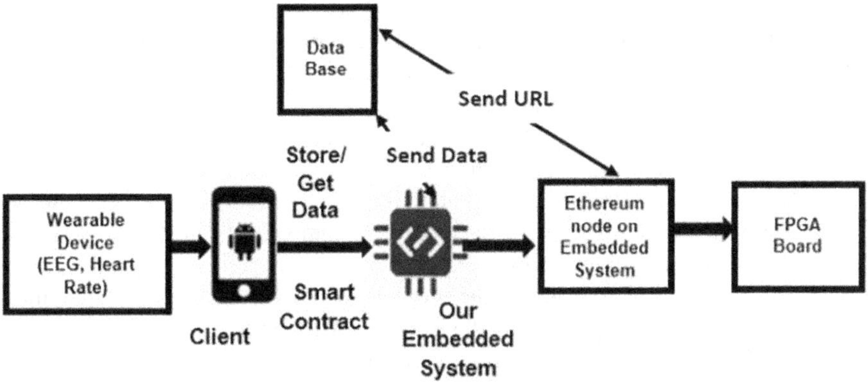

FIGURE 15.2 Block diagram of proposed system.

15.3.2 Proposed System

Wearable technology is used to collect data, such as heart rate. When the wearable system option is chosen, a smartphone serves as a relay to send data to "our system" via Bluetooth or Wi-Fi, and an embedded hardware with a powerful multiprocessor is used. Every bit of information that embedded hardware collects is stored locally in a database. Stored data is accessed via a web server that runs on the integrated hardware, making it simple and straightforward for a variety of users to access. Blockchain is used to store vast amount of data. Ethereum blockchain technology will make use of an embedded hardware design. The components are as follows (Figure 15.2):

1. Different nodes that facilitate the formation of encrypted transactions are operated by the embedded hardware, which is connected via Wi-Fi. Before being delivered, the data is encrypted and transferred via a connection to the FPGA.
2. The complex encryption and the blockchain mining functions are controlled by the FPGA chip. Using FPGA as a cryptoprocessing unit instead of embedded hardware-based software, encryption has one additional important advantage. With FPGA, data storage security can be increased. First off, the architecture can be greatly customized using specific Central Processing Unit (CPU) cores that are neither extensively used nor challenging to decode. The second step involves using encrypted bitstreams, which fortifies the decryption procedure even further.

15.3.3 Tools for Software

React is a front-end JavaScript library that is free and open-source and is used to create user interfaces using User Interface (UI) components. Meta and a group of independent developers and businesses maintain it.

Android Studio: The official Integrated Development Environment for creating Android applications is called Android Studio. It offers additional features that boost app developers' productivity. Its primary component is a Gradle-based build system that supports NDK and C++ and offers a quick and feature-rich emulator for app testing.

Flutter: With just one codebase, developers can create stunning natively compiled desktop, web, and mobile applications with Flutter, Google's portable UI toolkit. Flutter is a free and open-source software that integrates with existing code and is utilized by developers and organizations globally.

Ethereum virtual machine is a computation engine with millions of executable projects that functions similarly to a decentralized computer. With Remix, you can write Solidity contracts directly from your browser. It's an open-source tool.

Truffle, Ganache, and Drizzle are the three distinct tools that make up the Suite ecosystem for Web3 development. Truffle is a framework for testing smart contracts, an asset pipeline, and a development environment.

Ethereum is an open-source, decentralized blockchain that supports smart contracts. Ethereum is a decentralized blockchain platform that creates a peer-to-peer network for the safe execution and validation of smart contract application code. Without the need for a reliable central authority, participants can transact with one another using smart contracts. The proof-of-work consensus, which is currently used by the Ethereum blockchain, is carried out by miners competing to create new blocks that contain processed transactions. The victor receives some newly created Electronic Health Record (ETH) in addition to sharing the new block with the entire network. The computer that can complete a math problem the fastest wins the race and creates the cryptographic link between the current block and the previous block. It is the task of "proof-of-work" to solve this puzzle. The proof-of-work consensus mechanism, which requires more than 51% of the votes to approve a transaction, protects the network.

Visual Studio is an open-source software that is both lightweight and powerful, and it requires an extension to be developed further.

MetaMask is a free online multipurpose cryptocurrency wallet that lets users interact with the Ethereum blockchain, store and exchange virtual currencies, and host an evolving collection of decentralized applications (dApps).

Ganache is a personal Ethereum blockchain where you can develop apps, deploy contracts, run tests, and carry out other tasks all for free. It is also used to test smart contracts. To test your Solidity contracts, you can set up a personal Ethereum blockchain using Ganache. It is used to test smart contracts. The Ganache provides the local environment for the development of the blockchain which creates an emulated Ethereum network hosted over a local network, and it provides the Fake Ethereum for carrying out the transaction with actual cost calculation for each transaction which can be customized and can be directly pushed into the main Ethereum network after complete development.

NPM Node JS: Written entirely in JavaScript, Node Package Manager (NPM) is the default package manager for Node.js. NPM is a command-line client that manages all of the Node.js packages and modules. Along with Node.js installation, it is installed on the system. NPM is used to install the necessary packages and modules for the Node project. All the files required for a module are included in a package, and modules are JavaScript libraries that can be included in a Node project based on the project's specifications. The package.json file is not able to install all of a project's dependencies using NPM. Packages can also be updated and uninstalled. Using the semantic versioning scheme, each dependency in the package.json file can specify a range of acceptable versions, enabling developers to automatically update their packages without having to make unintentional breaking changes. An open-source development platform called Node.js (Node) is used to run JavaScript scripts server side. Node is helpful when creating apps that need a continuous connection between the browser and the server. Real-time apps like chat, news feeds, and web push notifications are common uses for Node. Node.js is designed to use a single thread with a single process at a time and to operate on a dedicated HTTP server. Applications built with Node.js are asynchronous and event based. On the Node platform, the code is not written in the conventional sequence of receive, process, send, wait, and receive. Rather, Node sends out short requests one after the other without waiting for a response, handling incoming requests in a continuous event stack.

15.3.4 Tools for Hardware

1. Smartwatch for medical data
2. FPGA board
3. An embedded hardware

Figure 15.3 shows the system peripherals such as EEG headset, smartwatch, and other sensors for health data collection. The FPGA board is connected to MSP430

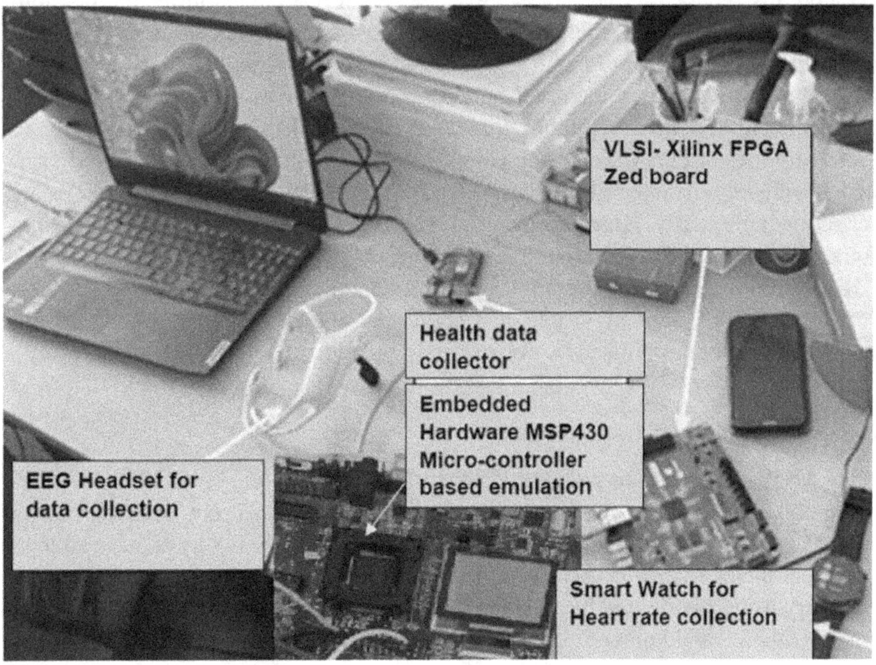

FIGURE 15.3 Hardware implementation of proposed system.

Embedded controller for the implementation of proof-of work (PoW) consensus Ethereum for our FPGA as explained in the next section.

15.3.5 Algorithms

Consensus methods in blockchain platforms can be classified as permissioned (such as Hyper Ledger and Ethereum Private) or permissionless (such as Ethereum). A permissioned blockchain has preselected nodes, as opposed to a permissionless blockchain where anyone can join. This makes it possible to use highly scalable and bandwidth-efficient consensus types. Proof-of-authority (PoA) consensus is a type of consensus that offers fault tolerance and high performance. Gavin Wood, a cofounder of Ethereum and Parity Technologies, coined the term in 2017.

Working of PoA: In PoA, nodes that have demonstrated their legitimacy are granted the ability to create new blocks. These nodes are known as "validators," and the software they run enables them to group transactions into blocks. Since the process is automated, validators do not need to keep a close eye on their computers, but they do need to keep them secure. PoA works well in distributed trust environments, such as public networks like PoA Network. Because the PoA consensus algorithm values identities, block validators are staking their reputations rather than coins. PoA is protected by confidence in the chosen identities.

Conditions for PoA Consensus: PoA consensus is generally applied under the following conditions, though it may vary depending on the specific implementation. Validators must attest to their true identities. A candidate needs to be prepared to risk both his reputation and money. Strict selection procedures lower the possibility of choosing dubious validators and encourage sustained dedication to the blockchain. Every candidate must be treated equally in the validator selection process. Verification of validators' identities is necessary to preserve the blockchain's integrity. There ought to be a procedure in place to choose trustworthy validators.

Advantages of PoA Consensus: High capacity for risk new blocks that are generated at a predictable interval that varies for PoW, provided that 51% of the nodes are not acting maliciously. **High Transaction Rate**: This is more robust than computationally intensive algorithms like PoW.

Limitations: PoA is an attempt to improve the efficiency of centralized systems; PoA validators are publicly visible. It is not decentralized. The identities of the validators could be compromised by outsider manipulation.

15.3.6 Module Implementations

IPFS MODULE: Peer-to-peer, or P2P, storage networks are what IPFS is. Peers can access content from anywhere in the globe by relaying, storing, or doing both at once. If you ask IPFS to locate something, it can do so by using its content address instead of its location. Understanding IPFS is based on three key concepts: content discovery using distributed hash tables (DHTs), content linking via directed acyclic graphs (DAGs), and unique identification via content addressing. The IPFS ecosystem is made possible by these three pillars that support one another. Let us begin by discussing content addressing and content unique identification. To store patient medical records, doctor-consulted reports, and other data, we have implemented an IPFS module. Physicians, insurance companies, and patients can all access the same records.

Transaction Module: We handled the transactions using a software cryptocurrency wallet called MetaMask, which communicates with the Ethereum blockchain. Through a mobile app or browser extension, it enables users to access their Ethereum wallet, which can be used to interact with decentralized applications. This module's implementation includes the MetaMask wallet, which keeps track of several user accounts and their corresponding cryptocurrency balances. Every transaction made through this wallet is mined by Ganache. Additionally, the transaction is completed using the money in the wallet. Starting with the front end, we used the JavaScript Library ReactJS and combined it with some styling libraries Tailwind CSS and Bootstrap. We implemented it using a template of boilerplate code generated by Truffle Suite which has the predefined connectivity with the Ethereum blockchain via Web3.js. We used a few tools to implement the blockchain as a backend: Ganache, which offers a private Ethereum blockchain environment that lets you emulate the Ethereum blockchain to interact with smart contracts, limited accounts, and cryptocurrency testing; and Solidity, an object-oriented programming language. We handled the transactions using a software cryptocurrency wallet called MetaMask, which communicates with the Ethereum blockchain. Through a mobile app or browser extension, it enables users to access their Ethereum wallet, which can be used to interact with decentralized applications. The majority of healthcare organizations today lack the infrastructure necessary to safeguard patient data from unauthorized access, which means that current EHRs might not satisfy patients' privacy demands. Healthcare 4.0, which makes use of blockchain, cloud computing, big data, and the IoT, has forced the industry to address security issues with processing medical data. To secure such data, several centralized cryptography solutions were introduced; however, they fell short of fully resolving the issues.

15.4 RESULTS

Figure 15.4 describes the admin page. Admin is the trusted center of authority; any participant in the network has to be approved by the Admin. Before making any transaction in the network, the participant must be added by the Admin. Figure 15.5 shows the Patient Dashboard where the patient has the option to grant or revoke access to viewing/adding his/her data onto the blockchain. Figure 15.6 shows the patient's home page where the patient can view the previous medical history and descriptions given by the doctors and also can add data to the existing blockchain. Figure 15.7 shows the doctor's home page where the doctor can view the patient's previous records. Also, a doctor has the option to add records to the patient's existing data. The doctor also has the option to add the consultations for the patient's future reference. When patient records are accessed in an EHR system, all such events must be documented in a log file for future access history audits. The log file can be viewed as a legal document and is used to reconstruct the previous state of medical records. As a result, we should, if at all possible, make the log file unchangeable and strictly guard against unauthorized access. They employ a centralized system that

FIGURE 15.4 Patient dashboard.

makes it possible for physicians and patients to look up a patient's dispersed medical records. One disadvantage of such a centralized architecture is that as the system gets bigger, it may run the risk of becoming a single point of failure and developing a data flow bottleneck. Patients fill out the information. The patient's information, including their heart rate, will be computed and entered into the patient list. The date of the patient's checkout will also be stored. Patient checkup (test) files can be kept in this patient list. Both the patient and the doctor have access to it. Additionally, the doctor can add files and consult with patients in the patient list. The doctor will consult with the patient during the additional consultation. In the patient list, they can also upload files. Figure 15.8 shows the appointments of the doctor, i.e., the patients who have given their access to add records to their block.

Figure 15.9 shows the hospital's home page. Hospitals can grant access to patient records to the insurance company by entering the public addresses of both parties. Figure 15.10 shows the insurance page; insurance companies can view all the

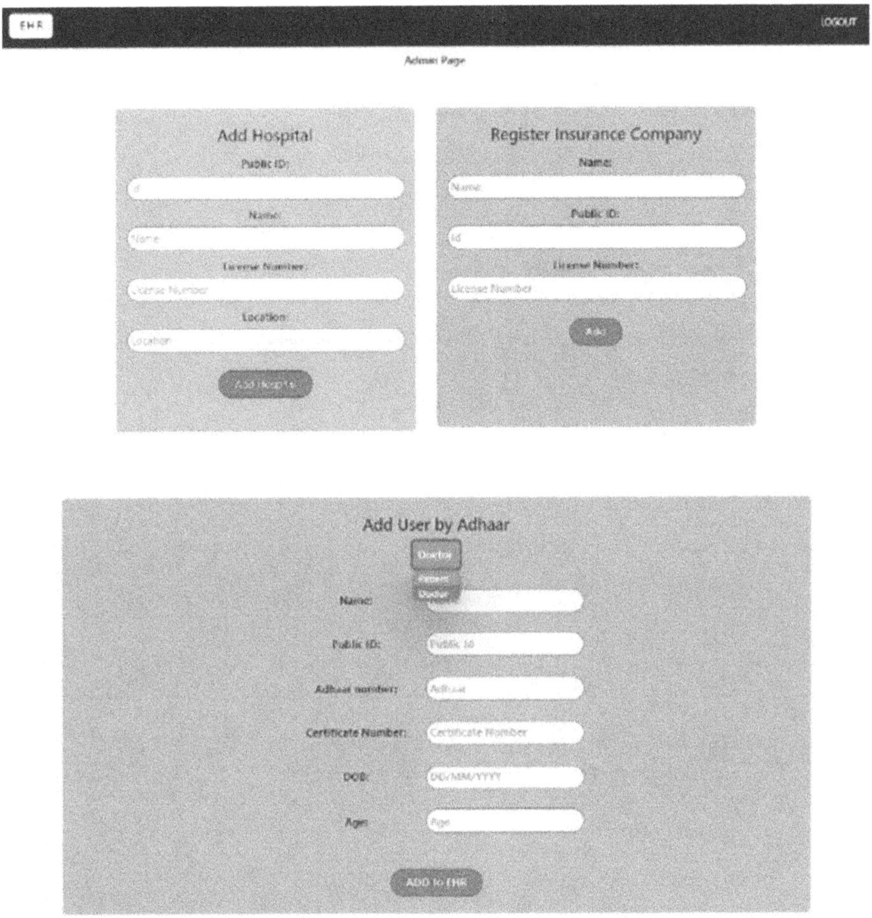

FIGURE 15.5 Admin page.

previous records of the patient. Insurance companies can only view the data; they cannot add the data. Figure 15.11 is the Flutter Login Screen that shows the login page of the Android application. Here, the user needs to successfully enter all his/her credentials to view the main screen, where the actual measuring process happens.

Figure 15.12 is the heartbeat measure that shows a person checking his heart rate by placing his finger on the camera. A person has to place his finger on the camera for a minute based on the intensity difference of the images taken by the application that will detect the heartbeat rate. Figure 15.13 is the heartbeat result that shows the sample heartbeat measure. This screen shows the measured output, i.e., Beats Per Minute (BPM) after sensing from the finger. The same result will be added to the EHR to the account of the user who is currently logged in. Figure 15.14 is the history of heartbeat that shows the history of heartbeat measures. In this screen, all the previous history of measured BPM will be listed with the date and time. This data may further be used to analyze the BPM record.

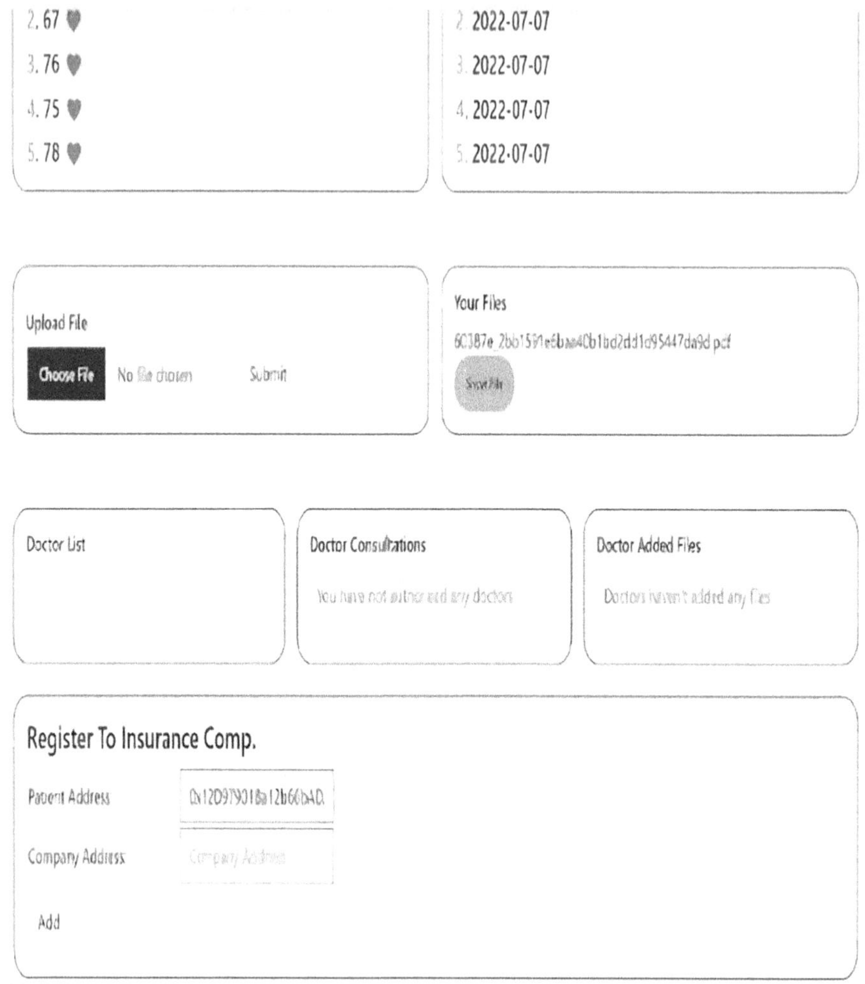

FIGURE 15.6 Doctor's homepage.

15.5 CONCLUSION

This work implemented a practical e-health platform based on VLSI systems and a proof of concept for blockchain-based e-health system. Since many hospitals and medical patients are linked through EHRs, robust security-based cryptography algorithms that are computationally efficient should be necessary to form the blockchain model. When performing medical data sharing, the blockchain model should take into account the possibility of various threats, including quantum threats. Scalability, computational load, and security performance efficiency are the three main requirements of the blockchain model. The model ought to have a low computation and a low power overhead. Our work also focuses on patients who will visit several hospitals in their lifetime. Since each hospital will have its own EHR and structure, it will

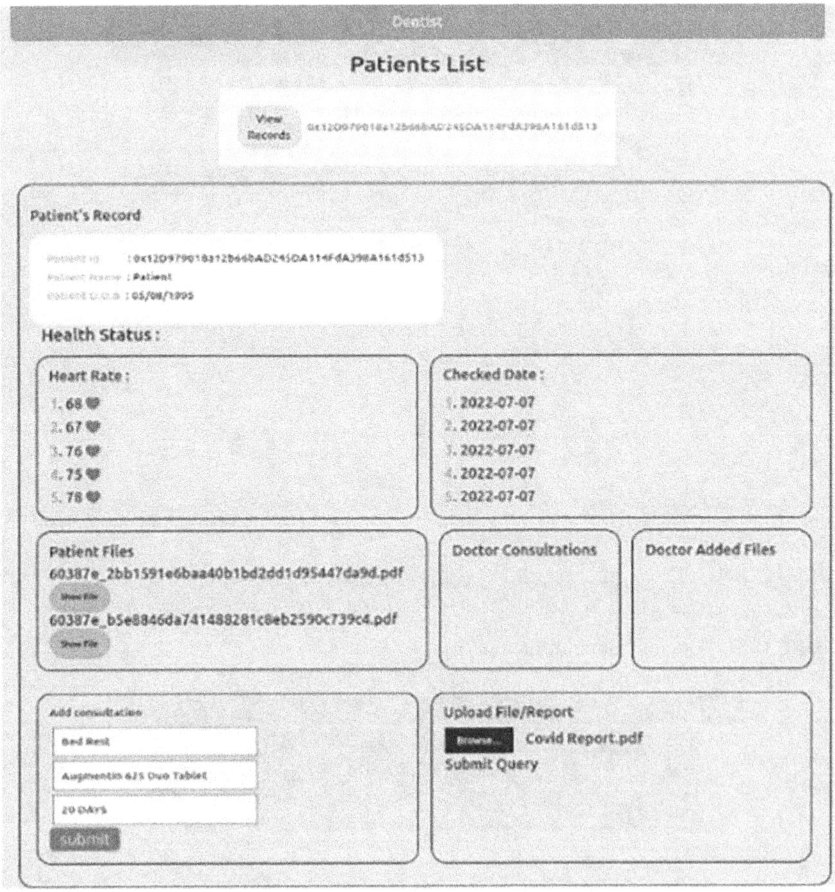

FIGURE 15.7 Patient homepage.

be difficult to transfer EHR within the same hospital, which will lead to loss of lots of patient data that could be used to analyze a patient's health history for better medication and health. Considering a current methodology where each hospital stores patient data in its central database, which poses a risk of a single point of failure, and which can be prevented by using a distributed structure of blockchain, the data is securely distributed across multiple nodes, ensuring redundancy and resilience. In future work, we intend to integrate a variety of health monitoring tools and enhance performance. Additionally, we intend to incorporate the payment module into the current framework. To do this, we must take a few factors into account. Specifically, we must determine the cost of a patient's doctor consultation on this decentralized blockchain system. To adhere to the values of the healthcare industry, we would also need to define specific policies and regulations. Millions of people's EHRs could be stored in an EHR. With time, the volume of data will increase exponentially.

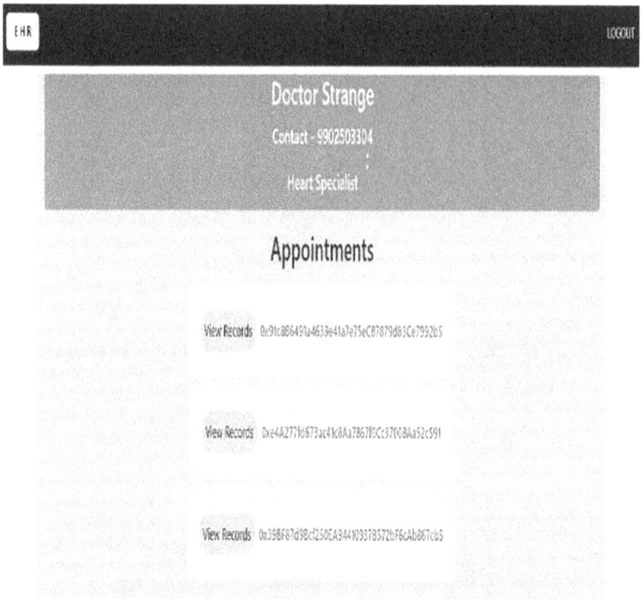

FIGURE 15.8 Doctor's appointment page.

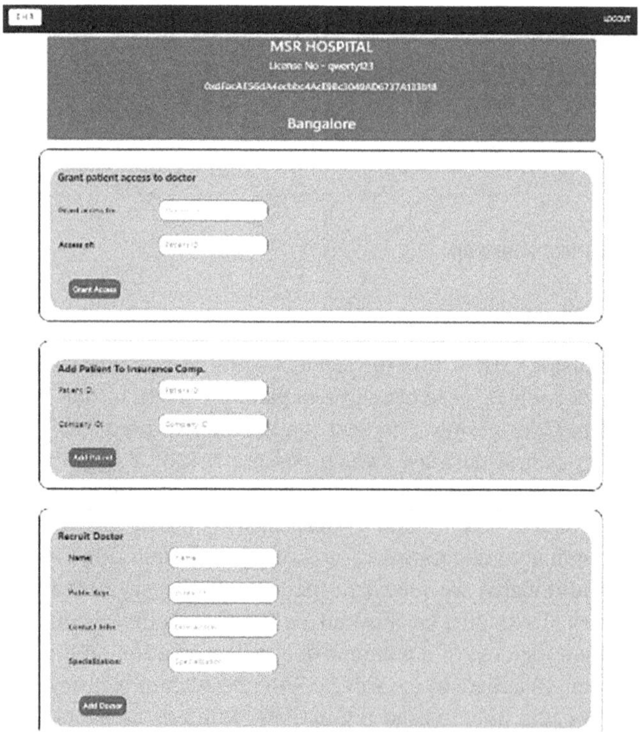

FIGURE 15.9 Hospital homepage.

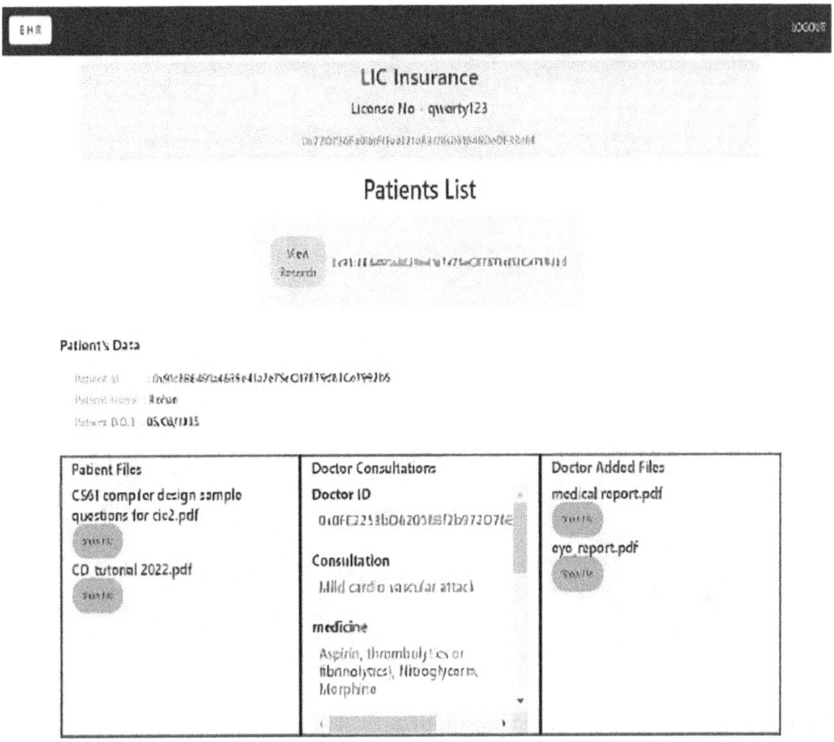

FIGURE 15.10 Insurance page.

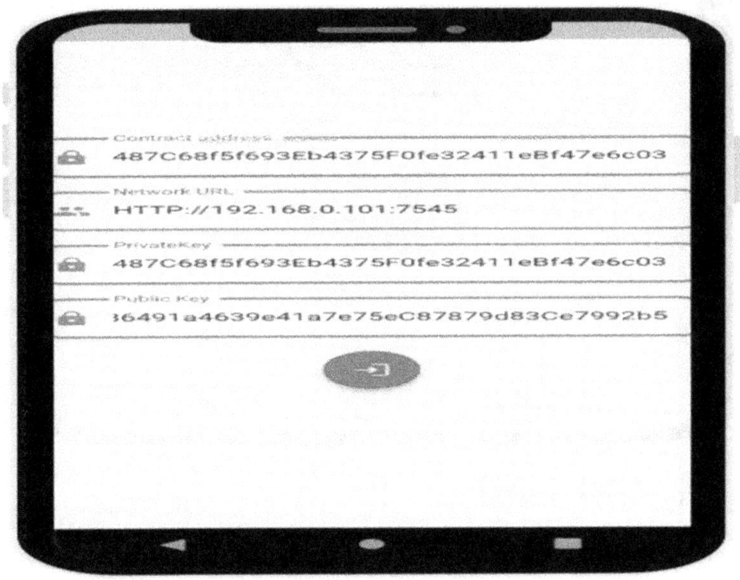

FIGURE 15.11 Flutter page.

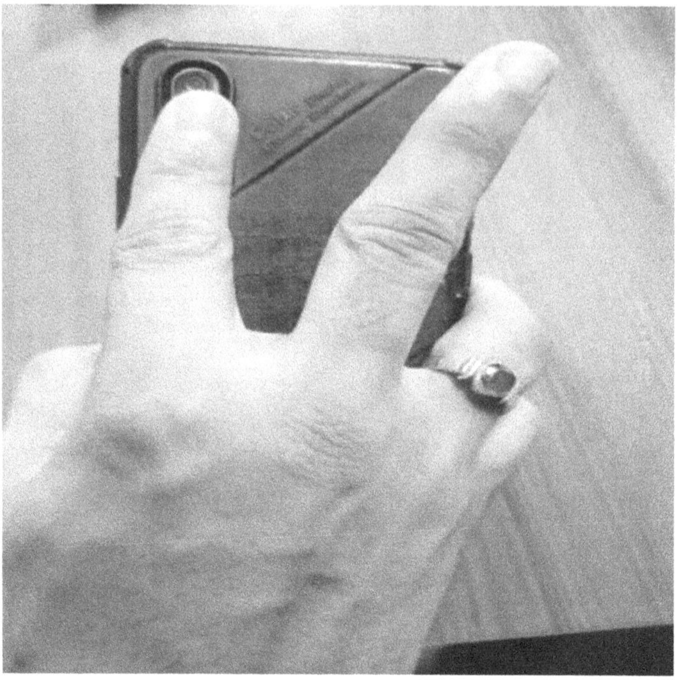

FIGURE 15.12 Individual hear rate monitoring.

FIGURE 15.13 Heartbeat rate result.

FIGURE 15.14 History of heartbeat rates.

For blockchain, a scalable system is required. The additional overhead brought on by the bandwidth resources needed to mine a new block and the broadcast communication to every network node is another issue with the suggested system. Dynamic reconfiguration could be added to this work to further improve IP usage and energy consumption.

REFERENCES

1. Shahnaz A., Qamar U., Khalid A. Using blockchain for electronic health records. *IEEE Access* (2019);7:147782–147795.
2. Jetley G., Zhang H. Electronic health records in IS research: Quality issues, essential thresholds, and remedial actions. *Decision Support Systems* (2019);126:113–137.
3. Cowie M.R., Blomster J.I., Curtis L.H., Duclaux S., Ford I., Fritz F., Goldman S., Janmohamed S., Kreuzer J., Leenay M., Michel A. Electronic health records to facilitate clinical research. *Clinical Research in Cardiology* (2017);106(1):1–9.
4. Mayer A.H., da Costa C.A.R., Righi R.D.R. Electronic health records in a block chain: A systematic review. *Health Informatics Journal* (2020);26(2):1273–1288.
5. Zheng Z., Xie S., Dai H., Chen X., Wang H. An overview of Blockchain technology: Architecture, consensus, and future trends. In: *2017 IEEE International Congress on Big Data (Bigdata Congress)*, Honolulu, Hawaii, USA, pp. 557–564 (2017). IEEE.
6. Sonkamble R.G., Phansalkar S.P., Potdar V.M., Bongale A.M. Survey of interoperability in electronic health records management and proposed blockchain-based framework: My Block EHR. *IEEE Access* (2021);9:58367–158401.

7. Dagher, G.G., Mohler, J., Milojkovic M., Marella, P.B. Ancile: Privacy-preserving framework for access control and interoperability of electronic health records using blockchain technology. *Sustainable Cities and Society* (2018);39:283–297. https://doi.org/10.1016/j.scs.2017.11.010.
8. Shuaib K., Abdella J., Sallabi F., Serhani M.A. Secure decentralized electronic health records sharing system based on blockchains. *Journal of King Saud University-Computer and Information Sciences* (2022);34(8):5045–5058.
9. Mahajan H.B., Rashid A.S., Junnarkar A.A., Uke N., Deshpande S.D., Futane P.R., Alkhayyat A., Alhayani B. Integration of healthcare 4.0 and Blockchain into secure cloud-based electronic health records systems. *Applied Nanoscience* (2023);13(3):2329–2342.
10. Hanumantharaju R., Shreenath K.N., Sowmya B.J., Srinivasa K.G. Fog-based smart healthcare: A machine learning paradigm for IoT sector. *Multimedia Tools and Applications* (2022);81(26):37299–37318.
11. Zīle K., Strazdiņa R. Blockchain use cases and their feasibility. *Applied Computer Systems* (2018);23(1):12–20.
12. Hanumantharaju R., Shreenath K.N., Sowmya B.J., Srinivasa K.G. Leveraging the fog-based machine learning model for ECG-based coronary disease prediction. *International Journal of Business Intelligence and Data Mining* (2022);21(3):328–353.
13. Lakhan A., Mohammed M.A., Rashid A.N., Kadry S., Panityakul T., Abdulkareem K.H., Thinnukool O. Smart-contract aware ethereum and client-fog-cloud healthcare system. *Sensors* (2021);21(12):4093.
14. Fatokun T., Nag A., Sharma S. Towards a Blockchain assisted patient owned system for electronic health records. *Electronics* (2021);10(5):580.
15. Prescient & Strategic Intelligence. *Blockchain in Healthcare Market by Application.* P&S Intelligence: New York, NY, USA (2018).
16. Simić Miloš M., Sladić Goran S., Milosavljević B. A Case Study IoT and Blockchain Powered. In: *Proceedings of the 8th PSU-UNS International Conference on Engineering,* Novi Sad, Serbia, pp. 8–10 (June 2017).
17. Dammak B., Turki M., Cheikhrouhou S., Baklouti M., Mars R., Dhahbi A. LoRaChainCare: an IoT architecture integrating blockchain and LoRa network for personal health care data monitoring. *Sensors* (2022); 22:1497.
18. Frikha T., Chaabane F., Aouinti N., Cheikhrouhou O., Ben Amor N., Kerrouche A. Implementation of blockchain consensus algorithm on embedded architecture. *Security and Communication Networks* (2021); 2021:9918697.
19. Tien Tuan Anh D., Ji W., Gang C., Rui L. Blockbench: A Framework for Analyzing Private Blockchains. In: *Proceedings of the 2017 ACM International Conference on Management of Data,* Chicago, IL, USA, pp. 1085–1100 (May 2017).

Index

Note: **Bold** page numbers refer to tables and *italic* page numbers refer to figures.

2D electron gas (2DEG) 18, 20, 28, 30
6T SRAM 191, 199, 206, 215, 229, 231
7T SRAM 215, *217*, 218, *221*, 222
9T SRAM 198, 199, 202, 206

accuracy 2, 40, 72, 99, 102–104, **107**, 107–111, 141, 150, 159, 162, 166, 167, 170, 173, 176, 190, 193
adaptive Iterative Learning Algorithm (ILA) 54
AlexNet 102, 103
AlGaN barrier 19–21, 24–27, 30
algorithm 1–3, 5, 10, 53, 54, 71–78, 84, 90, 103, 113, 117, 121–123, 134, 135, 139, 142, 145, 146, 150, 151, 153, 159, 161, 162, 165, 166, 174, 187–189, 233, 236, 240
amplifier 2, 3, 10, 14, 45, 52–56, 60, 63, **66**
architecture 2, 9, 44, 54, 71–78, 82, 84, 86, 102, 103, 106, 110, 127, 128, 130, 133, 140, 149, 153, 173, 191, 199, 215, 224, 226, 235, 237
area 2, 3, 5–11, **11**, 29, 38, 41, 44, 89, 91, 98, 131–134, 138, 145, 149, 150, 153, 156, 163, 170, 191, 199, 230, 234
artificial intelligence (AI) 40, 74, 106, 144, 185–187, 233
artificial neural networks (ANN) 1, 40, 161, 189, 193
ASIC 73–75, 84, 127, 128, 131–133, 138, 171, 174, 175
automation 1, 2, 54, 87, 93, 113, 129, 146, 149, 156, 174, 185, 192

backpropagation 105, 161, 175
bandgap discontinuity 15
Bayesian optimization 2, 53, 54
biomedical applications 2, 3
BIST 88, 96, 97, 154, 156
bitstream 76, 77, 130, 131, 237
blockchain 233–238, 240, 241, 244, 245, 249
boundary lock issues 38, **39**
BRATS 104, 110
buffers 21, 22, 78, 84, 198
butterfly curve 216, 218, *220–221*, 222

cache 88, 198, 199, 202
CAD 129, 130, 135, 137, 139, 145, 146, 185
Cadence Virtuoso 5, 10, 199, 218
capacitance 24, 27, 29, 30, 152, 154, 156, 161, 166, 167, 186, 209, 221
carrier mobility 17, 18, 224
channel length 5, 113, **115**, 184, 224
chip 36, 73, 79, 80, 86, 87, 89, 91–99, 134, 135, 142, 149–151, 153, 156, 161, 162, 164, 171, 184, 185, 192, 193, 198, 215, 237
circuit design 1, 5, 40, 41, **43**, 44, 54, 65, 75, 128, 134, 151, 153, 154, 160, 162, 163, 199, 202, 215, 224–226, 229, 230
circuit synthesis 1, 8, 9, 12
classification 102–104, **110**, 116, 139, 152, 157, 158, 161, 165, 187, *188*
clock gating 141, 149, 151
clock skew 91
clustering 104, 152, 157, 159, 188
CMOS 2, 36, 46, 52, 53, 56, *60*, *61*, 63–66, 184, 190, 198, 199, 216–218
CNN 103–106, **110**, 145, 191, 193
congestion 91, 135, 139, 141, 142, 150, 151, 161, 192
consensus algorithm 240
cryptography 176, 241, 244

data augmentation 146, 168
data cleaning 169
data privacy 234
data retention **43**, 44, 200
data synthesis 169
dataset 7, 9, 10, 102, 104, 106–108, 110, 111, 113–115, 117, 121, 122, 124–126, **126**, 137, 141, 158, 168, 170, 175, 185, 188, 190, 193
debugging 72, 130, 131, 149, 152, 153, 156
DenseNet 103, **107**
design automation 1, 2, 146, 149, 156
detection 98, 101, 104, 107, 110, **110**, 111, 142, 151, 152, 158, 161, 162, 164, 169, 176, 188, 189, 193
device variability 40
DFT 97, 185, 194
diagnosis 2, 101, 103, 154, 156, 161, 165, 176, 193
differential configuration **43**, 44–45
digital signal processing (DSP) 127
drain current 14, 17, 19, 21, 24–29, *29*, 205
dynamic power 172, 228

EDA 2, 54, 75–77, 146, 185, 192
efficiency 1–3, 9, 12, 14, 18, 22, 23, 44, 52–55, 66, 73, 103, 133, 144, 149, 170, 198, 224, 227

EHR 233–236, 241, 243–245
encryption 40, 127, 170, 237
energy efficiency 18, 44, 53, 73, 88, 149, 171, 173, 191, 200, 221, 224, 225, 227
evolutionary algorithms (EA) 1, 153, 173

fabrication 17, 19, 25, 74, 92–95, 128, 150, 151, 154, 167, 169, 176, 193, 224, 225, 229, 230
feature extraction 104, 110, 152
feature selection 159
FinFET 198–202, 213, 227
floorplanning 91
fog computing 233
folded cascode amplifier 2, 3, *3*
FPGA 71–80, **80**, 84, 127–137, 139–146, 173–175, 235–237, 239, 240

gain 5–7, 10, **11**, 18, 52, 55–58, 60, 63–66, 228
gallium nitride (GaN) 14, **30**, 56
gate dielectric **115**
gate leakage traps 18
gate voltage 27, 29, 30, 116, 205
gate-all-around (GAA) 224–226, 229, 230
gradient boosting (GB) 166, 188
gradient descent 161, 175

HDL 71, 75–77, 84, 88, 89, 128, 129, 131–133, 153, 185, 191
healthcare 86, 111, 145, 233–236, 241
HEMTs 14–16, 18, 22, 23, 56
heterojunction FETs 15
high-level synthesis (HLS) 131, 137, 138, *138*, 143, 145, 185, 191, 193
hold operation 201
hyper-FET 224–230
hyperparameters **117**, 117, 118

impedance matching 55, 56
inception 73, 86, 98, 103, 104, **107**, 110
inference 71, 74, 151, 171, 175, 191
integration 1, 2, 66, 73, 86, 93, 98, 110, 142, 171, 184, 198, 225, 229
interconnect 89, 91, 94, 97, 127, 134, 150, 151, 156, 166, 167, 191
interoperability 98, 234
IoT 87, 98, 99, 199, 221, 225, 226, 235, 236
IPFS 234, 236, 240, 241

lattice mismatch 19, 21, 24, 29
leakage current 18, 19, 21–24, 40, 199, 206, 208–210, 213, 224, 225, 227, 229
linear ionic drift model 38, 45, 46
lithography 91, 93, 150, 152, 176, 186, 193
logic blocks 72, 127
logic synthesis 76, 84, 98, 131–133, 137, 141, 153, 158, 160, 161, 194
low power 36, 56, 84, 199, 206, 215, 236, 244
LUTs **80**, **83**, 127, 133

machine learning (ML) 2, 40, 71–73, 83, 84, 102–104, 113, 114, 128, 135, 137, 138, *138*, 140–142, 144, 146, 149, 150, 157, 159, 165, 166, 174–176, 184–186
manufacturing 93, 95–97, *95*, 149, 150, 157, 176, 184, 192–194, 229, 230
materials 14, 17, 21, **30**, 37, 72, 93–95, 156, 173, 186, 222, 225, 226, 229, 230
matrix inverse 80–84
matrix multiplication 71–73, 83, 84, 102, 174, 191
mean squared error (MSE) 7, 125, 165
Memristive computing 173, 174
Memristor models 38, 40, 46
metaheuristic algorithms 2, 3
MetaMask 238, 241
MRI 101–104, 107, 109–111
multiply-accumulate 78, 79, 84

netlist 45, 76, 89, 91, 93, 130, 132, 133, 153, 154
neural network 1, 40, 71, 74, 103, 105, 113, 136, 137, 144, 153, 159, 161, 164, 168, 172–175, 189–191, 193
neuromorphic applications 40
noise margin 41, 202, *203–205*, 216, 221
nonlinear dopant drift 38, 45
NTFET 114, *115*, 120, 125, 126

optimization 1–3, **11**, 52, 54, 72, 73, 79, 80, 89, 91, 113, 132, 134–136, 141–143, 151, 160, 163, 173, 174, 202
optimization techniques 1, 54, 73, 89, 132

packaging 94, 95, 98, 156, 185
parallel computation 2
parallel processing 72, 74, 84, 127, 128, 144
parameters 1, 2, **6**, 7, 9, **11**, 17, 22, **39**, 41, **43**, 44, 54–56, 60, 61, *65*, **65**, **66**, 106, 107, 113–117, 122, 124–125, **126**, 133, 141, 152, 156, 162, 164, 172, 174, 175, 189, 191, 192, 194, 202, 218
particle swarm optimization (PSO) 3, 54
performance evaluation 9, 36, 185, 189, 213
performance metrics 1, 7, 86, 93, 114, **118**, **124**, 125, 159, 204
piezoelectric polarization 15, 19
placement 76, 90–92, 128, 129, 134–137, 139, 141, 143, 151, 160, 161, 167, 185, 192–194
power amplifier (PA) 52, 54, *56*, *57*, *60*, *63*, **66**
power consumption 36, 41, 72, 74, 75, 84, 89, 90, 98, 128, 132, 136–138, 141, 142, 149, 156, 166, 172–174, 176, 184, 191, 198, 202, 209, 210, 213, 215, 218, 221, 222, 224–227

power-added efficiency (PAE) 55, 58, *61*, 63–66
prediction 1, 2, 9, 12, 56, 113, 114, 116, 119–121, 123, 125, 126, 136, 139, 152, 158, 161, 165, 174, 187
process variation 151, 152, 161, 166, 168, 174, 185, 199, 209, 213
proof-of-work 238

quantization 72, 172, 173

random forest (RF) 104, 113, 145, 188, 190
read delay 41, **43**
read operation 41–45, 201, 203, 206, 213, 216, 218
reconfiguration 75, 175, 249
regression 7, *11*, 113, 116, **118**, *120*, 139, 152, 157, 161, 165–167, 187, 188, 190
reinforcement learning (RL) 137, 145, 153, 158, 163, 187, 189, 192
reliability 20, 23, 31, 44, 86, 94–98, 102, 111, 141, 142, 149, 150, 152, 156, 157, 160, 162–165, 168, 169, 172, 174, 176, 192, 194, 199, 201, 202, 209, 218
ResNet 102, 104–107, **107**, 109–111, 190
routing 72, 76, 84, 90, 91, 128, 129, 134, 136, 137, 139, 141, 143–145, 150, 151, 154, 156, 160, 161, 167, 186, 192, 194

scalability 38, 46, 73, 84, 168–170, 224, 229, 244
scan 96, 97, 103, 154, 156, 185
segmentation 103, 104, 110
sensitivity 108, 111, 184, 226
short channel effects (SCE) 17, 198
simulation 1, 2, 5, 7–11, **11**, 40, 41, 44, 45, 53, 54, 57, 72, 76, 77, 90, 92, 93, 113, *115*, 122, 125, 130, 131, 133, 136, 150, 152–154, 163, 168, 169, 185, 186, 189, 190, 193, 199, 213, 227
Simulink 77, 84
slew rate **4**, **6**, 7, **11**, 166
smart contracts 238, 241
sneak path current 40, **43**
S-parameters 56, 61
specificity 108, 111
spice simulations 2
spontaneous polarization 15
state drift 40, 41, 44
static noise margin 41, 202, *203–205*, 216, 221
static power 172, 202–206, 209, 225
static timing analysis (STA) 130, 131, 133, 166
supervised learning 136, 157, 158, 165–169, 187
surface traps 17, 20, *20*, 23, 24, 28–30
switching delay 228

technology node **5**, 46, 150, 168, 169, 186, 190
temperature 17, 18, 21, 24, 26, 30, 40, **42**, 94, 95, 98, 114, **115**, 135, 150, 161, 164, 172, 176, 207, 209–213, 226
testability 97, 99, 156, 185
testing 7, 10, 72, 74, 75, 86, 88, 92–99, 111, 114, 115, 117, 119, 121–125, 130, 152, 154, 156, 167, 169, 170, 186, 193, 194, 228, 238, 241
threshold voltage 14, 17, 18, 20, 21, 24, 30, 40, 114, 126, 215, 217
timing 76, 77, 88, 89, 91, 93, 128, 130, 132–136, 141, 149–151, 153, 160, 161, 165–167, 192, 194, 228
training 7, 9, 10, 71, 74, 102, 104, 106, 107, 114–119, *120*, 123–125, 128, 129, 137, 143, 144, 146, 159, 164, 167, 168, 170, 175, 187, 190, 191, 193
transconductance 3, 17, 26
transient analysis 218, *219*, *221*
transistor sizing 202, 218
trapping mechanisms 18, *26*
tumor 101–104, 107–111, *108*, *109*

unsupervised learning 157, 158, 187, 188

VGG 102–104, **110**
VLSI 86, 97, 149–153, 165, 167–170, 175, 184, 198, 233
voltage gain 55
voltage scaling 149, 151, 228
voltage TEAM (V-TEAM) 38–40

wearable 2, 226, 237
write delay 41, **46**
write operation 199, 201, 205, 207, 210, 213

YOLO 104, **110**

For Product Safety Concerns and Information please contact our EU representative GPSR@taylorandfrancis.com
Taylor & Francis Verlag GmbH, Kaufingerstraße 24, 80331 München, Germany

www.ingramcontent.com/pod-product-compliance
Lightning Source LLC
LaVergne TN
LVHW011122300926
849160LV00008B/481
9781032774299